# The
# Method of Moments
# in
# Electromagnetics
## Second Edition

# The Method of Moments in Electromagnetics

## Second Edition

**Walton C. Gibson**

Tripoint Industries Inc.
kalla@tripoint.org

**CRC Press**
Taylor & Francis Group
Boca Raton   London   New York

CRC Press is an imprint of the
Taylor & Francis Group, an **informa** business

A CHAPMAN & HALL BOOK

CRC Press
Taylor & Francis Group
6000 Broken Sound Parkway NW, Suite 300
Boca Raton, FL 33487-2742

© 2015 by Taylor & Francis Group, LLC
CRC Press is an imprint of Taylor & Francis Group, an Informa business

No claim to original U.S. Government works

Printed on acid-free paper
Version Date: 20140520

International Standard Book Number-13: 978-1-4822-3579-1 (Hardback)

**Visit the Taylor & Francis Web site at**
**http://www.taylorandfrancis.com**

**and the CRC Press Web site at**
**http://www.crcpress.com**

# Contents

# *Preface to the Second Edition*

The first edition of this book was well received, and since its publication I received a lot of feedback and questions about its content. When I began considering a second edition, I knew that a key element had to be the treatment of dielectric materials. After I started working on the new material, it became apparent that to accomplish this goal I would have to re-write a significant amount of the existing material. The result is a book that I believe significantly improves and expands upon the first edition in several key areas:

1. The first edition focused exclusively on integral equations for conducting problems. However, dielectric materials are an important aspect of practical electromagnetic devices, and the ability to include these in computational codes is of increasing importance. In this second edition, we will derive coupled surface integral equations that will treat conducting as well as composite conducting/dielectric objects. This treatment extends to objects having multiple dielectric regions with interfaces and junctions. Though many sections of this book have been changed or updated to reflect new material, we will continue to use the approach in the first edition, where we progress gradually from simple to more complex problems and topics.

2. As my background is in the design and programming of software codes for radar cross section (RCS) prediction, the first edition focused almost exclusively on far-zone radiated fields and RCS. In this second edition, I have added more material involving the calculation of near fields.

3. Technology continues to march ever onward, and I have attempted to keep the material in this edition as up to date as possible. As a result, existing material has been updated, and material deemed to be outdated or no longer relevant has been removed.

4. Known errors from the first edition have been corrected, and concepts that were unclear or poorly explained have been updated for clarity. Many equations have also been reformatted, and figures adjusted or regenerated to fix errors or improve their appearance.

# *Preface*

The Method of Moments (MOM) as applied to electromagnetic field problems was first described at length in Harrington's classic book,[1] and was no doubt in use before that. Since then, computing technology has grown at a staggering pace, and the field of computational electromagnetics (CEM) has followed closely behind. Though many dissertations and journal papers have been dedicated to the MOM in electromagnetics, few textbooks have been written for those who are unfamiliar with it. For graduate students who are just beginning their study of CEM, or seasoned professionals solving real-world problems, this sets the barrier to entry fairly high.

Thus, this author felt that a new book describing the Moment Method in electromagnetics was necessary. The first reason was because of the lack of a good introductory graduate-level text on the Moment Method. Though many universities offer courses in computational electromagnetics, the course material often comprises a disjoint collection of journal papers and copies of the instructor's personal notes. In the few books that do exist on the subject, the material is often presented at a high level with little to no attention paid to implementation details. Additionally, many of these often omit key information or refer to other papers (or worse, private communications), forcing the student to spend their time searching for missing information instead of focusing on course material. The second reason is the hope that a concise, up-to-date reference book will be of significant benefit to researchers and practicing professionals in the field of CEM. This book has several key features that set it apart from others of its kind:

1. *A straightforward, progressive introduction to field problems and the Moment Method.* This book begins by introducing the Method of Moments in the context of simple, electrostatic field problems. We then move to a review of frequency-domain electromagnetic theory, radiation equations, and the Green's functions for two- and three-dimensional objects. The surface equivalence theorem is then used to derive coupled surface integral equations of radiation and scattering for conducting and composite dielectric/conducting objects. Subsequent chapters are then dedicated to solving these integral equations for progressively more dif-

---

[1]R. F. Harrington *Field Computation by Moment Methods*, The Macmillan Company, 1968.

ficult problems involving thin wires, bodies of revolution, and two and three-dimensional bodies. With this material behind them, the student or researcher will be well-equipped for more advanced MOM topics encountered in the literature.

2. *A clear and concise summary of equations.* One of the fundamental problems encountered by this author is that clear expressions for the MOM matrix elements are almost never found in the literature. This book derives or summarizes the matrix elements used in every MOM problem, and all examples are computed using the expressions summarized in the text.

3. *A focus on radiation and scattering problems.* This book is primarily focused on scattering and radiation problems. Therefore, we will consider many practical examples such as antenna impedance calculation and radar cross section prediction, as well as calculation of near fields. Each example is presented in a straightforward manner with a careful explanation of the approach as well as explanation of the results.

4. *An up-to-date reference.* The material contained within is presented in the context of current-day computing technology, and includes up-to-date material such as the Fast Multipole Method that is now commonly used in CEM.

5. *"Show your work."* In this work we attempt to avoid a so-called "it can be shown" approach, which can be detrimental to students, and instead focus on step-by-step derivations in key areas that are often glossed over or omitted in other texts.

This book is intended for a one- or two- semester course in computational electromagnetics and a reference for the practicing engineer. It is expected that the reader will be familiar with time-harmonic electromagnetic fields and vector calculus, as well as differential/integral equations and linear algebra. The reader should also have some basic experience with computer programming in a language such as C, C++ or FORTRAN, or a mathematical environment such as MATLAB. Because some of the expressions in this book require the calculation of special functions, the reader should be aware of what they are and know how to calculate them.

This book comprises nine chapters:

Chapter 1 presents a very brief overview of computational electromagnetics and some of the commonly used numerical techniques in this field. This will show the reader how surface integral equations and the Method of Moments fits into the world of CEM algorithms.

Chapter 2 introduces the Method of Moments in the context of simple electrostatic field problems. We then formalize the MOM, discuss point matching and Galerkin's Method, and present some commonly used two-dimensional basis functions.

Chapter 3 begins with Maxwell's Equations, and then discusses electromagnetic boundary conditions, formulations for radiation, vector potentials, Green's functions, and near and far fields. It then uses the equivalence principle to derive a set of coupled integral equations for scattering by multi-region, composite conducting/dielectric objects. These equations are then discretized in general form using the Method of Moments, and are refined and adapted to specific problems in later chapters.

Chapter 4 discusses the solution of matrix equations, including topics such as Gaussian Elimination, LU Decomposition, matrix condition numbers, and iterative solvers.

Chapter 5 considers radiation and scattering by thin wires. We derive the thin wire kernel and the Hallén and Pocklington thin wire integral equations, and show how these are solved. We then apply the Moment Method to thin wires of arbitrary shape, and then solve practical thin wire problems involving antennas and feedlines.

Chapter 6 applies the Moment Method to two-dimensional problems. The integral equation formulations in Chapter 3 are adapted to general, two-dimensional boundaries, and at TM and TE polarizations. The first half of the chapter considers conducting geometries, and the latter half then moves into dielectric and composite geometries and problems.

Chapter 7 considers three-dimensional objects which can be described as bodies of revolution (BORs). Examples involving the radar cross section of conducting, dielectric, and composite objects are then presented. We will then discuss junctions between dielectric and conducting regions and their treatment, and consider some example problems having junctions.

Chapter 8 moves to three-dimensional (3D) surfaces of arbitrary shape. We discuss the modeling of surfaces using triangular facets, and devote significant effort to summarizing the expressions used to evaluate singular potential integrals over triangular elements. We will then present a method by which generalized junctions between regions can be treated in a systematic and straightforward manner. Next, we consider radar cross section problems involving conducting and composite objects, as well as some three-dimensional antenna problems.

Chapter 9 discusses the Fast Multipole Method (FMM) and how it is used to accelerate the matrix-vector product in the iterative solution of 3D problems. We will cover the addition theorem, wave translation, and single and multi-level fast multipole algorithms. The treatment is concise and contains all the information required to successfully implement the FMM for conducting or composite geometries in new or existing moment method codes. We will then consider examples of electrically large problems that could not be solved using a full-matrix approach, but become tractable using the FMM.

Chapter 10 discusses some methods of numerical integration including the trapezoidal and Simpson's rule, and then area coordinates and Gaussian quadrature over planar triangular elements.

Throughout this text, an $e^{jwt}$ time convention is assumed and suppressed throughout. We use SI units except in some examples where the test articles had dimensions in inches or feet. Numbers in parentheses ( ) refer to equations, and numbers in brackets [ ] are citations. Scalar quantities are written in italic font ($a$), vectors in bold lowercase font (**a**) with unit vectors denoted using the caret ($\hat{\mathbf{a}}$), and matrices in bold uppercase font (**A**).

# Acknowledgments

This book came about because of my experience as a student in electromagnetics and as a developer of computational electromagnetics codes. My walk down this path undoubtedly began very early with my interest in the power of the computer, as well as the mysteries behind radio waves and antennas. I got my first taste as an undergraduate student at Auburn University, where my professor of electromagnetics, Dr. Thomas Shumpert, encouraged me to pursue the subject at the graduate level. It progressed further during my study at the University of Illinois, where I learned many great things about electromagnetics from Dr. Weng Chew and my graduate advisor Dr. Jianming Jin. During my subsequent work on the code suite known as *lucernhammer*, I encountered many hard problems in mathematics and software algorithms, programming, and optimization. Without the help of friends and colleagues, it is likely I would not have completed this path at all. Therefore, I would like to thank the following people who provided me with useful assistance, advice, and information when it was needed:

> Jason Amos
> Michael J. Perry
> Dr. John L. Griffin
> Scott Kelley
> Paul Burns
> Dr. Stuart DeGraaf
> Jay Higley
> Dr. Andy Harrison
> Sally Colocho
> Dr. Enrico Poggio
> Dr. Bassem R. Mahafza

I would like to give extra special thanks to Stephen Robinson for proofreading the second edition manuscript prior to publication.

# *About the Author*

**Walton C. Gibson** was born in Birmingham, Alabama in 1975. He received a Bachelor of Science degree in electrical engineering from Auburn University in 1996, and a Master of Science degree from the University of Illinois Urbana-Champaign in 1998. His professional interests include electromagnetic theory, computational electromagnetics, radar cross section prediction, and computer graphics. He is the author of the industry-standard *lucernhammmer* electromagnetic signature prediction software suite. He is a licensed amateur radio operator (callsign K4LLA), with interests in antenna building and low-frequency HF propagation. He currently resides in Huntsville, Alabama.

# Chapter 1

## Computational Electromagnetics

In the beginning, the design and analysis of electromagnetic devices and structures was largely experimental. However, once the computer and numerical programming languages were developed, people immediately began using them to solve electromagnetic field problems of ever-increasing complexity. This pursuit of *computational electromagnetics* (CEM) has yielded many innovative, powerful analysis algorithms, and it now drives the development of electromagnetic devices people use every day. As the power of the computer continues to grow, so do the number of available algorithms as well as the size and complexity of the problems that can be solved. While the data gleaned from experimental measurements is invaluable, the entire process can be costly in terms of money and the manpower required to do the machine work, assembly, and measurements at the range. One of the fundamental drives behind reliable computational electromagnetics algorithms is the ability to simulate the behavior of devices and systems before they are actually built. This allows engineers to engage in levels of optimization that would be painstaking or even impossible if done experimentally. CEM also helps to provide fundamental insights into electromagnetic problems through the power of computation and computer visualization, making it one of the most important areas of engineering today.

## 1.1 CEM Algorithms

The range of electromagnetic problems is extensive, and this has led to the development of different classes of CEM algorithms, each with its own benefits and limitations. In the "early days" of CEM, many problems of practical size could not be solved unless some assumptions were made about the underlying physics and approximations made, usually under the asymptotic or high-frequency limit. These approximate algorithms are now commonly known as "high-frequency" methods. Algorithms that do not make these sorts of approximations are more demanding in terms of CPU and system memory, and historically have been limited to problems of small electrical size. These are usually referred to as "exact" or "low-frequency" algorithms. Both classes of algorithms can be further subdivided into time or frequency-domain algorithms. We will now summarize some of the most commonly used methods to provide some context in how the Moment Method fits in the CEM environment.

## 1.1.1   Low-Frequency Methods

Low-frequency (LF) methods are so-named because they solve Maxwell's Equations with no implicit approximations, and are typically limited to problems of small electrical size due to limitations of computation time and system memory. Though computers continue to grow more powerful and can solve problems of ever-increasing size, this nomenclature will likely remain common in the literature for the foreseeable future.

### 1.1.1.1   Finite Difference Time Domain Method

The Finite Difference Time Domain (FDTD) method [1, 2] uses finite differences to solve Maxwell's Equations in the time domain. The application of FDTD is usually very straightforward: the solution domain is discretized into small rectangular or curvilinear elements, and a "leap frog" in time is used to compute electric and magnetic fields from one another at discrete time steps. FDTD excels at analysis of inhomogeneous and nonlinear media, though its demands for system memory are high due to discretization of the entire solution domain, and it suffers from dispersion issues and the need to artificially truncate of the solution boundary. FDTD is typically applied in EM packaging and waveguide problems, as well as the study of wave propagation in complex (often composite) materials.

### 1.1.1.2   Finite Element Method

The Finite Element Method (FEM) [3, 4] is a method used to solve frequency-domain boundary-valued electromagnetic problems via variational techniques. It can be used with two- and three-dimensional canonical elements of differing shape, allowing for a highly accurate discretization of the solution domain. FEM is often used in the frequency domain for computing the field distribution in complex, closed regions such as cavities and waveguides. As in the FDTD, the solution domain must be discretized and truncated, making the FEM approach often unsuitable for radiation or scattering problems unless combined with a boundary integral equation approach such as the Method of Moments [3].

### 1.1.1.3 Method of Moments

The Method of Moments (MOM) is a technique used to solve electromagnetic surface[1] or volume integral equations in the frequency domain. MOM differs from FDTD and FEM as the electromagnetic sources (surface or volume currents) are the quantities of interest, and so only the surface or volume of the antenna or scatterer must be discretized. As a result, the MOM is widely used in solving radiation and scattering problems. In this book, we focus on the practical solution of surface integral equations of radiation and scattering using MOM.

## 1.1.2 High-Frequency Methods

Electromagnetic problems of large size have existed long before the computers that could solve them. Common examples of larger problems are those involving radar cross section prediction, or the calculation of an antenna's radiation pattern when in close proximity to a large structure. Many approximations have been made to the equations of radiation and scattering to make these problems tractable. Most of these treat the fields in their asymptotic limit, and employ ray-optics and edge diffraction. When the problem is electrically large, many asymptotic methods produce results that are accurate enough on their own, or can be used as a "first pass" before a more accurate though computationally demanding method is applied.

### 1.1.2.1 Geometrical Theory of Diffraction

The Geometrical Theory of Diffraction (GTD) [5, 6] uses ray-optics to determine electromagnetic wave propagation. The spreading and amplitude intensity and decay in a ray bundle is computed using Fermat's principle and the radius of curvature at the bounce points. GTD attempts to account for the fields diffracted by edges, allowing for a calculation of the fields in shadow regions. GTD is fast but the results are often fair to poor for complex geometries.

### 1.1.2.2 Physical Optics

Physical Optics (PO) [7] is a method for approximating the surface currents, allowing a boundary integration to be performed to obtain the fields. As we will see, PO and the MOM are closely related as they use the same equations to integrate the surface currents, however the MOM calculates the surface currents directly instead approximating them. While robust, the PO does not account for the fields diffracted by edges or those from multiple reflections, so supplemental corrections are typically added to it. The PO method is used

---

[1] In this context, it is also referred to as the Boundary Element Method (BEM).

extensively in high-frequency reflector antenna analysis, as well as many radar cross section prediction codes such as *lucernhammer MT*.

### 1.1.2.3   Physical Theory of Diffraction

The Physical Theory of Diffraction (PTD) [8, 9] is a means for supplementing the PO solution by adding to it the fields radiated by nonuniform currents along diffracting edges of an object. PTD is commonly used in high-frequency radar cross section and scattering analysis.

### 1.1.2.4   Shooting and Bouncing Rays

The Shooting and Bouncing Ray (SBR) method [10, 11] was developed to predict the multiple-bounce scattered fields from complex objects. It uses the ray-optics model to determine the path and amplitude of a ray bundle, but uses a physical optics-based scheme that integrates the surface currents induced by the ray at each bounce point. SBR is often used in scattering codes to account for multiple reflections on a surface or those inside a cavity, and as such it supplements the fields computed by PO and the PTD. SBR is also used to predict wave propagation and scattering in complex urban environments to determine the coverage for cellular telephone service.

---

# References

[1]  A. Taflove and S. C. Hagness, *Computational Electrodynamics: The Finite-Difference Time-Domain Method*. Artech House, third ed., 2005.

[2]  K. Kunz and R. Luebbers, *The Finite Difference Time Domain Method for Electromagnetics*. CRC Press, 1993.

[3]  J. Jin, *The Finite Element Method in Electromagnetics*. John Wiley and Sons, 1993.

[4]  J. L. Volakis, A. Chatterjee, and L. C. Kempel, *Finite Element Method for Electromagnetics*. IEEE Press, 1998.

[5]  J. B. Keller, "Geometrical theory of diffraction," *J. Opt. Soc. Amer.*, vol. 52, pp. 116–130, February 1962.

[6]  R. G. Kouyoumjian and P. H. Pathak, "A uniform geometrical theory of diffraction for an edge in a perfectly conducting surface," *Proc. IEEE*, vol. 62, pp. 1448–1461, November 1974.

[7] C. A. Balanis, *Advanced Engineering Electromagnetics*. John Wiley and Sons, 1989.

[8] P. Ufimtsev, "Approximate computation of the diffraction of plane electromagnetic waves at certain metal bodies (i and ii)," *Sov. Phys. Tech.*, vol. 27, pp. 1708–1718, August 1957.

[9] A. Michaeli, "Equivalent edge currents for arbitrary aspects of observation," *IEEE Trans. Antennas Propagat.*, vol. 23, pp. 252–258, March 1984.

[10] H. Ling, S. W. Lee, and R. Chou, "Shooting and bouncing rays: calculating the RCS of an arbitrarily shaped cavity," *IEEE Trans. Antennas Propagat.*, vol. 37, pp. 194–205, February 1989.

[11] H. Ling, S. W. Lee, and R. Chou, "High-frequency RCS of open cavities with rectangular and circular cross sections," *IEEE Trans. Antennas Propagat.*, vol. 37, pp. 648–652, May 1989.

# Chapter 2

## *The Method of Moments*

In this book, we are concerned with problems where we must solve surface integral equations. Because no analytic solutions exist for most of these problems, in practice we must apply numerical methods. In this chapter we will introduce the Method of Moments (MOM), a technique that is used to convert integral equations into a linear system that can then be solved numerically using the computer. To begin, we will analyze a few simple electrostatic problems to provide a bit of context, and then formally define the MOM. Next, we will discuss the expansion of an unknown function by a sum of weighted basis functions, and compare and contrast point matching and the Method of Galerkin. In the chapters that follow, we will then discuss electrodynamic problems and numerical methods for solving a general system of linear equations. With this material behind us, we will then be ready to apply the MOM to the more challenging integral equation problems in later chapters.

## 2.1 Electrostatic Problems

Because electrostatic problems are relatively simple compared to the electrodynamic case, they provide a good context for introducing the algorithms used to solve integral equations. Recall that the electric potential $\phi_e$ at a point $\mathbf{r}$ due to an electric charge density $q_e$ is given by the integral

$$\phi_e(\mathbf{r}) = \int_V \frac{q_e(\mathbf{r}')}{4\pi\epsilon|\mathbf{r} - \mathbf{r}'|} \, d\mathbf{r}' \tag{2.1}$$

If we know $q_e(\mathbf{r}')$, we can obtain the electric potential everywhere. If we instead know the electric potential but not the charge density, (2.1) becomes an integral equation for $q_e(\mathbf{r}')$. We will now solve this problem numerically for a pair of practical examples, the charged wire and plate.

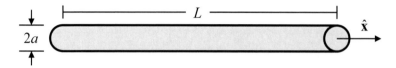

**FIGURE 2.1:** Thin Wire Dimensions

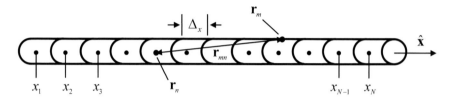

**FIGURE 2.2:** Thin Wire Segmentation

### 2.1.1   Charged Wire

Consider a thin, conducting wire of length $L$ and radius $a$ oriented along the $x$ axis, as shown in Figure 2.1. If the radius of the wire is very small compared to the length ($a << L$), the electric potential on the wire can be expressed via the integral

$$\phi_e(\mathbf{r}) = \int_0^L \frac{q_e(x')}{4\pi\epsilon|\mathbf{r}-\mathbf{r'}|}\, dx' \tag{2.2}$$

where

$$|\mathbf{r}-\mathbf{r'}| = \sqrt{(x-x')^2 + (y-y')^2} \tag{2.3}$$

We intend to convert (2.2) into a linear system of equations, so we will first subdivide the wire into $N$ subsegments, each having length $\Delta_x$, as shown in Figure 2.2. Within each subsegment, we assume that the charge density has a constant value such that $q_e(x')$ is piecewise constant over the length of the wire. Mathematically, this can be written as

$$q_e(x') = \sum_{n=1}^N a_n f_n(x') \tag{2.4}$$

where $a_n$ are unknown coefficients, and $f_n(x')$ is a pulse function having a constant value on segment $n$ but zero on all others, i.e.

$$f_n(x') = \begin{cases} 0 & x' < (n-1)\Delta_x \\ 1 & (n-1)\Delta_x \leq x' \leq n\Delta_x \\ 0 & x' > n\Delta_x \end{cases} \tag{2.5}$$

Let us now set the potential of the wire to a value of $\phi_e = 1\mathrm{V}$. Substituting

(2.4) into (2.2) then yields

$$1 = \int_0^L \sum_{n=1}^N a_n f_n(x') \frac{1}{4\pi\epsilon|\mathbf{r}-\mathbf{r}'|} \, dx' \tag{2.6}$$

Using the previous definition of the pulse function, we can rewrite this as

$$1 = \frac{1}{4\pi\epsilon} \sum_{n=1}^N a_n \int_{(n-1)\Delta_x}^{n\Delta_x} \frac{1}{|\mathbf{r}-\mathbf{r}'|} \, dx' \tag{2.7}$$

where we now have a sum of integrals, each over the domain of a single pulse function. Let us now fix the source points so that they are on the wire axis, and the field point to be on the wire's surface. This ensures that we do not encounter a singularity in the integrand. The denominator of the integrand now becomes

$$|\mathbf{r}-\mathbf{r}'| = \sqrt{(x-x')+a^2} \tag{2.8}$$

and (2.7) can be written as

$$4\pi\epsilon = a_1 \int_0^{\Delta_x} \frac{1}{\sqrt{(x-x')+a^2}} \, dx' + a_2 \int_{\Delta_x}^{2\Delta_x} \frac{1}{\sqrt{(x-x')+a^2}} \, dx' + \cdots$$
$$+a_{N-1} \int_{(N-2)\Delta_x}^{(N-1)\Delta_x} \frac{1}{\sqrt{(x-x')+a^2}} \, dx' + a_N \int_{(N-1)\Delta_x}^{N\Delta_x} \frac{1}{\sqrt{(x-x')+a^2}} \, dx' \tag{2.9}$$

which comprises one equation in $N$ unknowns. If we can now somehow convert this equation to $N$ equations in $N$ unknowns, we can solve it by common matrix algebra routines. To do so, let us choose $N$ independent field points $x_m$ on the surface of the wire, each at the center of a wire segment. Doing so yields

$$4\pi\epsilon = a_1 \int_0^{\Delta_x} \frac{1}{\sqrt{(x_1-x')+a^2}} \, dx' + \cdots + a_N \int_{(N-1)\Delta_x}^{N\Delta_x} \frac{1}{\sqrt{(x_1-x')+a^2}} \, dx'$$

$$\vdots$$

$$4\pi\epsilon = a_1 \int_0^{\Delta_x} \frac{1}{\sqrt{(x_N-x')+a^2}} \, dx' + \cdots + a_N \int_{(N-1)\Delta_x}^{N\Delta_x} \frac{1}{\sqrt{(x_N-x')+a^2}} \, dx' \tag{2.10}$$

which comprises a matrix system of the form

$$\begin{bmatrix} Z_{11} & Z_{12} & Z_{13} & \cdots & Z_{1N} \\ Z_{21} & Z_{22} & Z_{23} & \cdots & Z_{2N} \\ Z_{31} & Z_{32} & Z_{33} & \cdots & Z_{3N} \\ \vdots & \vdots & \vdots & \ddots & \vdots \\ Z_{N1} & Z_{N2} & Z_{N3} & \cdots & Z_{NN} \end{bmatrix} \begin{bmatrix} a_1 \\ a_2 \\ a_3 \\ \vdots \\ a_N \end{bmatrix} = \begin{bmatrix} b_1 \\ b_2 \\ b_3 \\ \vdots \\ b_N \end{bmatrix} \tag{2.11}$$

where the matrix elements $Z_{mn}$ are

$$Z_{mn} = \int_{(n-1)\Delta_x}^{n\Delta_x} \frac{1}{\sqrt{(x_m - x') + a^2}} \, dx' \tag{2.12}$$

and the right-hand side (RHS) vector elements $b_m$ are

$$b_m = 4\pi\epsilon \tag{2.13}$$

### 2.1.1.1   Matrix Element Evaluation

The integral in (2.12) can be evaluated in closed form. Performing this integration yields [1] (200.01)

$$Z_{mn} = \log\left[\frac{(x_b - x_m) + \sqrt{(x_b - x_m)^2 - a^2}}{(x_a - x_m) + \sqrt{(x_a - x_m)^2 - a^2}}\right] \tag{2.14}$$

where $x_b = n\Delta_x$ and $x_a = (n-1)\Delta_x$. Note that the linear geometry of this problem yields a matrix that has a symmetric Toeplitz structure of the form

$$\mathbf{Z} = \begin{bmatrix} Z_1 & Z_2 & Z_3 & \dots & Z_N \\ Z_2 & Z_1 & Z_2 & \dots & Z_{N-1} \\ Z_3 & Z_2 & Z_1 & \dots & Z_{N-2} \\ \vdots & \vdots & \vdots & \ddots & \vdots \\ Z_N & Z_{N-1} & Z_{N-2} & \dots & Z_1 \end{bmatrix} \tag{2.15}$$

where the elements in the first row can be used to populate the entire matrix.

### 2.1.1.2   Solution

In Figures 2.3a and 2.3b is plotted the computed charge density on the wire using 15 and 100 segments, respectively. The representation of the charge at the lower level of discretization is somewhat crude, as expected. The increase to 100 unknowns greatly increases the fidelity of the result. Using the computed charge density, we then compute the potential at 100 points along the wire using (2.2). The potential using 15 charge segments is shown in Figure 2.4a. While the voltage is near the expected value of 1V, it is not of constant value, particularly near the ends of the wire. Figure 2.4b shows the potential obtained using 100 segments. The voltage is now nearly constant across the entire wire, except at the endpoints. Because we used a uniform segment size for the wire, the charge density tends to be somewhat oversampled in the middle of the wire and undersampled near the ends. As a result, the variation of the charge near the ends of the wire is not represented as accurately as in the center, and the

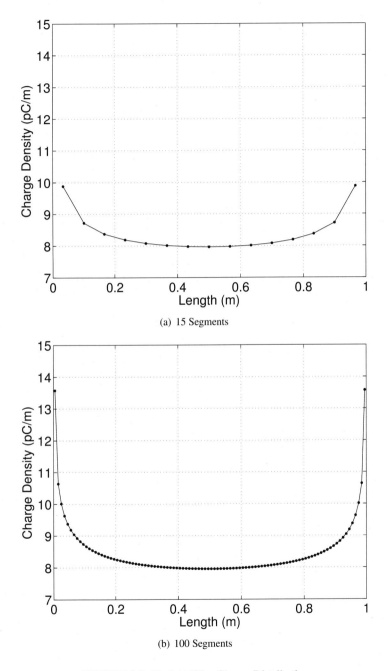

(a) 15 Segments

(b) 100 Segments

**FIGURE 2.3:** Straight Wire Charge Distribution

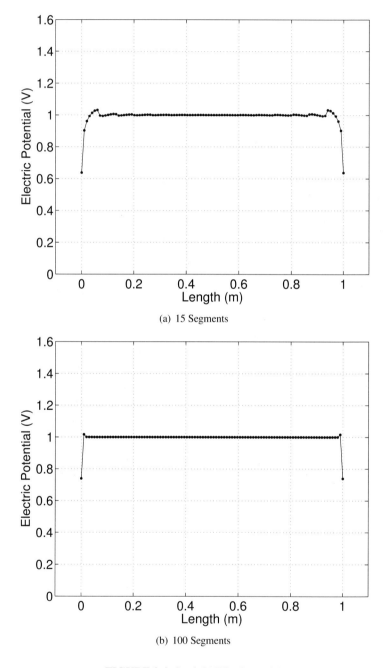

(a) 15 Segments

(b) 100 Segments

**FIGURE 2.4:** Straight Wire Potential

computed voltage tends to diverge from the true value. Many realistic shapes have irregular surface features such as cracks, gaps, and corners that give rise to a more rapid variation in the solution at those points. In an attempt to increase accuracy, it is often advantageous to employ a denser level of discretization in areas where the most variation is expected.

## 2.1.2 Charged Plate

We next consider a similar problem involving a thin, charged conducting square plate of side length $L$, as shown in Figure 2.5. The potential on the plate can be written as

$$\phi_e(\mathbf{r}) = \int_{-\frac{L}{2}}^{\frac{L}{2}} \int_{-\frac{L}{2}}^{\frac{L}{2}} \frac{q_e(x', y')}{4\pi\epsilon|\mathbf{r} - \mathbf{r}'|} \, dx' \, dy' \qquad (2.16)$$

Fixing the plate to a potential of 1V as before, (2.16) becomes

$$4\pi\epsilon = \int_{-\frac{L}{2}}^{\frac{L}{2}} \int_{-\frac{L}{2}}^{\frac{L}{2}} \frac{q_e(x', y')}{\sqrt{(x - x')^2 + (y - y')^2}} \, dx' \, dy' \qquad (2.17)$$

We now subdivide the plate into $N$ square patches having edge length $2a$ and area $\Delta_s = 4a^2$, and assume the charge to be of constant value within each patch. We then choose $N$ independent field points, each located at the center $(x_m, y_m)$ of a patch. Doing so yields a matrix equation with elements $Z_{mn}$ given by

$$Z_{mn} = \int_{S_n} \frac{1}{\sqrt{(x_m - x')^2 + (y_m - y')^2}} \, dx' \, dy' \qquad (2.18)$$

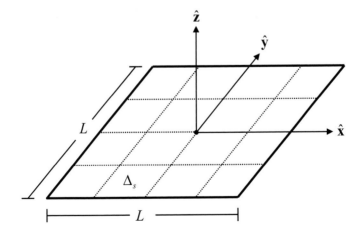

**FIGURE 2.5:** Thin Charged Plate Dimensions

where $S_n$ comprises the extents of patch $n$. Right-hand side vector elements remain the same as in (2.13).

### 2.1.2.1   Matrix Element Evaluation

When the observation and source patches are the same ($m = n$), the integrand has a singularity and the integral must be evaluated analytically. These matrix elements are called *self terms*, and represent the most dominant interactions between elements. We will devote significant attention to the evaluation of self-term matrix elements in this book. The self-term integral for the charged plate is

$$Z_{mm} = \int_{-a}^{a} \int_{-a}^{a} \frac{1}{\sqrt{(x')^2 + (y')^2}} \, dx' \, dy' \tag{2.19}$$

Performing the innermost integration yields [1] (200.01)

$$Z_{mm} = \int_{-a}^{a} \log \left[ \frac{\sqrt{a^2 + (y')^2} + a}{\sqrt{a^2 + (y')^2} - a} \right] dy' \tag{2.20}$$

and performing the outermost integration then yields [2]

$$Z_{mm} = 2a \log \left[ y + \sqrt{a^2 + y^2} \right] + y \log \left[ \frac{y^2 + 2a(a + \sqrt{a^2 + y^2})}{y^2} \right] \Bigg|_{-a}^{a} \tag{2.21}$$

which reduces to

$$Z_{mm} = \frac{2a}{\pi \epsilon} \log(1 + \sqrt{2}) \tag{2.22}$$

For patches that do not overlap ($m \neq n$) we will use a simple centroidal approximation to the integral, resulting in

$$Z_{mn} = \frac{\Delta_s}{\sqrt{(x_m - x_n)^2 + (y_m - y_n)^2}} \tag{2.23}$$

where $x_n$ and $y_n$ are at the center of the source patch. This approximation is not very accurate for elements that are near to one other, though it serves to illustrate the problem. In such cases, an analytic or adaptive numerical integration should be used instead. Matrix elements involving source and field points that do not overlap but are still sufficiently close are called *near terms*, and are also considered in greater detail in this book.

### 2.1.2.2   Solution

Figures 2.6a and 2.6b show the computed surface charge densities obtained using 15 and 35 patches in the $x$ and $y$ directions, which comprises 225 and 1225 unknowns, respectively. Figures 2.7a and 2.7b show the surface charge

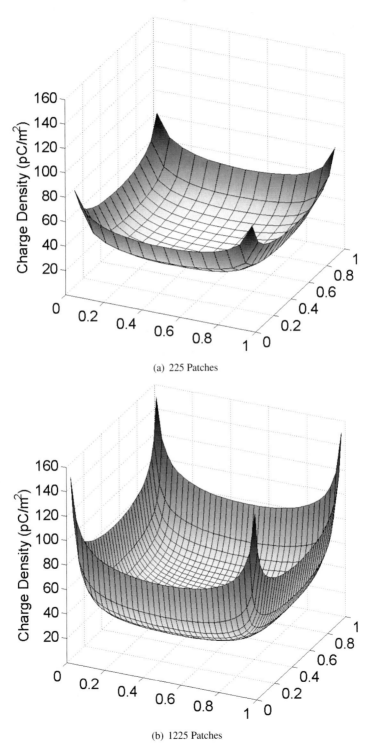

(a) 225 Patches

(b) 1225 Patches

**FIGURE 2.6:** Charge Distribution on Square Plate

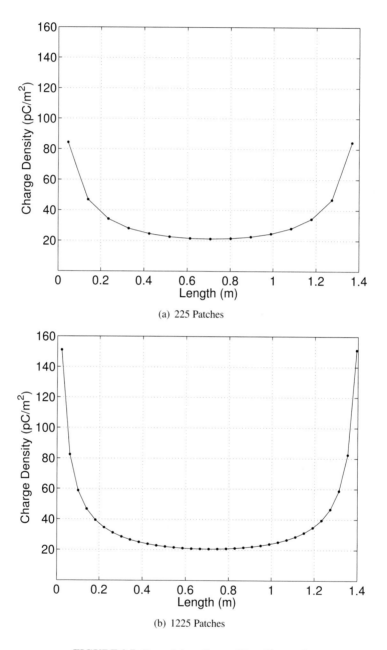

(a) 225 Patches

(b) 1225 Patches

**FIGURE 2.7:** Potential on Square Plate Diagonal

density on the patches along the diagonal of the plate. As with the thin wire, the electric charge accumulates near the corners and the edges of the plate, and our solution would likely benefit from additional discretization density in those areas.

---

## 2.2    The Method of Moments

In the previous section, we considered the problem of computing an unknown charge distribution on a thin wire or plate at a known potential. Our basic approach was to expand the unknown quantity using a set of known functions with unknown coefficients. We then converted the resulting equation into a linear system of equations by enforcing the boundary conditions, in this case the electric potential, at a number of field points on the object. The resulting linear system was then solved numerically for the unknown coefficients. Let us formalize this process by introducing a method of weighted residuals known as the Method of Moments (MOM). Consider the generalized problem

$$L(f) = g \tag{2.24}$$

where $L$ is a linear operator, $g$ is a known forcing function, and $f$ is unknown [3]. In electromagnetic problems, $L$ is typically an integro-differential operator such as $\mathcal{L}$ (3.77) and $\mathcal{K}$ (3.78), $f$ is the unknown function (current) and $g$ is a known driving function (incident field). Let us now expand $f$ into a sum of $N$ weighted *basis functions*

$$f = \sum_{n=1}^{N} a_n f_n \tag{2.25}$$

where $a_n$ are unknown coefficients. Because $L$ is linear, substitution of the above into (2.24) yields

$$\sum_{n=1}^{N} a_n L(f_n) \approx g \tag{2.26}$$

where the residual is

$$R = g - \sum_{n=1}^{N} a_n L(f_n) \tag{2.27}$$

The basis functions are chosen so they model the expected behavior of the unknown function throughout its domain, and they may be scalars or vectors depending on the problem. If the basis functions have local support in the domain, they are called *local* or *subsectional* basis functions. If their support spans the entire problem domain, they are called *global* or *entire-domain* basis functions. In this book we focus almost exclusively on local basis functions.

Let us now generalize the method by which the boundary conditions were previously enforced. We define an inner product or *moment* between a basis function $f_n(\mathbf{r}')$ and a *testing* or *weighting* function $f_m(\mathbf{r})$

$$\langle f_m, f_n \rangle = \int_{f_m} f_m(\mathbf{r}) \cdot \int_{f_n} f_n(\mathbf{r}') \, d\mathbf{r}' d\mathbf{r} \qquad (2.28)$$

where the integrals can be line, surface, or volume integrals depending on the basis and testing functions. Requiring the inner product of each testing function with the residual function to be zero yields

$$\sum_{n=1}^{N} a_n \langle f_m, L(f_n) \rangle = \langle f_m, g \rangle \qquad (2.29)$$

which results in the $N \times N$ matrix equation $\mathbf{Za} = \mathbf{b}$ with matrix elements

$$Z_{mn} = \langle f_m, L(f_n) \rangle \qquad (2.30)$$

and right-hand side vector elements

$$b_m = \langle f_m, g \rangle \qquad (2.31)$$

In electromagnetic problems, a basis function will interact with all others via the Green's function and the resulting system matrix is full. This can be compared to other algorithms such as the Finite Element Method, where the matrix is typically sparse, symmetric, and banded, and many elements of each matrix row are zero [4].

## 2.2.1 Point Matching

In Sections 2.1.1 and 2.1.2, we enforced the boundary conditions by testing the integral equation at a set of discrete points on the object. This is equivalent to using a delta function as the testing function in (2.28):

$$f_m(\mathbf{r}) = \delta(\mathbf{r}) \qquad (2.32)$$

This method is referred to as *point matching* or *point collocation*. There are advantages as well as disadvantages to this. One benefit is that in evaluating the matrix elements, no integral is required over the range of the testing function, only that of the source function, which may make evaluation of the matrix elements easier. The primary disadvantage is that the boundary conditions are matched only at discrete locations throughout the solution domain. Regardless, the results are often still reasonable, and we use this method for a few problems in this book for comparative purposes.

### 2.2.2 Galerkin's Method

For testing, we are free to use whatever functions we wish, however for most problems the choice of testing function is crucial to obtaining a good solution. One of the most commonly used is the *Method of Galerkin*, where the basis functions are used as the testing functions. This has the advantage of enforcing the boundary conditions throughout the solution domain, instead of at discrete points as with point matching. Most of the MOM problems encountered in the literature use Galerkin testing as the standard method (with its use often assumed or implied), and we will solve most problems in this book using it as well.

---

## 2.3 Common 2D Basis Functions

The most important characteristic of a basis function is that it will reasonably represent the behavior of the unknown function throughout its domain. If the solution has a high level of variation throughout a particular region, pulse basis functions may not be as good a choice as a linear or higher-order function. The choice of basis function also determines the level of difficulty in evaluating the MOM matrix elements, which in some cases may be quite high. We will now briefly consider some two-dimensional local basis functions commonly used in Moment Method problems, as well as entire-domain functions.

### 2.3.1 Pulse Functions

A set of pulse basis functions is depicted in Figure 2.8, where the domain has been divided into $N$ points with $N - 1$ subsegments/pulses. In our figure the segments all have equal lengths, however this is not required. The pulse

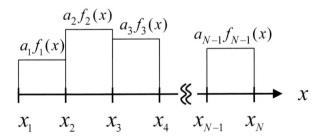

**FIGURE 2.8:** Pulse Functions

function is defined as

$$f_n(x) = 1 \qquad x_n \leq x \leq x_{n+1} \tag{2.33}$$

$$f_n(x) = 0 \qquad \text{elsewhere} \tag{2.34}$$

Pulse functions comprise a simple and crude approximation to the solution over each segment, but can greatly simplify the evaluation of MOM matrix elements. Note that since the derivative of pulse functions is impulsive, they cannot be used when the integral operator contains a derivative with respect to $x$ [5].

## 2.3.2   Piecewise Triangular Functions

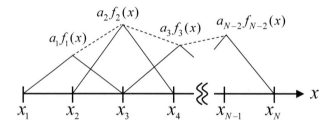

**FIGURE 2.9:** Triangle Functions (End Condition 1)

Where pulse functions are constant on a single segment, a triangle function spans two segments and varies from zero at the outer points to unity at the center. A set of triangle functions is shown in Figure 2.9. The domain has been divided into $N$ points and $N - 1$ subsegments, resulting in $N - 2$ basis functions. Though we have again shown the segments as having equal length, this is not required. Since adjacent functions overlap by one segment, triangles provide a piecewise linear variation of the solution between segments. The triangle function is defined as

$$f_n(x) = \frac{x - x_{n-1}}{x_n - x_{n-1}} \qquad x_{n-1} \leq x \leq x_n \tag{2.35}$$

$$f_n(x) = \frac{x_{n+1} - x}{x_{n+1} - x_n} \qquad x_n \leq x \leq x_{n+1} \tag{2.36}$$

Triangle functions can be used when the integral operator contains a derivative with respect to $x$, which is useful in redistributing vector derivatives.

Note that the configuration of Figure 2.9 forces the solution to zero at $x_1$ and $x_N$. This configuration may be desirable when the value of the solution at the ends of the domain is known to be zero *a priori*, however it should not

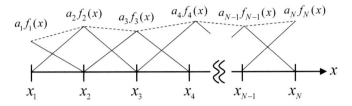

**FIGURE 2.10:** Triangle Functions (End Condition 2)

be used when the solution might be nonzero. In this case, we can assign a *half triangle* to the first and last segments, which no longer forces the solution to be zero. This is illustrated in Figure 2.10, where we now have a total of $N$ basis functions.

### 2.3.3 Piecewise Sinusoidal Functions

Piecewise sinusoid functions are similar to triangle functions, as illustrated in Figure 2.11. They are sometimes used in the analysis of wire antennas because of their ability to represent sinusoidal current distributions. These functions are defined as

$$f_n(x) = \frac{\sin k(x - x_{n-1})}{\sin k(x_n - x_{n-1})} \qquad x_{n-1} \le x \le x_n \qquad (2.37)$$

$$f_n(x) = \frac{\sin k(x_{n+1} - x)}{\sin k(x_{n+1} - x_n)} \qquad x_n \le x \le x_{n+1} \qquad (2.38)$$

where $k$ is the wavenumber, and the length of the segments are generally much less than the period of the sinusoid.

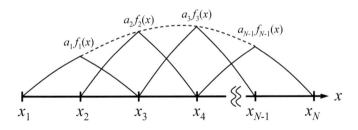

**FIGURE 2.11:** Piecewise Sinusoidal Functions

### 2.3.4 Entire-Domain Functions

Unlike local basis functions, entire-domain functions exist everywhere throughout the problem domain. One might have reason to use entire-domain functions if certain information about the solution is known *a priori*. For example, the solution may be known to comprise a sum of weighted polynomials or sines and cosines. As an example, consider the current $I(x)$ on a thin dipole antenna of length $L$. We might represent this current by the sum [6]

$$I(x) = \sum_{n=1}^{N} a_n (1 - |x/L|)^m \qquad (2.39)$$

where the testing procedure is still performed as before. After solving the resulting matrix equation, only the first few coefficients $a_n$ may be needed to accurately represent the current in the above sum. A disadvantage of entire-domain functions is that it may not be feasible to apply them to a geometry of arbitrary shape. As a result, entire-domain functions are not often encountered in the literature, and we do not consider them further in this book.

### 2.3.5 Number of Basis Functions

For a given problem, the number of basis functions (unknowns) must be chosen so they can adequately represent local variation in the solution. Because we are concerned with time-harmonic problems, we must model amplitude as well as phase behavior. For linear basis functions such as the triangle of Section 2.3.2, a common "rule of thumb" is that at least ten unknowns per wavelength should be used to represent a sinusoid. This number should increase for areas where the solution may vary significantly, such as gaps, cracks, and edges on a surface. Note also that in problems with multiple dielectric regions, the wavelength in each material will differ and so must the basis function density on the different interfaces. Higher-order basis functions exist that may reduce the unknown count at the expense of increased complexity in the geometry description as well as the MOM formulation. Such functions are covered in detail in advanced references on basis functions in electromagnetics.

Though the number of unknowns $N$ always increases with the size of the problem, the rate of the increase depends on whether the problem domain is a line, surface, or volume. In the MOM, $N$ increases linearly for one-dimensional problems and exponentially for the surface problems considered in Chapter 8. The resulting number may be as small as a few hundred, or may grow into the thousands or even millions for very large problems. This has significant consequences in terms of the system memory, as it requires the storage of an $N \times N$ matrix, as well as the compute time to solve it.

# References

[1] H. B. Dwight, *Tables of Integrals and Other Mathematical Data.* The Macmillan Company, 1961.

[2] *The Integrator.* Wolfram Research, http://integrals.wolfram.com.

[3] R. F. Harrington, *Field Computation by Moment Methods.* The Macmillan Company, 1968.

[4] J. Jin, *The Finite Element Method in Electromagnetics.* John Wiley and Sons, 1993.

[5] N. Morita, N. Kumagai, and J. R. Mautz, *Integral Equation Methods for Electromagnetics.* Artech House, 1990.

[6] B. D. Popovich, "Polynomial approximation of current along thin symmetrical cylindrical dipoles," *Proc. IEE*, vol. 117, pp. 873–878, May 1970.

# Chapter 3

## Radiation and Scattering

In this chapter we will review Maxwell's Equations and the electromagnetic boundary conditions, derive the integral equations for radiation and scattering, and then derive the Green's functions needed to solve those equations. We will next derive the electric and magnetic vector potential and expressions for the radiated electric and magnetic near and far fields. We will then consider the solution of radiation problems by way of the equivalence principle, and derive coupled surface integral equations that can be used to treat multi-region conducting, dielectric, and hybrid conducting/dielectric geometries. We will then discretize these equations in a general sense via the Method of Moments, which will be applied to more specific two- and three-dimensional problems of increasing complexity in later chapters.

## 3.1 Maxwell's Equations

In a homogeneous dielectric region having constituent parameters $\epsilon$ and $\mu$, the electric and magnetic fields must satisfy Maxwell's Equations, which in the frequency domain are

$$\nabla \times \mathbf{E} = -\mathbf{M} - j\omega\mu\mathbf{H} \tag{3.1}$$

$$\nabla \times \mathbf{H} = \mathbf{J} + j\omega\epsilon\mathbf{E} \tag{3.2}$$

$$\nabla \cdot \mathbf{D} = q_e \tag{3.3}$$

$$\nabla \cdot \mathbf{B} = q_m \tag{3.4}$$

where $\mathbf{D} = \epsilon\mathbf{E}$, $\mathbf{B} = \mu\mathbf{H}$, and the time dependence $e^{jwt}$ is assumed and suppressed throughout this text. Though the magnetic current $\mathbf{M}$ and charge $q_m$ are not physically realizable quantities, they are useful mathematical tools in solving radiation and scattering problems, as we will show in Section 3.6.1.

## 3.2    Electromagnetic Boundary Conditions

At the interface between two regions $R_1$ and $R_2$, each having different dielectric parameters, the generalized electromagnetic boundary conditions can be written as

$$\hat{\mathbf{n}}_1 \times (\mathbf{E}_1 - \mathbf{E}_2) = -\mathbf{M} \tag{3.5}$$

$$\hat{\mathbf{n}}_1 \times (\mathbf{H}_1 - \mathbf{H}_2) = \mathbf{J} \tag{3.6}$$

$$\hat{\mathbf{n}}_1 \cdot (\mathbf{D}_1 - \mathbf{D}_2) = q_e \tag{3.7}$$

$$\hat{\mathbf{n}}_1 \cdot (\mathbf{B}_1 - \mathbf{B}_2) = q_m \tag{3.8}$$

where any of $\mathbf{M}, \mathbf{J}, q_e$, or $q_m$ may be present, and $\hat{\mathbf{n}}_1$ is the normal vector on the interface between $R_1$ and $R_2$ that points into $R_1$. At the interface between a dielectric ($R_1$) and a perfect electric conductor (PEC, $R_2$), the boundary conditions are

$$-\hat{\mathbf{n}}_1 \times \mathbf{E}_1 = 0 \tag{3.9}$$

$$\hat{\mathbf{n}}_1 \times \mathbf{H}_1 = \mathbf{J} \tag{3.10}$$

$$\hat{\mathbf{n}}_1 \cdot \mathbf{D}_1 = q_e \tag{3.11}$$

$$\hat{\mathbf{n}}_1 \cdot \mathbf{B}_1 = 0 \tag{3.12}$$

and between a dielectric ($R_1$) and perfect magnetic conductor (PMC, $R_2$) they are

$$-\hat{\mathbf{n}}_1 \times \mathbf{E}_1 = \mathbf{M} \tag{3.13}$$

$$\hat{\mathbf{n}}_1 \times \mathbf{H}_1 = 0 \tag{3.14}$$

$$\hat{\mathbf{n}}_1 \cdot \mathbf{D}_1 = 0 \tag{3.15}$$

$$\hat{\mathbf{n}}_1 \cdot \mathbf{B}_1 = q_m \tag{3.16}$$

In this book we will consider dielectric and PEC regions only.

## 3.3    Formulations for Radiation

The electromagnetic radiation problem involves the calculation of fields everywhere in space due to a set of electric and magnetic currents. Scattering problems can be treated as radiation problems where these currents are generated by other currents or impressed fields. Let us start from first principles and derive the expression for the electric field in the presence of the electric and magnetic currents $\mathbf{J}$ and $\mathbf{M}$. To begin, we take the curl of (3.1) and combine with (3.2) to get

$$\nabla \times \nabla \times \mathbf{E} = -j\omega\mu\nabla \times \mathbf{H} = \omega^2\mu\epsilon\mathbf{E} - j\omega\mu\mathbf{J} - \nabla \times \mathbf{M} \qquad (3.17)$$

or

$$\nabla \times \nabla \times \mathbf{E} - \omega^2\mu\epsilon\mathbf{E} = -j\omega\mu\mathbf{J} - \nabla \times \mathbf{M} \qquad (3.18)$$

Using the vector identity

$$\nabla \times \nabla \times \mathbf{E} = \nabla(\nabla \cdot \mathbf{E}) - \nabla^2\mathbf{E} \qquad (3.19)$$

we write this as

$$\nabla(\nabla \cdot \mathbf{E}) - \nabla^2\mathbf{E} - k^2\mathbf{E} = -j\omega\mu\mathbf{J} - \nabla \times \mathbf{M} \qquad (3.20)$$

where $k = w\sqrt{\mu\epsilon} = 2\pi/\lambda$ is the wavenumber. Substituting (3.3) into the above we get

$$\nabla^2\mathbf{E} + k^2\mathbf{E} = j\omega\mu\mathbf{J} + \frac{\nabla q_e}{\epsilon} - \nabla \times \mathbf{M} \qquad (3.21)$$

Employing the equation of continuity

$$\nabla \cdot \mathbf{J} = -j\omega q_e \qquad (3.22)$$

allows us to obtain

$$\nabla^2\mathbf{E} + k^2\mathbf{E} = j\omega\mu\mathbf{J} - \frac{1}{j\omega\epsilon}\nabla(\nabla \cdot \mathbf{J}) - \nabla \times \mathbf{M} \qquad (3.23)$$

Since Maxwell's Equations are linear, we can consider $\mathbf{J}$ and $\mathbf{M}$ to be a superposition of point sources distributed over some volume. Therefore, if we know the response of a point source, we can solve the original problem by integrating that response over the volume. We now make use of this idea to convert (3.23) into an integral equation. To do so, we introduce the *Green's function* $G(\mathbf{r}, \mathbf{r}')$, which satisfies the scalar Helmholtz equation [1]

$$\nabla^2 G(\mathbf{r}, \mathbf{r}') + k^2 G(\mathbf{r}, \mathbf{r}') = -\delta(\mathbf{r}, \mathbf{r}') \qquad (3.24)$$

Assuming that we know $G(\mathbf{r}, \mathbf{r}')$, the radiated electric field is then

$$\mathbf{E}(\mathbf{r}) = -j\omega\mu \int_V G(\mathbf{r}, \mathbf{r}') \left[1 + \frac{1}{k^2}\nabla'\nabla'\cdot\right]\mathbf{J}(\mathbf{r}')\,d\mathbf{r}' - \int_V G(\mathbf{r}, \mathbf{r}')\nabla'\times\mathbf{M}(\mathbf{r}')\,d\mathbf{r}' \qquad (3.25)$$

and by a similar derivation, the radiated magnetic field is

$$\mathbf{H}(\mathbf{r}) = -j\omega\epsilon \int_V G(\mathbf{r}, \mathbf{r}') \left[1 + \frac{1}{k^2}\nabla'\nabla'\cdot\right]\mathbf{M}(\mathbf{r}')\,d\mathbf{r}' + \int_V G(\mathbf{r}, \mathbf{r}')\nabla'\times\mathbf{J}(\mathbf{r}')\,d\mathbf{r}' \qquad (3.26)$$

We will next solve (3.24) for $G(\mathbf{r}, \mathbf{r}')$ in two and three dimensions. To do so, we will first consider the homogeneous version, and then match the boundary conditions of the inhomogeneous case to obtain a unique solution.

### 3.3.1   Three-Dimensional Green's Function

Since $G(\mathbf{r}, \mathbf{r}')$ is the solution for a point electromagnetic source, it must exhibit spherical symmetry in three dimensions. Therefore, we will retain only the radial term of the Laplacian and write

$$\nabla^2 G = \frac{1}{r^2}\frac{d}{dr}\left(r^2\frac{dG}{dr}\right) = \frac{d^2G}{dr^2} + \frac{2}{r}\frac{dG}{dr} \tag{3.27}$$

Next, we observe that

$$\frac{1}{r}\frac{d^2(rG)}{dr^2} = \frac{1}{r}\frac{d}{dr}\left[r\frac{dG}{dr} + G\right] = \frac{d^2G}{dr^2} + \frac{2}{r}\frac{dG}{dr} \tag{3.28}$$

and if we substitute in the function $rG$ for $r > 0$, we can write

$$\frac{d^2(rG)}{dr^2} + k^2(rG) = 0 \tag{3.29}$$

The solution to the above is

$$G(r) = A\frac{e^{-jkr}}{r} + B\frac{e^{jkr}}{r} \tag{3.30}$$

where $r = |\mathbf{r} - \mathbf{r}'|$. Noting that the solution must contain only outgoing waves, we retain only the first term in (3.30), yielding

$$G(r) = A\frac{e^{-jkr}}{r} \tag{3.31}$$

It should be pointed out that the phase convention of the exponential in the Green's function is not standard throughout the literature. The phase as defined in (3.31) is standard in most electrical engineering references, however other references [2, 3] define the Green's function with a positive exponent, assuming the time harmonic term $e^{-j\omega t}$. Similarly, the two-dimensional Green's function in (3.49) would instead use the Hankel function of the first kind.

We must now match the boundary conditions to determine a unique solution. Since the wave must decay to zero with increasing $r$, this requires that $G(r) \to 0$ as $r \to \infty$, which it is seen to do via inspection. We must now match it at the location of the point source ($r = 0$), which will allow us to determine $A$. To do so, let us integrate (3.24) over a spherical volume of radius $a$ around the source. Substituting (3.31) for $G(r)$ we get

$$A\int_V\left[\nabla\cdot\nabla\left(\frac{e^{-jkr}}{r}\right) + k^2\frac{e^{-jkr}}{r}\right]dV = -1 \tag{3.32}$$

To evaluate the first term, we use the divergence theorem to write

$$\int_V\nabla\cdot\nabla\left(\frac{e^{-jkr}}{r}\right)dV = \int_S\hat{\mathbf{n}}\cdot\nabla\left(\frac{e^{-jkr}}{r}\right)dS \tag{3.33}$$

On the sphere, $\hat{\mathbf{n}} = \hat{\mathbf{r}}$, therefore

$$\int_S \hat{\mathbf{r}} \cdot \nabla (\frac{e^{-jkr}}{r}) \, dS = \int_S \frac{\partial}{\partial r} (\frac{e^{-jkr}}{r}) \, dS \tag{3.34}$$

which is

$$4\pi a^2 \left[ \frac{\partial}{\partial r} \left( \frac{e^{-jkr}}{r} \right) \right]_{r=a} \tag{3.35}$$

Taking the limit as $a \to 0$ we get

$$\lim_{a \to 0} 4\pi a^2 \left[ \frac{\partial}{\partial r} \left( \frac{e^{-jkr}}{r} \right) \right]_{r=a} = -4\pi \tag{3.36}$$

The second term is

$$k^2 \int_0^a \frac{e^{-jkr}}{r} 4\pi r^2 \, dr = 4\pi k^2 \int_0^a r e^{-jkr} \, dr \tag{3.37}$$

By inspection, this integral tends to zero as $a \to 0$. Therefore,

$$A = \frac{1}{4\pi} \tag{3.38}$$

and

$$G(r) = \frac{e^{-jkr}}{4\pi r} \tag{3.39}$$

which is the electrodynamic Green's function in three dimensions.

### 3.3.2 Two-Dimensional Green's Function

Two-dimensional problems are those where the currents, surfaces, and fields extend to $\pm\infty$ along the $\hat{\mathbf{z}}$ axis. As such, there is no wave propagation in the $\hat{\mathbf{z}}$ direction. For these problems we will work the cylindrical coordinate system $(\rho, \phi, z)$ where $\boldsymbol{\rho} = \rho \cos\phi \, \hat{\mathbf{x}} + \rho \sin\phi \, \hat{\mathbf{y}}$, and surface integrals are performed in the $(x, y)$ plane $(z = 0)$. In two dimensions, (3.24) can be written as

$$\nabla^2 G(\boldsymbol{\rho}, \boldsymbol{\rho}') + k^2 G(\boldsymbol{\rho}, \boldsymbol{\rho}') = -\delta(\boldsymbol{\rho}, \boldsymbol{\rho}') \tag{3.40}$$

The solutions to the above for the homogeneous case are the Hankel functions of the first and second kinds of order zero. Since we know the solution to contain outgoing waves only, we then write

$$G(\rho) = AH_0^{(2)}(k\rho) \tag{3.41}$$

where $\rho = |\boldsymbol{\rho} - \boldsymbol{\rho}'|$. To obtain $A$, we use the small-argument approximation to the Hankel function [4]

$$H_0^{(2)}(k\rho) \approx 1 - j\frac{2}{\pi}\log\left(\frac{\gamma k\rho}{2}\right) \qquad \rho \to 0 \tag{3.42}$$

and integrate (3.40) over a very small circle of radius $a$ centered at the origin, yielding

$$A\int_S \left[\nabla^2 + k^2\right]\left[1 - j\frac{2}{\pi}\log\left(\frac{\gamma k\rho}{2}\right)\right]dS = -1 \tag{3.43}$$

Noting that $\nabla^2 f = \nabla \cdot (\nabla f)$, the first term can be converted to a line integral using the divergence theorem, yielding

$$-j\frac{2}{\pi}\int_0^{2\pi} \nabla\left[\log\left(\frac{\gamma k\rho}{2}\right)\right]\rho\,d\phi = -4j \tag{3.44}$$

The second term is

$$k^2\int_0^a \left[1 - j\frac{2}{\pi}\log\left(\frac{\gamma k\rho}{2}\right)\right]2\pi\rho\,d\rho \tag{3.45}$$

The first part of (3.45) goes to zero as $a \to 0$. Integrating the second part yields

$$4jk^2\int_0^a \log\left(\frac{\gamma k\rho}{2}\right)\rho\,d\rho = \left[\frac{\rho^2}{2}\log\left(\frac{\gamma k\rho}{2}\right) - \frac{\rho^2}{4}\right]\Bigg|_0^a \tag{3.46}$$

The above goes to zero since

$$\lim_{\rho\to 0} \rho^2\log\rho = 0 \tag{3.47}$$

Therefore,

$$A = -\frac{j}{4} \tag{3.48}$$

and

$$G(\rho) = -\frac{j}{4}H_0^{(2)}(k\rho) \tag{3.49}$$

which is the electrodynamic Green's function in two dimensions.

## 3.4 Vector Potentials

In Section 3.3 we derived expressions that allow us to determine the radiated field everywhere in space from electric and magnetic current distributions. In this section we will derive alternative expressions for the radiated fields in terms of vector potentials, also commonly used in radiation and scattering problems. We will then address the differences between the formulations in the following section.

### 3.4.1 Magnetic Vector Potential

We will first derive the magnetic vector potential for a homogeneous, source-free region. We begin by observing that since the magnetic field $\mathbf{H}$ is always solenoidal, it can be written as the curl of another vector $\mathbf{A}$, which is

$$\mathbf{H} = \frac{1}{\mu}\nabla \times \mathbf{A} \tag{3.50}$$

Substituting the above into (3.1), we get

$$\nabla \times \mathbf{E} = -j\omega\nabla \times \mathbf{A} \tag{3.51}$$

which can be written as

$$\nabla \times (\mathbf{E} + j\omega\mathbf{A}) = 0 \tag{3.52}$$

We next use the identity

$$\nabla \times (-\nabla\Phi_e) = 0 \tag{3.53}$$

to write

$$\mathbf{E} = -j\omega\mathbf{A} - \nabla\Phi_e \tag{3.54}$$

where $\Phi_e$ is an arbitrary electric *scalar potential*. We next use the identity

$$\nabla \times \nabla \times \mathbf{A} = \nabla\nabla \cdot \mathbf{A} - \nabla^2\mathbf{A} \tag{3.55}$$

and take the curl of both sides of (3.50) allowing us to write

$$\mu\nabla \times \mathbf{H} = \nabla\nabla \cdot \mathbf{A} - \nabla^2\mathbf{A} \tag{3.56}$$

and combining this with (3.2) leads to

$$\mu\mathbf{J} + j\omega\mu\epsilon\mathbf{E} = \nabla\nabla \cdot \mathbf{A} - \nabla^2\mathbf{A} \tag{3.57}$$

Substituting (3.54) into the above leads to

$$\mu\mathbf{J} + j\omega\mu\epsilon(-j\omega\mathbf{A} - \nabla\Phi_e) = \nabla\nabla \cdot \mathbf{A} - \nabla^2\mathbf{A} \tag{3.58}$$

which is

$$\nabla^2 \mathbf{A} + k^2 \mathbf{A} = -\mu \mathbf{J} + \nabla(\nabla \cdot \mathbf{A} + j\omega\mu\epsilon\Phi_e) \qquad (3.59)$$

We have already defined the curl of $\mathbf{A}$ in (3.50), however we have not yet defined its divergence. We are therefore free to make it whatever we wish as long as we remain consistent with that definition elsewhere. Therefore, if we make the following assignment

$$\Phi_e = -\frac{1}{j\omega\mu\epsilon}\nabla \cdot \mathbf{A} \qquad (3.60)$$

(3.59) simplifies to

$$\nabla^2 \mathbf{A} + k^2 \mathbf{A} = -\mu \mathbf{J} \qquad (3.61)$$

which is an inhomogeneous vector Helmholtz equation for $\mathbf{A}$. Using the Green's function, $\mathbf{A}(\mathbf{r})$ can be written as

$$\mathbf{A}(\mathbf{r}) = \mu \int_V G(\mathbf{r}, \mathbf{r}') \, \mathbf{J}(\mathbf{r}') \, d\mathbf{r}' \qquad (3.62)$$

and the corresponding electric field is then

$$\mathbf{E}(\mathbf{r}) = -j\omega\left[1 + \frac{1}{k^2}\nabla\nabla \cdot \right]\mathbf{A}(\mathbf{r}) \qquad (3.63)$$

### 3.4.1.1    Three-Dimensional Magnetic Vector Potential

Substitution of (3.39) into (3.62) yields the magnetic vector potential in three dimensions, which is

$$\mathbf{A}(\mathbf{r}) = \mu \int_V \mathbf{J}(\mathbf{r}') \frac{e^{-jk|\mathbf{r}-\mathbf{r}'|}}{4\pi|\mathbf{r}-\mathbf{r}'|} \, d\mathbf{r}' \qquad (3.64)$$

### 3.4.1.2    Two-Dimensional Magnetic Vector Potential

The corresponding magnetic vector potential in two dimensions is obtained via (3.49), and is

$$\mathbf{A}(\boldsymbol{\rho}) = -j\frac{\mu}{4} \int_S \mathbf{J}(\boldsymbol{\rho}')H_0^{(2)}(k|\boldsymbol{\rho}-\boldsymbol{\rho}'|) \, d\boldsymbol{\rho}' \qquad (3.65)$$

## 3.4.2    Electric Vector Potential

Using the symmetry of Maxwell's Equations, we can derive similar expressions for the electric vector potential $\mathbf{F}$. We will not repeat the derivation, and simply summarize these expressions below:

$$\mathbf{E} = -\frac{1}{\epsilon}\nabla \times \mathbf{F} \qquad (3.66)$$

$$\Phi_m = -\frac{1}{j\omega\mu\epsilon}\nabla\cdot\mathbf{F} \tag{3.67}$$

$$\nabla^2\mathbf{F} + k^2\mathbf{F} = -\epsilon\mathbf{M} \tag{3.68}$$

$$\mathbf{H}(\mathbf{r}) = -j\omega\left[1 + \frac{1}{k^2}\nabla\nabla\cdot\right]\mathbf{F}(\mathbf{r}) \tag{3.69}$$

$$\mathbf{F}(\mathbf{r}) = \epsilon\int_V \mathbf{M}(\mathbf{r}')\,G(\mathbf{r},\mathbf{r}')\,d\mathbf{r}' \tag{3.70}$$

### 3.4.2.1 Three-Dimensional Electric Vector Potential

Substitution of (3.39) into (3.70) yields the electric vector potential in three dimensions, which is

$$\mathbf{F}(\mathbf{r}) = \epsilon\int_V \mathbf{M}(\mathbf{r}')\,\frac{e^{-jk|\mathbf{r}-\mathbf{r}'|}}{4\pi|\mathbf{r}-\mathbf{r}'|}\,d\mathbf{r}' \tag{3.71}$$

### 3.4.2.2 Two-Dimensional Electric Vector Potential

Similarly, the two-dimensional electric vector potential is obtained via (3.49), and is

$$\mathbf{F}(\boldsymbol{\rho}) = -j\frac{\epsilon}{4}\int_S \mathbf{M}(\boldsymbol{\rho}')H_0^{(2)}(k|\boldsymbol{\rho}-\boldsymbol{\rho}'|)\,d\boldsymbol{\rho}' \tag{3.72}$$

## 3.4.3 Total Fields

Using the results from Sections 3.4.1 and 3.4.2, we can write the total electric field $\mathbf{E}(\mathbf{r})$ as

$$\mathbf{E}(\mathbf{r}) = -j\omega\left[\mathbf{A}(\mathbf{r}) + \frac{1}{k^2}\nabla\nabla\cdot\mathbf{A}(\mathbf{r})\right] - \frac{1}{\epsilon}\nabla\times\mathbf{F}(\mathbf{r}) \tag{3.73}$$

and the magnetic field $\mathbf{H}(\mathbf{r})$ as

$$\mathbf{H}(\mathbf{r}) = -j\omega\left[\mathbf{F}(\mathbf{r}) + \frac{1}{k^2}\nabla\nabla\cdot\mathbf{F}(\mathbf{r})\right] + \frac{1}{\mu}\nabla\times\mathbf{A}(\mathbf{r}) \tag{3.74}$$

In *operator notation*, we can rewrite (3.73) and (3.74) as

$$\mathbf{E}(\mathbf{r}) = -j\omega\mu(\mathcal{L}\mathbf{J})(\mathbf{r}) - (\mathcal{K}\mathbf{M})(\mathbf{r}) \tag{3.75}$$

and

$$\mathbf{H}(\mathbf{r}) = -j\omega\epsilon(\mathcal{L}\mathbf{M})(\mathbf{r}) + (\mathcal{K}\mathbf{J})(\mathbf{r}) \tag{3.76}$$

where the operators $\mathcal{L}$ and $\mathcal{K}$ are

$$(\mathcal{L}\mathbf{X})(\mathbf{r}) = \left[ 1 + \frac{1}{k^2}\nabla\nabla \cdot \right] \int_V G(\mathbf{r}, \mathbf{r}')\mathbf{X}(\mathbf{r}') \, d\mathbf{r}' \qquad (3.77)$$

and

$$(\mathcal{K}\mathbf{X})(\mathbf{r}) = \nabla \times \int_V G(\mathbf{r}, \mathbf{r}')\mathbf{X}(\mathbf{r}') \, d\mathbf{r}' \qquad (3.78)$$

### 3.4.4   Comparison of Radiation Formulas

The expressions for the electric and magnetic fields in (3.73) and (3.74) are almost identical to those of (3.25) and (3.26) except that the differential operators are distributed differently. It is important that we show their equivalence, as the reader will likely encounter both forms in the literature, as well as expressions where the operators are distributed differently than shown in this book. We will also use techniques from this section to redistribute differential operators in cases where it simplifies evaluation of the MOM matrix elements. To demonstrate this equivalence, we must show that

$$\mathbf{I}_1(\mathbf{r}) = \nabla\nabla \cdot \int_V G(\mathbf{r}, \mathbf{r}')\, \mathbf{X}(\mathbf{r}') \, d\mathbf{r}' = \int_V G(\mathbf{r}, \mathbf{r}')\nabla'\nabla' \cdot \mathbf{X}(\mathbf{r}') \, d\mathbf{r}' \quad (3.79)$$

and

$$\mathbf{I}_2(\mathbf{r}) = \nabla \times \int_V G(\mathbf{r}, \mathbf{r}')\, \mathbf{X}(\mathbf{r}') \, d\mathbf{r}' = \int_V G(\mathbf{r}, \mathbf{r}')\nabla' \times \mathbf{X}(\mathbf{r}') \, d\mathbf{r}' \quad (3.80)$$

Let us begin with $\mathbf{I}_1(\mathbf{r})$. Using the vector identity

$$\nabla \cdot \left[ G(\mathbf{r}, \mathbf{r}')\, \mathbf{X}(\mathbf{r}') \right] = \left[ \nabla G(\mathbf{r}, \mathbf{r}') \right] \cdot \mathbf{X}(\mathbf{r}') + G(\mathbf{r}, \mathbf{r}')\nabla \cdot \mathbf{X}(\mathbf{r}') \qquad (3.81)$$

and the fact that $\mathbf{X}(\mathbf{r}')$ is not a function of the unprimed coordinates, we can write

$$\mathbf{I}_1(\mathbf{r}) = \nabla\nabla \cdot \int_V G(\mathbf{r}, \mathbf{r}')\, \mathbf{X}(\mathbf{r}') \, d\mathbf{r}' = \nabla \int_V \left[ \nabla G(\mathbf{r}, \mathbf{r}') \right] \cdot \mathbf{X}(\mathbf{r}') \, d\mathbf{r}' \quad (3.82)$$

Using the symmetry of the Green's function,

$$\nabla G(\mathbf{r}, \mathbf{r}') = -\nabla' G(\mathbf{r}, \mathbf{r}') \qquad (3.83)$$

the above becomes

$$\mathbf{I}_1(\mathbf{r}) = -\nabla \int_V \left[ \nabla' G(\mathbf{r}, \mathbf{r}') \right] \mathbf{X}(\mathbf{r}') \, d\mathbf{r}' \qquad (3.84)$$

Using the previous vector identity we can write the above as a sum of two integrals

$$\mathbf{I}_1(\mathbf{r}) = \nabla \int_V G(\mathbf{r}, \mathbf{r}') \nabla' \cdot \mathbf{X}(\mathbf{r}') \, d\mathbf{r}' - \nabla \int_V \nabla' \cdot \left[ G(\mathbf{r}, \mathbf{r}') \, \mathbf{X}(\mathbf{r}') \right] \, d\mathbf{r}' \quad (3.85)$$

We make use of the divergence theorem to write the second term as

$$\mathbf{I}_1(\mathbf{r}) = \nabla \int_V G(\mathbf{r}, \mathbf{r}') \nabla' \cdot \mathbf{X}(\mathbf{r}') \, d\mathbf{r}' - \nabla \oint_S \hat{\mathbf{n}} \cdot \left[ G(\mathbf{r}, \mathbf{r}') \, \mathbf{X}(\mathbf{r}') \right] \, d\mathbf{r}' \quad (3.86)$$

Assuming $\mathbf{X}(\mathbf{r}')$ is of finite extent and completely inside $V$, it will be zero on $S$. Thus, the second integral is zero and we are left with

$$\mathbf{I}_1(\mathbf{r}) = \nabla \int_V G(\mathbf{r}, \mathbf{r}') \nabla' \cdot \mathbf{X}(\mathbf{r}') \, d\mathbf{r}' \quad (3.87)$$

We must now move the remaining gradient operator under the integral sign, and doing so results in

$$\mathbf{I}_1(\mathbf{r}) = - \int_V \nabla' G(\mathbf{r}, \mathbf{r}') \nabla' \cdot \mathbf{X}(\mathbf{r}') \, d\mathbf{r}' \quad (3.88)$$

where we have again employed the symmetry of the Green's function. We must now move the gradient so that it operates on the $\nabla' \cdot \mathbf{X}(\mathbf{r}')$ term. To do so, we use the vector identity

$$\nabla' G(\mathbf{r}, \mathbf{r}') \nabla' \cdot \mathbf{X}(\mathbf{r}') = \nabla' \left[ G(\mathbf{r}, \mathbf{r}') \nabla' \cdot \mathbf{X}(\mathbf{r}') \right] - G(\mathbf{r}, \mathbf{r}') \nabla' \nabla' \cdot \mathbf{X}(\mathbf{r}') \quad (3.89)$$

to write

$$\mathbf{I}_1(\mathbf{r}) = \int_V G(\mathbf{r}, \mathbf{r}') \nabla' \nabla' \cdot \mathbf{X}(\mathbf{r}') \, d\mathbf{r}' - \int_V \nabla' \left[ G(\mathbf{r}, \mathbf{r}') \nabla' \cdot \mathbf{X}(\mathbf{r}') \right] \, d\mathbf{r}' \quad (3.90)$$

To show that the second term is zero, we apply the following result of the divergence theorem

$$\int_V \nabla f(\mathbf{r}) \, dV = \oint_S f(\mathbf{r}) \, d\mathbf{S} \quad (3.91)$$

to write the above as

$$\mathbf{I}_1(\mathbf{r}) = \int_V G(\mathbf{r}, \mathbf{r}') \nabla' \nabla' \cdot \mathbf{X}(\mathbf{r}') \, d\mathbf{r}' - \oint_S G(\mathbf{r}, \mathbf{r}') \nabla' \cdot \mathbf{X}(\mathbf{r}') \, d\mathbf{r}' \quad (3.92)$$

Again making $V$ large enough such that $\mathbf{X}(\mathbf{r}')$ has zero value on $S$, we are left with

$$\mathbf{I}_1(\mathbf{r}) = \int_V G(\mathbf{r}, \mathbf{r}') \nabla' \nabla' \cdot \mathbf{X}(\mathbf{r}') \, d\mathbf{r}' \quad (3.93)$$

which is the expected result. We next focus our attention on $\mathbf{I}_2(\mathbf{r})$. Using the vector identity

$$\nabla \times \left[ G(\mathbf{r}, \mathbf{r}') \, \mathbf{X}(\mathbf{r}') \right] = G(\mathbf{r}, \mathbf{r}') \nabla \times \mathbf{X}(\mathbf{r}') + \nabla G(\mathbf{r}, \mathbf{r}') \times \mathbf{X}(\mathbf{r}') \quad (3.94)$$

we can write

$$\nabla \times \int_V G(\mathbf{r}, \mathbf{r}') \, \mathbf{X}(\mathbf{r}') \, d\mathbf{r}' = \int_V G(\mathbf{r}, \mathbf{r}') \nabla \times \mathbf{X}(\mathbf{r}') \, d\mathbf{r}' + \int_V \nabla G(\mathbf{r}, \mathbf{r}') \times \mathbf{X}(\mathbf{r}') \, d\mathbf{r}'$$
$$(3.95)$$

Noting that $\nabla G(\mathbf{r}, \mathbf{r}') = -\nabla' G(\mathbf{r}, \mathbf{r}')$ and $\nabla \times \mathbf{X}(\mathbf{r}') = 0$, we write the above as

$$\nabla \times \int_V G(\mathbf{r}, \mathbf{r}') \, \mathbf{X} \, d\mathbf{r}' = - \int_V \nabla' G(\mathbf{r}, \mathbf{r}') \times \mathbf{X}(\mathbf{r}') \, d\mathbf{r}' \quad (3.96)$$

Again using the identity (3.94) but on primed coordinates, we write

$$-\int_V \nabla' G(\mathbf{r}, \mathbf{r}') \times \mathbf{X}(\mathbf{r}') \, d\mathbf{r}' = \int_V G(\mathbf{r}, \mathbf{r}') \nabla' \times \mathbf{X}(\mathbf{r}') \, d\mathbf{r}'$$
$$-\int_V \nabla' \times \left[ G(\mathbf{r}, \mathbf{r}') \, \mathbf{X}(\mathbf{r}') \right] d\mathbf{r}' \quad (3.97)$$

and using the following result of the divergence theorem

$$\int_V \nabla \times \mathbf{X}(\mathbf{r}) \, dV = \oint_S \hat{\mathbf{n}}(\mathbf{r}) \times \mathbf{X}(\mathbf{r}) \, dS \quad (3.98)$$

we can write (3.4.4) as

$$-\int_V \nabla' G(\mathbf{r}, \mathbf{r}') \times \mathbf{X}(\mathbf{r}') \, d\mathbf{r}' = \int_V G(\mathbf{r}, \mathbf{r}') \nabla' \times \mathbf{X}(\mathbf{r}') \, d\mathbf{r}'$$
$$-\oint_S \hat{\mathbf{n}} \times \left[ G(\mathbf{r}, \mathbf{r}') \, \mathbf{X}(\mathbf{r}') \right] d\mathbf{r}' \quad (3.99)$$

Making $V$ large enough such that $\mathbf{X}(\mathbf{r}')$ has zero value on $S$, the second term on the right hand side of (3.4.4) goes to zero and we are left with

$$\nabla \times \int_V G(\mathbf{r}, \mathbf{r}') \, \mathbf{X} \, d\mathbf{r}' = \int_V G(\mathbf{r}, \mathbf{r}') \nabla' \times \mathbf{X}(\mathbf{r}') \, d\mathbf{r}' \quad (3.100)$$

which is again the expected result.

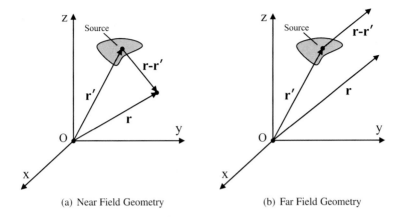

(a) Near Field Geometry          (b) Far Field Geometry

**FIGURE 3.1:** Near and Far Field Geometry

## 3.5 Near and Far Field

The radiation equations presented thus far are general formulas that can be used to compute the electric and magnetic fields anywhere. Usually though, one is interested in the fields in one of two regions: the *near field* region or the *far field* region. The near field region is located close to the source, and the fields there comprise a superposition of stored (reactive) as well as radiated power. The near field region is of interest in applications such as antenna design, electromagnetic interference (EMI), and packaging of electronic components. The far field region is that which is far away from the source, where the fields have transitioned into purely radiative plane waves. In this section we will derive expressions for the near and far fields in two and three dimensions.

### 3.5.1 Three-Dimensional Near Field

Let us obtain expressions for the three-dimensional near fields at a point close to a source, as illustrated in Figure 3.1a. Recall that the magnetic field radiated by an electric current can be obtained from (3.50) and (3.62) and is

$$\mathbf{H}_A(\mathbf{r}) = \frac{1}{\mu} \nabla \times \mathbf{A}(\mathbf{r}) = \nabla \times \int_V \mathbf{J}(\mathbf{r}') \, G(\mathbf{r}, \mathbf{r}') \, d\mathbf{r}' \qquad (3.101)$$

Moving the curl operator under the integral sign and using the vector identity (3.94) and the fact that $\nabla \times \mathbf{J}(\mathbf{r}') = 0$, we write

$$\mathbf{H}_A(\mathbf{r}) = \int_V \nabla G(\mathbf{r}, \mathbf{r}') \times \mathbf{J}(\mathbf{r}') \, d\mathbf{r}' \qquad (3.102)$$

Taking the gradient of the 3D Green's function (3.39) yields

$$\nabla\left(\frac{e^{-jkr}}{4\pi r}\right) = -(\mathbf{r} - \mathbf{r}')\left(\frac{1+jkr}{4\pi r^3}\right)e^{-jkr} \tag{3.103}$$

and using this result we can then write the magnetic field as

$$\mathbf{H}_A(\mathbf{r}) = -\frac{1}{4\pi}\int_V \left[(\mathbf{r}-\mathbf{r}') \times \mathbf{J}(\mathbf{r}')\right]\frac{1+jkr}{r^3}e^{-jkr}\,d\mathbf{r}' \tag{3.104}$$

Expanding into Cartesian components, the above can be written as

$$H_{A,x}(\mathbf{r}) = \frac{1}{4\pi}\int_V \left[\Delta_z J_y - \Delta_y J_z\right]\frac{1+jkr}{r^3}e^{-jkr}\,d\mathbf{r}'$$

$$H_{A,y}(\mathbf{r}) = \frac{1}{4\pi}\int_V \left[\Delta_x J_z - \Delta_z J_x\right]\frac{1+jkr}{r^3}e^{-jkr}\,d\mathbf{r}' \tag{3.105}$$

$$H_{A,z}(\mathbf{r}) = \frac{1}{4\pi}\int_V \left[\Delta_y J_x - \Delta_x J_y\right]\frac{1+jkr}{r^3}e^{-jkr}\,d\mathbf{r}'$$

where

$$\Delta_x = x - x' \qquad \Delta_y = y - y' \qquad \Delta_z = z - z' \tag{3.106}$$

and

$$r = |\mathbf{r} - \mathbf{r}'| = \sqrt{(x-x')^2 + (y-y')^2 + (z-z')^2} \tag{3.107}$$

Using (3.2), the corresponding electric field is

$$E_{A,x}(\mathbf{r}) = -\frac{j}{4\pi\omega\epsilon}\int_V \left(2C_1 J_x + C_2\left[\Delta_x(\Delta_z J_z + \Delta_y J_y) - J_x(\Delta_y^2 + \Delta_z^2)\right]\right)d\mathbf{r}'$$

$$E_{A,y}(\mathbf{r}) = -\frac{j}{4\pi\omega\epsilon}\int_V \left(2C_1 J_y + C_2\left[\Delta_y(\Delta_x J_x + \Delta_z J_z) - J_y(\Delta_z^2 + \Delta_x^2)\right]\right)d\mathbf{r}'$$

$$E_{A,z}(\mathbf{r}) = -\frac{j}{4\pi\omega\epsilon}\int_V \left(2C_1 J_z + C_2\left[\Delta_z(\Delta_y J_y + \Delta_x J_x) - J_z(\Delta_x^2 + \Delta_y^2)\right]\right)d\mathbf{r}'$$

$$\tag{3.108}$$

where

$$C_1 = \frac{1+1jkr}{r^3}e^{-jkr} \tag{3.109}$$

and

$$C_2 = \frac{3+3jkr-k^2r^2}{r^5}e^{-jkr} \tag{3.110}$$

Similarly, the electric field radiated by a magnetic current is

$$\mathbf{E}_F(\mathbf{r}) = -\frac{1}{\epsilon}\nabla \times \mathbf{F}(\mathbf{r}) = -\nabla \times \int_V \mathbf{M}(\mathbf{r}')\,G(\mathbf{r},\mathbf{r}')\,d\mathbf{r}' \tag{3.111}$$

and its rectangular components are

$$E_{F,x}(\mathbf{r}) = -\frac{1}{4\pi} \int_V \left[\Delta_z M_y - \Delta_y M_z\right] \frac{1 + jkr}{r^3} e^{-jkr} \, d\mathbf{r}'$$

$$E_{F,y}(\mathbf{r}) = -\frac{1}{4\pi} \int_V \left[\Delta_x M_z - \Delta_z M_x\right] \frac{1 + jkr}{r^3} e^{-jkr} \, d\mathbf{r}' \qquad (3.112)$$

$$E_{F,z}(\mathbf{r}) = -\frac{1}{4\pi} \int_V \left[\Delta_y M_x - \Delta_x M_y\right] \frac{1 + jkr}{r^3} e^{-jkr} \, d\mathbf{r}'$$

and using (3.1), the corresponding magnetic field is

$$H_{F,x}(\mathbf{r}) = -\frac{j}{4\pi\omega\mu} \int_V \left(C_1 M_x + C_2\left[\Delta_x(\Delta_z M_z + \Delta_y M_y) - M_x(\Delta_y^2 + \Delta_z^2)\right]\right) d\mathbf{r}'$$

$$H_{F,y}(\mathbf{r}) = -\frac{j}{4\pi\omega\mu} \int_V \left(C_1 M_y + C_2\left[\Delta_y(\Delta_x M_x + \Delta_z M_z) - M_y(\Delta_z^2 + \Delta_x^2)\right]\right) d\mathbf{r}'$$

$$H_{F,z}(\mathbf{r}) = -\frac{j}{4\pi\omega\mu} \int_V \left(C_1 M_z + C_2\left[\Delta_z(\Delta_y M_y + \Delta_x M_x) - M_z(\Delta_x^2 + \Delta_y^2)\right]\right) d\mathbf{r}'$$

$$(3.113)$$

Note that terms $C_1$ and $C_2$ in these equations become singular for field points located very close to the source, and numerical integration algorithms alone may yield inaccurate results. In such cases, a more advanced singularity treatment may be needed, such as those outlined in [5] and [6] for current distributions on planar triangles.

### 3.5.2 Two-Dimensional Near Field

Following (3.101) and using the two-dimensional Green's function (3.49), the magnetic field $\mathbf{H}(\boldsymbol{\rho})$ can be written as

$$\mathbf{H}_A(\boldsymbol{\rho}) = \frac{1}{\mu} \nabla \times \mathbf{A}(\boldsymbol{\rho}) = -\frac{j}{4} \int_S \nabla H_0^{(2)}(k\rho) \times \mathbf{J}(\boldsymbol{\rho}') \, d\boldsymbol{\rho}' \qquad (3.114)$$

The gradient of the Hankel function in cylindrical coordinates is [4]

$$\nabla H_0^{(2)}(k\rho) = -(\boldsymbol{\rho} - \boldsymbol{\rho}')\frac{k}{\rho} H_1^{(2)}(k\rho) \qquad (3.115)$$

which results in

$$\mathbf{H}_A(\boldsymbol{\rho}) = j\frac{k}{4} \int_S (\boldsymbol{\rho} - \boldsymbol{\rho}') \times \mathbf{J}(\boldsymbol{\rho}') \frac{H_1^{(2)}(k\rho)}{\rho} \, d\boldsymbol{\rho}' \qquad (3.116)$$

Expanding into Cartesian components, the above can be written as

$$H_{A,x}(\boldsymbol{\rho}) = j\frac{k}{4}\int_S \left[\Delta_y J_z\right]\frac{H_1^{(2)}(k\rho)}{\rho}\,d\boldsymbol{\rho}'$$

$$H_{A,y}(\boldsymbol{\rho}) = j\frac{k}{4}\int_S \left[-\Delta_x J_z\right]\frac{H_1^{(2)}(k\rho)}{\rho}\,d\boldsymbol{\rho}' \qquad (3.117)$$

$$H_{A,z}(\boldsymbol{\rho}) = j\frac{k}{4}\int_S \left[\Delta_x J_y - \Delta_y J_x\right]\frac{H_1^{(2)}(k\rho)}{\rho}\,d\boldsymbol{\rho}'$$

where

$$\Delta_x = x - x' \qquad \Delta_y = y - y' \qquad (3.118)$$

and

$$\rho = |\boldsymbol{\rho} - \boldsymbol{\rho}'| = \sqrt{(x - x')^2 + (y - y')^2} \qquad (3.119)$$

Using (3.2), the corresponding electric field is

$$E_{A,x}(\boldsymbol{\rho}) = \frac{k^2}{8\omega\epsilon}\int_S \left[\left([\Delta_y^2 - \Delta_x^2]J_x - 2\Delta_x\Delta_y J_y\right)\frac{H_2^{(2)}(k\rho)}{\rho^2} - J_x H_0^{(2)}(k\rho)\right]d\boldsymbol{\rho}'$$

$$E_{A,y}(\boldsymbol{\rho}) = \frac{k^2}{8\omega\epsilon}\int_S \left[\left([\Delta_x^2 - \Delta_y^2]J_y - 2\Delta_x\Delta_y J_x\right)\frac{H_2^{(2)}(k\rho)}{\rho^2} - J_y H_0^{(2)}(k\rho)\right]d\boldsymbol{\rho}'$$

$$E_{A,z}(\boldsymbol{\rho}) = -\frac{k^2}{4\omega\epsilon}\int_S J_z H_0^{(2)}(k\rho)\,d\boldsymbol{\rho}'$$

$$(3.120)$$

Similarly, the electric field radiated by a magnetic current is

$$\mathbf{E}_F(\boldsymbol{\rho}) = -j\frac{k}{4}\int_S (\boldsymbol{\rho} - \boldsymbol{\rho}') \times \mathbf{M}(\boldsymbol{\rho}')\frac{H_1^{(2)}(k\rho)}{\rho}\,d\boldsymbol{\rho}' \qquad (3.121)$$

and its rectangular components are

$$E_{F,x}(\boldsymbol{\rho}) = -j\frac{k}{4}\int_S \left[\Delta_y M_z\right]\frac{H_1^{(2)}(k\rho)}{\rho}\,d\boldsymbol{\rho}'$$

$$E_{F,y}(\boldsymbol{\rho}) = -j\frac{k}{4}\int_S \left[-\Delta_x M_z\right]\frac{H_1^{(2)}(k\rho)}{\rho}\,d\boldsymbol{\rho}' \qquad (3.122)$$

$$E_{F,z}(\boldsymbol{\rho}) = -j\frac{k}{4}\int_S \left[\Delta_x M_y - \Delta_y M_x\right]\frac{H_1^{(2)}(k\rho)}{\rho}\,d\boldsymbol{\rho}'$$

The corresponding magnetic field is

$$H_{A,x}(\boldsymbol{\rho}) = \frac{k^2}{8\omega\mu} \int_S \left[ \left( [\Delta_y^2 - \Delta_x^2]M_x - 2\Delta_x\Delta_y M_y \right) \frac{H_2^{(2)}(k\rho)}{\rho^2} - M_x H_0^{(2)}(k\rho) \right] d\boldsymbol{\rho}'$$

$$H_{A,y}(\boldsymbol{\rho}) = \frac{k^2}{8\omega\mu} \int_S \left[ \left( [\Delta_x^2 - \Delta_y^2]M_y - 2\Delta_x\Delta_y M_x \right) \frac{H_2^{(2)}(k\rho)}{\rho^2} - M_y H_0^{(2)}(k\rho) \right] d\boldsymbol{\rho}'$$

$$H_{A,z}(\boldsymbol{\rho}) = -\frac{k^2}{4\omega\mu} \int_S M_z H_0^{(2)}(k\rho) \, d\boldsymbol{\rho}'$$

$$(3.123)$$

### 3.5.3 Three-Dimensional Far Field

When the field point is located very far away from the source, the vectors $\mathbf{r}$ and $\mathbf{r} - \mathbf{r}'$ are virtually parallel, as shown in Figure 3.1b. As a general rule of thumb, this is typically satisfied when $kr \gg 1$. In this case, the fields are virtually planar, and approximations are made that simplify their calculation. Under this assumption, we make the following approximations for $r$ in the far field region [1]

$$r = \begin{cases} r - \mathbf{r}' \cdot \hat{\mathbf{r}} & \text{for phase variations} \\ r & \text{for amplitude variations} \end{cases} \qquad (3.124)$$

Consider the total electric field as expressed in (3.73). We first focus on the components of $\mathbf{E}(\mathbf{r})$ due to $\mathbf{A}(\mathbf{r})$, which are

$$\mathbf{E}_A(\mathbf{r}) = -j\omega \left[ \mathbf{A}(\mathbf{r}) + \frac{1}{k^2}\nabla\nabla \cdot \mathbf{A}(\mathbf{r}) \right] \qquad (3.125)$$

The first term on the right-hand side of (3.125) will result in fields that vary according to $1/r$, and the second term in fields that vary according to $1/r^2, 1/r^3$, etc., due to the differential operators. In the far field, we know that only those fields that vary according to $1/r$ are of significant amplitude, and that these components are planar with no components along the direction of propagation. The far electric field $\mathbf{E}_A(\mathbf{r})$ is therefore

$$\mathbf{E}_A(\mathbf{r}) = -j\omega\mathbf{A}(\mathbf{r}) \qquad (3.126)$$

and the far magnetic field $\mathbf{H}_F(\mathbf{r})$ is

$$\mathbf{H}_F(\mathbf{r}) = -j\omega\mathbf{F}(\mathbf{r}) \qquad (3.127)$$

where $\mathbf{E}_A(\mathbf{r})$ and $\mathbf{H}_F(\mathbf{r})$ can be written in terms of their $\hat{\theta}$ and $\hat{\phi}$ components as

$$\mathbf{E}_A(\mathbf{r}) = -j\omega \left[ A_\theta(\mathbf{r})\hat{\theta} + A_\phi(\mathbf{r})\hat{\phi} \right] \qquad (3.128)$$

$$\mathbf{H}_F(\mathbf{r}) = -j\omega\left[F_\theta(\mathbf{r})\hat{\theta} + F_\phi(\mathbf{r})\hat{\phi}\right] \tag{3.129}$$

To obtain the far electric field $\mathbf{E}_F(\mathbf{r})$ due to $\mathbf{F}(\mathbf{r})$ we can use (3.66), but instead we will use the fact that for plane waves, the electric and magnetic field are related via the relationship

$$\mathbf{H}(\mathbf{r}) = \frac{1}{\eta}\,\hat{\mathbf{r}} \times \mathbf{E}(\mathbf{r}) \tag{3.130}$$

where $\hat{\mathbf{r}}$ is the direction of propagation and $\eta = \sqrt{\mu/\epsilon}$ is the impedance of the dielectric medium. Using this relationship we can then write

$$\mathbf{E}_F(\mathbf{r}) = -j\omega\eta\left[F_\phi(\mathbf{r})\hat{\theta} - F_\theta(\mathbf{r})\hat{\phi}\right] \tag{3.131}$$

and

$$\mathbf{H}_A(\mathbf{r}) = -j\frac{\omega}{\eta}\left[-A_\phi(\mathbf{r})\hat{\theta} + A_\theta(\mathbf{r})\hat{\phi}\right] \tag{3.132}$$

The total far fields are then

$$\mathbf{E}(\mathbf{r}) = -j\omega\left(\left[A_\theta(\mathbf{r}) + \eta F_\phi(\mathbf{r})\right]\hat{\theta} + \left[A_\phi(\mathbf{r}) - \eta F_\theta(\mathbf{r})\right]\hat{\phi}\right) \tag{3.133}$$

and

$$\mathbf{H}(\mathbf{r}) = -j\frac{\omega}{\eta}\left(\left[-A_\phi(\mathbf{r}) + \eta F_\theta(\mathbf{r})\right]\hat{\theta} + \left[A_\theta(\mathbf{r}) + \eta F_\phi(\mathbf{r})\right]\hat{\phi}\right) \tag{3.134}$$

where in the far field, $\mathbf{A}(\mathbf{r})$ and $\mathbf{F}(\mathbf{r})$ can be written as

$$\mathbf{A}(\mathbf{r}) = \mu\frac{e^{-jkr}}{4\pi r}\int_V \mathbf{J}(\mathbf{r}')e^{jkr'\cdot\hat{\mathbf{r}}}\,d\mathbf{r}' \tag{3.135}$$

and

$$\mathbf{F}(\mathbf{r}) = \epsilon\frac{e^{-jkr}}{4\pi r}\int_V \mathbf{M}(\mathbf{r}')e^{jkr'\cdot\hat{\mathbf{r}}}\,d\mathbf{r}' \tag{3.136}$$

For scattering problems, where the currents $\mathbf{J}$ and $\mathbf{M}$ induced by the incident fields $\mathbf{E}^i$ and $\mathbf{H}^i$ generate the scattered far fields $\mathbf{E}^s$ and $\mathbf{H}^s$, the three-dimensional radar cross section $\sigma_{3D}$ is defined as [1]

$$\sigma_{3D} = 4\pi r^2 \frac{|\mathbf{E}^s|^2}{|\mathbf{E}^i|^2} \tag{3.137}$$

It is common in the literature to see $\hat{\theta}$ polarization referred to as *vertical* polarization and $\hat{\phi}$ polarization as *horizontal* polarization. We will use this nomenclature when analyzing the RCS of three-dimensional objects in later chapters.

### 3.5.4 Two-Dimensional Far Field

To determine the far field expressions for $\mathbf{A}(\boldsymbol{\rho})$ and $\mathbf{F}(\boldsymbol{\rho})$, we use the large-argument approximation of the Hankel function [4]

$$H_n^{(2)}(k\rho) \approx \sqrt{\frac{2j}{\pi k \rho}}\, j^n e^{-jk\rho} \qquad \rho \to \infty \qquad (3.138)$$

and the two-dimensional version of (3.124) to write

$$H_0^{(2)}(k\rho) = \sqrt{\frac{2j}{\pi k \rho}}\, e^{-jk\rho} e^{jk\boldsymbol{\rho}' \cdot \hat{\rho}} \qquad (3.139)$$

which yields

$$\mathbf{A}(\boldsymbol{\rho}) = -j\mu \sqrt{\frac{j}{8\pi k}}\, \frac{e^{-jk\rho}}{\sqrt{\rho}} \int_S \mathbf{J}(\boldsymbol{\rho}')\, e^{jk\boldsymbol{\rho}' \cdot \hat{\rho}}\, d\boldsymbol{\rho}' \qquad (3.140)$$

and

$$\mathbf{F}(\boldsymbol{\rho}) = -j\epsilon \sqrt{\frac{j}{8\pi k}}\, \frac{e^{-jk\rho}}{\sqrt{\rho}} \int_S \mathbf{M}(\boldsymbol{\rho}')\, e^{jk\boldsymbol{\rho}' \cdot \hat{\rho}}\, d\boldsymbol{\rho}' \qquad (3.141)$$

In the two-dimensional case, we know that the total fields $\mathbf{E}(\boldsymbol{\rho})$ and $\mathbf{H}(\boldsymbol{\rho})$ comprise $\hat{\mathbf{z}}$ and $\hat{\phi}$ components only. Following similar reasoning as in the three-dimensional case, we can then write the far electric and magnetic fields as

$$\mathbf{E}(\boldsymbol{\rho}) = -j\omega \left( \left[ A_\phi(\boldsymbol{\rho}) + \eta F_z(\boldsymbol{\rho}) \right] \hat{\phi} + \left[ A_z(\boldsymbol{\rho}) - \eta F_\phi(\boldsymbol{\rho}) \right] \hat{\mathbf{z}} \right) \qquad (3.142)$$

and

$$\mathbf{H}(\boldsymbol{\rho}) = -j\frac{\omega}{\eta} \left( \left[ -A_z(\boldsymbol{\rho}) + \eta F_\phi(\boldsymbol{\rho}) \right] \hat{\phi} + \left[ A_\phi(\boldsymbol{\rho}) + \eta F_z(\boldsymbol{\rho}) \right] \hat{\mathbf{z}} \right) \qquad (3.143)$$

Similarly, the two-dimensional radar cross section $\sigma_{2D}$ is defined as [1]

$$\sigma_{2D} = 2\pi\rho \frac{|\mathbf{E}^s|^2}{|\mathbf{E}^i|^2} \qquad (3.144)$$

## 3.6 Formulations for Scattering

Scattering problems can be treated as radiation problems where the radiating currents are themselves generated by other currents or fields. For example, to determine the radiation pattern of an antenna, we compute the radiated field due to the current distribution on the antenna. These currents, however, are generated by other sources, such as a voltage applied at the antenna's feedpoint or the fields radiated by other antennas nearby. Computation of an object's radar cross section (RCS) is another common example of this problem. In this case, the scattering object (or *target*) is in the far field of an external antenna, and induced currents on the scatterer radiate a secondary or *scattered* field.

Strictly speaking, radiation problems require the integration of known currents $\mathbf{J}$ and $\mathbf{M}$ to obtain the fields by way of either (3.25) and (3.26) or (3.73) and (3.74). Though in some cases the expressions for the currents may be known *a priori*, most practical problems of interest have no analytic solution and the currents must be determined numerically. Thus, these problems usually involve three steps:

1. Formulating an integral equation for the unknown currents $\mathbf{J}$ and $\mathbf{M}$ generated by known fields $\mathbf{E}^i$ or $\mathbf{H}^i$

2. Solving the integral equation

3. Integrating the currents $\mathbf{J}$ and $\mathbf{M}$ to obtain the radiated or scattered fields $\mathbf{E}^s$ and $\mathbf{H}^s$.

The last step is achieved using the methods outlined in Section 3.5 and is relatively straightforward once $\mathbf{J}$ or $\mathbf{M}$ are known. However, the first two steps must be considered carefully, and the remainder of this book is dedicated to this task. We will first review the equivalence principle in Section 3.6.1 and use it to reformulate the scattering problem using equivalent surface currents. We will then develop surface integral equations of scattering in Section 3.6.2. We will then apply these integral equations to problems of interest in subsequent chapters, which will be solved numerically using the Method of Moments.

### 3.6.1 Surface Equivalent

The expressions for the radiated fields in (3.73) and (3.74) assume that the currents $\mathbf{J}$ and $\mathbf{M}$ radiate in a homogeneous dielectric region with parameters $(\mu, \epsilon)$. If we now introduce an obstacle into this region with different material properties, we can no longer use (3.73) and (3.74) to determine the radiated fields. This is a serious problem, as virtually all problems of practical interest involve currents radiating in the presence of an obstacle. Fortunately, it is possible to reformulate problems like these in terms of an *equivalent* problem that is more convenient to solve in the region of interest. By introducing certain fictitious electric and magnetic currents, we can generate a set of coupled integral equations that effectively eliminate the obstacle. Once the equations are solved for these currents, it is then easy to determine the field in each region. As we treat conducting, homogeneous dielectric and composite conducting/dielectric objects in this book, we will formulate our equivalent problem in terms of surface currents that exist on the interfaces between dielectric and conducting regions.

The Surface Equivalence Theorem (also called *Huygen's Principle*) states that every point on an advancing wavefront is itself a source of radiated waves. Using this theorem, a radiating source can be replaced by a fictitious set of different but *equivalent* sources that lie on an arbitrary closed surface around the original source. By matching the appropriate boundary conditions, these currents will generate the *same* radiated field outside the closed surface as the original sources. The theorem allows us to determine the far fields radiated by a structure if the near fields are known, or to formulate a surface integral equation that can be solved for the currents induced on an object by an externally incident field.

To introduce the concept, we will develop the integral equations for a small, conceptual scattering problem, and generalize them to general problems later. To begin, consider the geometry of Figure 3.2a, where the electric and magnetic currents $\mathbf{J}$ and $\mathbf{M}$ generate known *incident* fields $\mathbf{E}_1^i$ and $\mathbf{H}_1^i$ in an unbounded region $R_1$ with dielectric parameters $(\mu_1, \epsilon_1)$. A small, bounded region $R_2$ with the same parameters is placed inside $R_1$, where the interface between the two regions is the surface $S$ with normal $\hat{\mathbf{n}}_1$ that points into $R_1$. Because $R_1$ and $R_2$ have the same dielectric parameters, $\mathbf{E}_1^i$ and $\mathbf{H}_1^i$ comprise the field everywhere. If we now change the parameters of $R_2$ to $(\mu_2, \epsilon_2)$ as in Figure 3.2b, it becomes an obstacle inside $R_1$ and the fields are no longer the same. The fields in $R_1$ now comprise $\mathbf{E}_1$ and $\mathbf{H}_1$, which are

$$\mathbf{E}_1 = \mathbf{E}_1^i + \mathbf{E}_1^s \tag{3.145}$$

$$\mathbf{H}_1 = \mathbf{H}_1^i + \mathbf{H}_1^s \tag{3.146}$$

where $\mathbf{E}_1^s$ and $\mathbf{H}_1^s$ are the *scattered* or *perturbed* fields generated by the obstacle, and the fields $\mathbf{E}_2$ and $\mathbf{H}_2$ inside $R_2$ are referred to as the *transmitted* fields. Let us now remove $\mathbf{J}$ and $\mathbf{M}$ and replace them by fictitious but *equivalent* surface currents $\mathbf{J}_1$ and $\mathbf{M}_1$ on the outside of $S$, which we refer to as $S^+$. For the fields in $R_1$ and $R_2$ to remain unchanged, the boundary conditions (3.2) on $S^+$ must be satisfied as follows

$$\mathbf{M}_1 = -\hat{\mathbf{n}} \times (\mathbf{E}_1 - \mathbf{E}_2) \tag{3.147}$$

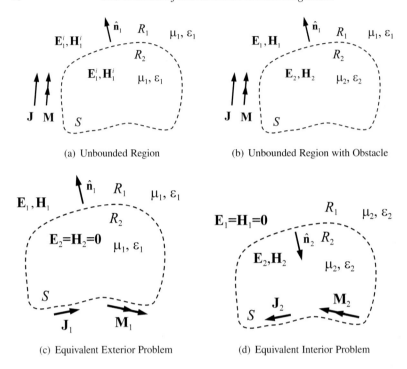

**FIGURE 3.2:** Surface Equivalent

$$\mathbf{J}_1 = \hat{\mathbf{n}} \times (\mathbf{H}_1 - \mathbf{H}_2) \tag{3.148}$$

Note that because the obstacle is still present, we cannot use (3.73) and (3.74) in either (3.147) or (3.148). However, the quantities of interest in scattering problems are usually the fields $\mathbf{E}_1^s$ and $\mathbf{H}_1^s$, not the fields inside the obstacle. Therefore, we will apply the *extinction theorem* [2] and simply assume that those fields are zero. Doing so, we can now assign to $R_2$ the same dielectric parameters as $R_1$, and the boundary conditions (3.2) on $S^+$ become

$$\mathbf{M}_1 = -\hat{\mathbf{n}} \times \mathbf{E}_1 = -\hat{\mathbf{n}} \times (\mathbf{E}_1^s + \mathbf{E}_1^i) \tag{3.149}$$

$$\mathbf{J}_1 = \hat{\mathbf{n}} \times \mathbf{H}_1 = \hat{\mathbf{n}} \times (\mathbf{H}_1^s + \mathbf{H}_1^i) \tag{3.150}$$

We now have an equivalent *exterior* problem, as illustrated in Figure 3.2c, where the fields outside $S$ remain unchanged. We can likewise formulate an equivalent *interior* problem inside $R_2$. To do so, we place the surface currents $\mathbf{J}_2$ and $\mathbf{M}_2$ on the inside of $S$, which we refer to as $S^-$. As before, if we assume that the fields in $R_1$ are zero, the boundary conditions (3.2) on $S^-$ become

$$\mathbf{M}_2 = -\hat{\mathbf{n}}_2 \times \mathbf{E}_2 \tag{3.151}$$

$$\mathbf{J}_2 = \hat{\mathbf{n}}_2 \times \mathbf{H}_2 \tag{3.152}$$

where $\hat{\mathbf{n}}_2$ points into $R_2$.

Let us focus again on the fields in $R_1$. Substituting (3.75) and (3.76) into (3.149) and (3.150) yields

$$\mathbf{M}_1(\mathbf{r}) + \hat{\mathbf{n}}_1(\mathbf{r}) \times \left[ -j\omega\mu(\mathcal{L}\mathbf{J}_1)(\mathbf{r}) - (\mathcal{K}\mathbf{M}_1)(\mathbf{r}) \right] = -\hat{\mathbf{n}}_1(\mathbf{r}) \times \mathbf{E}_1^i(\mathbf{r}) \tag{3.153}$$

and

$$\mathbf{J}_1(\mathbf{r}) - \hat{\mathbf{n}}_1(\mathbf{r}) \times \left[ -j\omega\epsilon(\mathcal{L}\mathbf{M}_1)(\mathbf{r}) + (\mathcal{K}\mathbf{J}_1)(\mathbf{r}) \right] = \hat{\mathbf{n}}_1(\mathbf{r}) \times \mathbf{H}_1^i(\mathbf{r}) \tag{3.154}$$

As $\mathbf{r}$ approaches $\mathbf{r}'$ on $S^+$, the $\mathcal{L}$ operator is well behaved, however the $\mathcal{K}$ operator contains a singularity which must be addressed [2]. To do so, we must determine the value of

$$\lim_{\mathbf{r} \to \mathbf{r}'} \left[ \hat{\mathbf{n}}(\mathbf{r}) \times (\mathcal{K}\mathbf{X})(\mathbf{r}) \right] \tag{3.155}$$

Using (3.78), this is

$$\lim_{\mathbf{r} \to \mathbf{r}'} \left[ \hat{\mathbf{n}}(\mathbf{r}) \times \nabla \times \int_S G(\mathbf{r}, \mathbf{r}') \mathbf{X}(\mathbf{r}') \, d\mathbf{r}' \right] \tag{3.156}$$

and following (3.96), we can move the gradient inside, which yields

$$\lim_{\mathbf{r} \to \mathbf{r}'} \left[ -\hat{\mathbf{n}}(\mathbf{r}) \times \int_S \mathbf{X}(\mathbf{r}') \times \nabla G(\mathbf{r}, \mathbf{r}') \, d\mathbf{r}' \right] \tag{3.157}$$

We now break the integral into two parts following [7]

$$-\hat{\mathbf{n}}(\mathbf{r}) \times \left[ \int_{S-\delta S} \mathbf{X}(\mathbf{r}') \times \nabla G(\mathbf{r}, \mathbf{r}') \, d\mathbf{r}' + \int_{\delta S} \mathbf{X}(\mathbf{r}') \times \nabla G(\mathbf{r}, \mathbf{r}') \, d\mathbf{r}' \right] \tag{3.158}$$

where $\delta S$ is a very small, circular region of radius $a$ in $S$ about $\mathbf{r}$. If we now make the center of $\delta S$ the origin of a local cylindrical coordinate system, we can write

$$|\mathbf{r} - \mathbf{r}'| = \sqrt{(\rho')^2 + z^2} \tag{3.159}$$

and for $|\mathbf{r} - \mathbf{r}'| \ll 1$, the Green's function inside $\delta S$ can be written as

$$G(\mathbf{r}, \mathbf{r}') = \frac{e^{-jk|\mathbf{r}-\mathbf{r}'|}}{4\pi|\mathbf{r}-\mathbf{r}'|} \approx \frac{1}{4\pi\sqrt{(\rho')^2 + (z-z')^2}} \tag{3.160}$$

The gradient in cylindrical coordinates is

$$\nabla = \frac{\partial}{\partial \rho}\hat{\rho} + \frac{1}{\rho}\frac{\partial}{\partial \phi}\hat{\phi} + \frac{\partial}{\partial z}\hat{z} \tag{3.161}$$

and because $\hat{n}(\mathbf{r}) = \hat{z}$ on $\delta S$ and $\mathbf{X}(\mathbf{r}')$ is everywhere tangent to $\delta S$, we can write

$$\hat{z} \times \mathbf{X}(\mathbf{r}') \times \nabla G(\mathbf{r}, \mathbf{r}') = \mathbf{X}(\mathbf{r}')\left[\frac{\partial}{\partial z}G(\mathbf{r}, \mathbf{r}')\right] \tag{3.162}$$

which is

$$\mathbf{X}(\mathbf{r}')\left[\frac{\partial}{\partial z}G(\mathbf{r}, \mathbf{r}')\right] = \mathbf{X}(\mathbf{r}')\frac{-z}{4\pi\left[(\rho')^2 + z^2\right]^{3/2}} \tag{3.163}$$

Because $\delta S$ is very small, we will assume that $\mathbf{X}(\mathbf{r}')$ is approximately equal to $\mathbf{X}(\mathbf{r})$. We now write the integral over $\delta S$ as

$$-\hat{n}(\mathbf{r}) \times \int_{\delta S} \mathbf{X}(\mathbf{r}') \times \nabla G(\mathbf{r}, \mathbf{r}')\, d\mathbf{r}' = \frac{\mathbf{X}(\mathbf{r})}{2}\int_0^a \frac{z\rho'}{\left[(\rho')^2 + z^2\right]^{3/2}}\, d\rho' \tag{3.164}$$

Integrating this expression we get

$$\frac{\mathbf{X}(\mathbf{r})}{2}\left[\frac{z}{|z|} - \frac{z}{\sqrt{a^2 + z^2}}\right] \tag{3.165}$$

and taking the limit as $z$ approaches 0 from above yields

$$\lim_{z \to 0^+} \frac{\mathbf{X}(\mathbf{r})}{2}\left[\frac{z}{|z|} - \frac{z}{\sqrt{a^2 + z^2}}\right] = \frac{\mathbf{X}(\mathbf{r})}{2} \tag{3.166}$$

Thus,

$$\lim_{\mathbf{r} \to \mathbf{r}'} \left[\hat{n}_1(\mathbf{r}) \times (\mathcal{K}\mathbf{X})(\mathbf{r})\right] = \frac{\mathbf{X}(\mathbf{r})}{2} \tag{3.167}$$

Note that in cases where the small area $\delta S$ is not locally planar, such as the apex of a cone or an edge shared by two planar facets, a modification of (3.167) is required. It is shown in [7] that the result of (3.167) should be adjusted to take into account the exterior solid angle $\Omega_0$, which results in

$$\lim_{\mathbf{r} \to \mathbf{r}'} \left[\hat{n}_1(\mathbf{r}) \times (\mathcal{K}\mathbf{X})(\mathbf{r})\right] = \frac{\Omega_0}{4\pi}\mathbf{X}(\mathbf{r}) \tag{3.168}$$

For smooth geometries, the exterior solid angle is $\Omega_0 = 2\pi$ at every observation point and (3.168) reduces to (3.167). When an object has sharp corners or edges the solid angle can be adjusted, as is done in [8] for three-dimensional objects (see Chapter 8). However, in this book we will use (3.167) with no modifications, which is a common approach in the literature. Thus, (3.153) and (3.154) now become

$$\frac{1}{2}\mathbf{M}_1(\mathbf{r}) + \hat{\mathbf{n}}_1(\mathbf{r}) \times \left[ -j\omega\mu(\mathcal{L}\mathbf{J}_1)(\mathbf{r}) - (\mathcal{K}\mathbf{M}_1)(\mathbf{r}) \right] = -\hat{\mathbf{n}}_1(\mathbf{r}) \times \mathbf{E}_1^i(\mathbf{r}) \quad (3.169)$$

and

$$\frac{1}{2}\mathbf{J}_1(\mathbf{r}) - \hat{\mathbf{n}}_1(\mathbf{r}) \times \left[ -j\omega\epsilon(\mathcal{L}\mathbf{M}_1)(\mathbf{r}) + (\mathcal{K}\mathbf{J}_1)(\mathbf{r}) \right] = \hat{\mathbf{n}}_1(\mathbf{r}) \times \mathbf{H}_1^i(\mathbf{r}) \quad (3.170)$$

for $\mathbf{r}$ on $S^+$, where $(\mathcal{K}\mathbf{X})(\mathbf{r})$ is excluded from points very close to $\mathbf{r}'$. Similarly, the fields in $R_2$ can be written as

$$\frac{1}{2}\mathbf{M}_2(\mathbf{r}) + \hat{\mathbf{n}}_2(\mathbf{r}) \times \left[ -j\omega\mu(\mathcal{L}\mathbf{J}_2)(\mathbf{r}) - (\mathcal{K}\mathbf{M}_2)(\mathbf{r}) \right] = 0 \quad (3.171)$$

and

$$\frac{1}{2}\mathbf{J}_2(\mathbf{r}) - \hat{\mathbf{n}}_2(\mathbf{r}) \times \left[ -j\omega\epsilon(\mathcal{L}\mathbf{M}_2)(\mathbf{r}) + (\mathcal{K}\mathbf{J}_2)(\mathbf{r}) \right] = 0 \quad (3.172)$$

for $\mathbf{r}$ on $S^-$. Note that since a limiting argument was used to determine the value of $(\mathcal{K}\mathbf{X})(\mathbf{r})$ as $\mathbf{r} \to \mathbf{r}'$ on $S^\pm$, it is only valid for surfaces that are closed and cannot be used on open surfaces or those that are very thin.

We can now state the surface integral equations for the concept problem outlined in Section 3.6.1. Using the fact that on $S$,

$$\hat{\mathbf{n}} \times \hat{\mathbf{n}} \times \mathbf{A} = (\hat{\mathbf{n}} \cdot \mathbf{A})\hat{\mathbf{n}} - \mathbf{A} = -\mathbf{A}_{tan} \quad (3.173)$$

we can write (3.169) and (3.170) as

$$\left[ j\omega\mu(\mathcal{L}\mathbf{J}_1)(\mathbf{r}) + (\mathcal{K}\mathbf{M}_1)(\mathbf{r}) \right]_{tan} + \frac{1}{2}\hat{\mathbf{n}}_1(\mathbf{r}) \times \mathbf{M}_1(\mathbf{r}) = \left[ \mathbf{E}_1^i(\mathbf{r}) \right]_{tan} \quad (3.174)$$

and

$$\left[ j\omega\epsilon(\mathcal{L}\mathbf{M}_1)(\mathbf{r}) - (\mathcal{K}\mathbf{J}_1)(\mathbf{r}) \right]_{tan} - \frac{1}{2}\hat{\mathbf{n}}_1(\mathbf{r}) \times \mathbf{J}_1(\mathbf{r}) = \left[ \mathbf{H}_1^i(\mathbf{r}) \right]_{tan} \quad (3.175)$$

and (3.171) and (3.172) as

$$\left[ j\omega\mu(\mathcal{L}\mathbf{J}_2)(\mathbf{r}) + (\mathcal{K}\mathbf{M}_2)(\mathbf{r}) \right]_{tan} + \frac{1}{2}\hat{\mathbf{n}}_2(\mathbf{r}) \times \mathbf{M}_2(\mathbf{r}) = 0 \quad (3.176)$$

and

$$\left[ j\omega\epsilon(\mathcal{L}\mathbf{M}_2)(\mathbf{r}) - (\mathcal{K}\mathbf{J}_2)(\mathbf{r}) \right]_{tan} - \frac{1}{2}\hat{\mathbf{n}}_2(\mathbf{r}) \times \mathbf{J}_2(\mathbf{r}) = 0 \quad (3.177)$$

Equations (3.170 – 3.174) comprise a set of integral equations for the unknown currents $\mathbf{J}_1, \mathbf{M}_1, \mathbf{J}_2$ and $\mathbf{M}_2$, in terms of the known incident fields $\mathbf{E}_1^i$ and $\mathbf{H}_1^i$ in $R_1$. Once the currents are found, the fields everywhere in $R_1$ and $R_2$ can be found using (3.75) and (3.76). As written, these equations do not appear coupled, as they have no quantities in common. However, the currents on each side of $S$ are related via the boundary conditions. We will consider this in more detail in the next section.

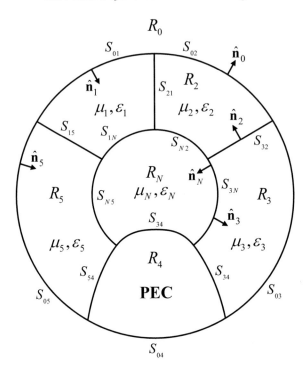

**FIGURE 3.3:** Composite Multi-Region Geometry

### 3.6.2   Surface Integral Equations

Let us now replace the simple scatterer from Section 3.6.1 with an object of a more general configuration, as shown in Figure 3.3. This object comprises multiple, piecewise homogeneous dielectric (penetrable) or conducting (PEC) regions $R_1, \ldots, R_N$. The material parameters of dielectric region $R_l$ is $(\mu_l, \epsilon_l)$, where either $\mu$ or $\epsilon$ may be complex (denoting a lossy material) and $R_0$ is the unbounded external region to which we assign the free space parameters $(\mu_0, \epsilon_0)$. The boundary of $R_l$ we denote as the surface $S_l$, and the interface between regions $R_l$ and $R_m$ is the surface $S_{lm}$ (or $S_{ml}$), which may comprise a portion or all of $S_l$. The normal vector $\hat{\mathbf{n}}_l$ points into $R_l$ on each of its interfaces. Conductors may be closed regions with finite volume ($R_4$ in Figure 3.3), or open, infinitely thin sheets. For each non-PEC region $R_l$, the surface integral equations on $S_l$ are

$$\left[ j\omega\mu(\mathcal{L}\mathbf{J}_l)(\mathbf{r}) + (\mathcal{K}\mathbf{M}_l)(\mathbf{r}) \right]_{tan} + \frac{1}{2}\hat{\mathbf{n}}_l(\mathbf{r}) \times \mathbf{M}_l(\mathbf{r}) = \left[ \mathbf{E}_l^i(\mathbf{r}) \right]_{tan} \quad (3.178)$$

and

$$\left[ j\omega\epsilon(\mathcal{L}\mathbf{M}_l)(\mathbf{r}) - (\mathcal{K}\mathbf{J}_l)(\mathbf{r}) \right]_{tan} - \frac{1}{2}\hat{\mathbf{n}}_l(\mathbf{r}) \times \mathbf{J}_l(\mathbf{r}) = \left[ \mathbf{H}_l^i(\mathbf{r}) \right]_{tan} \qquad (3.179)$$

where for scattering problems, $\mathbf{E}_l^i(\mathbf{r})$ and $\mathbf{H}_l^i(\mathbf{r})$ are zero everywhere except in $R_0$. For antenna and radiation problems, $\mathbf{E}_l^i(\mathbf{r})$ and $\mathbf{H}_l^i(\mathbf{r})$ may be nonzero in one or more regions, depending on where the excitation is applied. We refer to (3.178) as the Electric Field Integral Equation (EFIE), and to (3.179) as the Magnetic Field Integral Equation (MFIE).

### 3.6.2.1 Interior Resonance Problem

Applying (3.178) and (3.179) to dielectrics will yield a result free of spurious solutions [9], however on closed conductors the EFIE and MFIE cannot produce a unique solution for all frequencies [9, 10, 11, 12, 13]. This is because there are homogeneous solutions to these equations that satisfy the boundary conditions with zero incident field. These spurious solutions correspond to interior cavity (resonant) modes of the object itself, and radiate no field outside the object. One method for dealing with this problem is through the application of an extended boundary condition, where the original integral equations are modified or augmented such that the value of the field inside the object is explicitly forced to be zero [7, 14]. Dual-surface electric and magnetic integral equations (DSEFIE, DSMFIE) [15] are an example of such an extended boundary condition. In such a formulation, a second surface is placed just inside the original surface. The appropriate boundary condition for the internal fields on this surface is used to generate an additional integral equation. The combination of this new equation with the original results in a combined equation that produces a unique solution at all frequencies. This method, however, requires additional effort in generating the secondary surface, and the number of currents is doubled, which drives up the demand for CPU time and system memory.

*Combined Field Integral Equation*

To date, the most commonly used method in handling the resonance problem is through a linear combination of the EFIE and a modified MFIE. This enforces the boundary conditions on the electric and magnetic ield simultaneously and is free of spurious solutions, as the null spaces of the EFIE and the modified MFIE differ. To create this modified MFIE, we form the cross product of $\hat{\mathbf{n}}_l$ and (3.179) to yield

$$\hat{\mathbf{n}}_l \times \left[ j\omega\epsilon(\mathcal{L}\mathbf{M}_l)(\mathbf{r}) - (\mathcal{K}\mathbf{J}_l)(\mathbf{r}) \right]_{tan} + \frac{1}{2}\mathbf{J}_l(\mathbf{r}) = \hat{\mathbf{n}}_l \times \left[ \mathbf{H}_l^i(\mathbf{r}) \right]_{tan} \qquad (3.180)$$

where there are now $\hat{\mathbf{n}} \times \mathcal{L}$ and $\hat{\mathbf{n}} \times \mathcal{K}$ operators. We refer to (3.180) as the nM-FIE, which when combined with the EFIE yields the Combined Field Integral Equation (CFIE) given by [9]

$$\alpha\text{EFIE} + (1 - \alpha)\eta_l\text{nMFIE} \tag{3.181}$$

where $\eta_l = \sqrt{\mu_l/\epsilon_l}$ and $0 \leq \alpha \leq 1$, with $\alpha = 0.5$ commonly used. We apply the CFIE to closed, conducting volumes in this book. For open, thin conductors, we will use the EFIE only ($\alpha = 1$).

*A Note on Nomenclature*

Many other books and journal papers that discuss surface integral equations are concerned with conducting (PEC) objects only. The equations in these references are typically the EFIE (3.178) and the nMFIE of (3.180) with no magnetic currents, except that the nMFIE is often referred to as simply the MFIE. References that treat conducting as well as dielectric objects will more often use the nomenclature outlined in this book.

### 3.6.2.2 Discretization and Testing

We will now us the Method of Moments to convert (3.178), (3.179), and (3.180) into a matrix system using Galerkin-type testing in each region. Expanding the electric and magnetic currents in $R_l$ using a sum of weighted basis functions yields

$$\mathbf{J}_l(\mathbf{r}) = \sum_{n=1}^{N_J^{(l)}} I_n^{J(l)} \mathbf{f}_n(\mathbf{r}) \tag{3.182}$$

and

$$\mathbf{M}_l(\mathbf{r}) = \sum_{n=1}^{N_M^{(l)}} I_n^{M(l)} \mathbf{g}_n(\mathbf{r}) \tag{3.183}$$

where $N_J^{(l)}$ and $N_M^{(l)}$ are the number of electric and magnetic basis functions in $R_l$, respectively, and the basis functions $\mathbf{f}_n(\mathbf{r})$ and $\mathbf{g}_n(\mathbf{r})$ will remain undefined for now. We next test the EFIE and nMFIE with the electric testing functions $\mathbf{f}_m(\mathbf{r})$, and the MFIE with the magnetic testing functions $\mathbf{g}_m(\mathbf{r})$ as in [9]. This yields in $R_l$ the matrix system

$$\begin{bmatrix} Z^{EJ(l)} & Z^{EM(l)} \\ Z^{HJ(l)} & Z^{HM(l)} \\ Z^{nHJ(l)} & Z^{nHM(l)} \end{bmatrix} \begin{bmatrix} I^{J(l)} \\ I^{M(l)} \end{bmatrix} = \begin{bmatrix} V^{E(l)} \\ V^{H(l)} \\ V^{nH(l)} \end{bmatrix} \tag{3.184}$$

with matrix elements given by

$$
\begin{aligned}
Z_{mn}^{EJ(l)} = {}& j\omega\mu_l \int_{\mathbf{f}_m} \mathbf{f}_m(\mathbf{r}) \cdot \int_{\mathbf{f}_n} \mathbf{f}_n(\mathbf{r}') \, G(\mathbf{r},\mathbf{r}') \, d\mathbf{r}' \, d\mathbf{r} \\
& + \frac{j}{\omega\epsilon_l} \int_{\mathbf{f}_m} \mathbf{f}_m(\mathbf{r}) \cdot \left[ \nabla\nabla \cdot \int_{\mathbf{f}_n} \mathbf{f}_n(\mathbf{r}') \, G(\mathbf{r},\mathbf{r}') \, d\mathbf{r}' \right] d\mathbf{r} \qquad (3.185)
\end{aligned}
$$

$$
\begin{aligned}
Z_{mn}^{EM(l)} = {}& \int_{\mathbf{f}_m} \mathbf{f}_m(\mathbf{r}) \cdot \int_{\mathbf{g}_n} \nabla G(\mathbf{r},\mathbf{r}') \times \mathbf{g}_n(\mathbf{r}') \, d\mathbf{r}' \, d\mathbf{r} \\
& + \frac{1}{2} \int_{\mathbf{f}_m,\mathbf{g}_n} \mathbf{f}_m(\mathbf{r}) \cdot \left[ \hat{\mathbf{n}}_l(\mathbf{r}) \times \mathbf{g}_n(\mathbf{r}) \right] d\mathbf{r} \qquad (3.186)
\end{aligned}
$$

$$
\begin{aligned}
Z_{mn}^{HJ(l)} = {}& - \int_{\mathbf{g}_m} \mathbf{g}_m(\mathbf{r}) \cdot \int_{\mathbf{f}_n} \nabla G(\mathbf{r},\mathbf{r}') \times \mathbf{f}_n(\mathbf{r}') \, d\mathbf{r}' \, d\mathbf{r} \\
& - \frac{1}{2} \int_{\mathbf{g}_m,\mathbf{f}_n} \mathbf{g}_m(\mathbf{r}) \cdot \left[ \hat{\mathbf{n}}_l(\mathbf{r}) \times \mathbf{f}_n(\mathbf{r}) \right] d\mathbf{r} \qquad (3.187)
\end{aligned}
$$

$$
\begin{aligned}
Z_{mn}^{HM(l)} = {}& j\omega\epsilon_l \int_{\mathbf{g}_m} \mathbf{g}_m(\mathbf{r}) \cdot \int_{\mathbf{g}_n} \mathbf{g}_n(\mathbf{r}') \, G(\mathbf{r},\mathbf{r}') \, d\mathbf{r}' \, d\mathbf{r} \\
& + \frac{j}{\omega\mu_l} \int_{\mathbf{g}_m} \mathbf{g}_m(\mathbf{r}) \cdot \left[ \nabla\nabla \cdot \int_{\mathbf{g}_n} \mathbf{g}_n(\mathbf{r}') \, G(\mathbf{r},\mathbf{r}') \, d\mathbf{r}' \right] d\mathbf{r} \qquad (3.188)
\end{aligned}
$$

$$
\begin{aligned}
Z_{mn}^{nHJ(l)} = {}& - \int_{\mathbf{f}_m} \mathbf{f}_m(\mathbf{r}) \cdot \left[ \hat{\mathbf{n}}_l(\mathbf{r}) \times \int_{\mathbf{f}_n} \nabla G(\mathbf{r},\mathbf{r}') \times \mathbf{f}_n(\mathbf{r}') \, d\mathbf{r}' \right] d\mathbf{r} \\
& + \frac{1}{2} \int_{\mathbf{f}_m,\mathbf{f}_n} \mathbf{f}_m(\mathbf{r}) \cdot \mathbf{f}_n(\mathbf{r}) \, d\mathbf{r} \qquad (3.189)
\end{aligned}
$$

$$
\begin{aligned}
Z_{mn}^{nHM(l)} = {}& j\omega\epsilon_l \int_{\mathbf{f}_m} \mathbf{f}_m(\mathbf{r}) \cdot \left[ \hat{\mathbf{n}}_l(\mathbf{r}) \times \int_{\mathbf{g}_n} \mathbf{g}_n(\mathbf{r}') \, G(\mathbf{r},\mathbf{r}') \, d\mathbf{r}' \right] d\mathbf{r} \\
& + \frac{j}{\omega\mu_l} \int_{\mathbf{f}_m} \mathbf{f}_m(\mathbf{r}) \cdot \left[ \hat{\mathbf{n}}_l(\mathbf{r}) \times \nabla\nabla \cdot \int_{\mathbf{g}_n} \mathbf{g}_n(\mathbf{r}') \, G(\mathbf{r},\mathbf{r}') \, d\mathbf{r}' \right] d\mathbf{r} \\
& \qquad\qquad\qquad\qquad\qquad\qquad\qquad\qquad\qquad\qquad (3.190)
\end{aligned}
$$

where the $\mathcal{K}$ and $\hat{\mathbf{n}} \times \mathcal{K}$ operators were modified following (3.96). The elements of the right-hand side vectors are

$$
V_m^{E(l)} = \int_{\mathbf{f}_m} \mathbf{f}_m(\mathbf{r}) \cdot \mathbf{E}_i^i(\mathbf{r}) \, d\mathbf{r} \qquad (3.191)
$$

$$V_m^{H(l)} = \int_{\mathbf{f}_m} \mathbf{f}_m(\mathbf{r}) \cdot \mathbf{H}_l^i(\mathbf{r}) \, d\mathbf{r} \tag{3.192}$$

$$V_m^{nH(l)} = \int_{\mathbf{f}_m} \mathbf{f}_m(\mathbf{r}) \cdot \left[\hat{\mathbf{n}}_l \times \mathbf{H}_l^i(\mathbf{r})\right] d\mathbf{r} \tag{3.193}$$

Note that all blocks of the matrix in (3.184) are square. If we now combine the matrix equations for all regions, we obtain a block-diagonal system of the form

$$\begin{bmatrix} Z^{(1)} & 0 & \cdots & 0 \\ 0 & Z^{(2)} & \cdots & 0 \\ \vdots & \vdots & \ddots & 0 \\ 0 & 0 & \cdots & Z^{(N)} \end{bmatrix} \begin{bmatrix} I^{(1)} \\ I^{(2)} \\ \vdots \\ I^{(N)} \end{bmatrix} = \begin{bmatrix} V^{(1)} \\ V^{(2)} \\ \vdots \\ V^{(N)} \end{bmatrix} \tag{3.194}$$

where

$$Z^{(l)} = \begin{bmatrix} Z^{EJ(l)} & Z^{EM(l)} \\ Z^{HJ(l)} & Z^{HM(l)} \\ Z^{nHJ(l)} & Z^{nHM(l)} \end{bmatrix} \tag{3.195}$$

$$I^{(l)} = \begin{bmatrix} I^{J(l)} \\ I^{M(l)} \end{bmatrix} \tag{3.196}$$

and

$$V^{(l)} = \begin{bmatrix} V^{E(l)} \\ V^{H(l)} \\ V^{nH(l)} \end{bmatrix} \tag{3.197}$$

### 3.6.2.3 Modification of Matrix Elements

If the basis and testing functions have a well-behaved divergence, it is possible to re-distribute the vector derivatives in the $\mathcal{L}$ operator so that they do not operate on the Green's function. This is advantageous as they would otherwise compound the $1/r$ singularity. To do so, let us concentrate on the expression having the form

$$\int_{\mathbf{f}_m} \mathbf{f}_m(\mathbf{r}) \cdot \left[\nabla\nabla \cdot \int_{\mathbf{f}_n} \mathbf{f}_n(\mathbf{r}')G(\mathbf{r},\mathbf{r}') \, d\mathbf{r}'\right] d\mathbf{r} \tag{3.198}$$

Following Section 3.4.4, this can be rewritten as

$$\int_{\mathbf{f}_m} \mathbf{f}_m(\mathbf{r}) \cdot \nabla S(\mathbf{r}) \, d\mathbf{r} = \int_{\mathbf{f}_m} \mathbf{f}_m(\mathbf{r}) \cdot \left[\nabla \int_{\mathbf{f}_n} \nabla' \cdot \mathbf{f}(\mathbf{r}') \, G(\mathbf{r},\mathbf{r}') \, d\mathbf{r}'\right] d\mathbf{r} \tag{3.199}$$

Using the vector identity

$$\mathbf{f}(\mathbf{r}) \cdot \nabla S(\mathbf{r}) = \nabla \cdot \left[\mathbf{f}(\mathbf{r})S(\mathbf{r})\right] - \left[\nabla \cdot \mathbf{f}(\mathbf{r})\right] S(\mathbf{r}) \qquad (3.200)$$

we can write the above as

$$\int_{\mathbf{f}_m} \mathbf{f}_m(\mathbf{r}) \cdot \nabla S(\mathbf{r}) \, d\mathbf{r} = \int_{\mathbf{f}_m} \nabla \cdot \left[\mathbf{f}_m(\mathbf{r})S(\mathbf{r})\right] \, d\mathbf{r} - \int_{\mathbf{f}_m} \left[\nabla \cdot \mathbf{f}_m(\mathbf{r})\right] S(\mathbf{r}) \, d\mathbf{r}$$
$$(3.201)$$

We convert the first term on the right-hand side to a surface integral using the divergence theorem

$$\int_V \nabla \cdot \left[\mathbf{f}_m(\mathbf{r})S(\mathbf{r})\right] \, d\mathbf{r} = \int_S \hat{\mathbf{n}} \cdot \left[\mathbf{f}_m(\mathbf{r})S(\mathbf{r})\right] \, d\mathbf{r} \qquad (3.202)$$

Since the bounding surface can be made large enough that $\mathbf{f}_m(\mathbf{r})$ vanishes, the above goes to zero leaving only the second term, which is

$$\int_{\mathbf{f}_m} \mathbf{f}_m(\mathbf{r}) \cdot \nabla S(\mathbf{r}) \, d\mathbf{r} = - \int_{\mathbf{f}_m} \nabla \cdot \mathbf{f}_m(\mathbf{r}) \int_n \nabla' \cdot \mathbf{f}(\mathbf{r}') \, G(\mathbf{r}, \mathbf{r}') \, d\mathbf{r}' \, d\mathbf{r} \quad (3.203)$$

Substitution of the above into (3.185) yields

$$Z_{mn}^{EJ(l)} = j\omega\mu_l \int_{\mathbf{f}_m} \mathbf{f}_m(\mathbf{r}) \cdot \int_{\mathbf{f}_n} \mathbf{f}_n(\mathbf{r}') \, G(\mathbf{r}, \mathbf{r}') \, d\mathbf{r}' \, d\mathbf{r}$$
$$- \frac{j}{\omega\epsilon_l} \int_{\mathbf{f}_m} \nabla \cdot \mathbf{f}_m(\mathbf{r}) \int_{\mathbf{f}_n} \nabla' \cdot \mathbf{f}_n(\mathbf{r}') \, G(\mathbf{r}, \mathbf{r}') \, d\mathbf{r}' \, d\mathbf{r} \qquad (3.204)$$

and a similar substitution can also be made in (3.188).

Using the vector identity $\mathbf{a} \cdot (\mathbf{b} \times \mathbf{c}) = (\mathbf{a} \times \mathbf{b}) \cdot \mathbf{c}$, we can also write the second integral in (3.190) as

$$- \int_{\mathbf{f}_m} \nabla \cdot \left[\mathbf{f}_m(\mathbf{r}) \times \hat{\mathbf{n}}_l(\mathbf{r})\right] \int_{\mathbf{g}_n} \nabla' \cdot \mathbf{g}_n(\mathbf{r}') \, G(\mathbf{r}, \mathbf{r}') \, d\mathbf{r}' \, d\mathbf{r} \qquad (3.205)$$

Using the divergence theorem, the outermost integral can be converted to a contour integral, which yields

$$- \oint_{\delta\mathbf{f}_m} \hat{\mathbf{c}}_m(\mathbf{r}) \cdot \left[\mathbf{f}_m(\mathbf{r}) \times \hat{\mathbf{n}}_l(\mathbf{r})\right] \int_{\mathbf{g}_n} \nabla' \cdot \mathbf{g}_n(\mathbf{r}') \, G(\mathbf{r}, \mathbf{r}') \, d\mathbf{r}' \, d\mathbf{r} \qquad (3.206)$$

This integral is now performed along the boundary of the support of $\mathbf{f}_m(\mathbf{r})$, where $\hat{\mathbf{c}}_m(\mathbf{r})$ is the outward-facing normal to that boundary. The choice of (3.205) or (3.206) in the implementation of (3.190) depends on the geometry and testing functions, and we will use both forms in this book.

### 3.6.3    Enforcement of Boundary Conditions

The block-diagonal system in (3.194) was defined region-wise without regard to the boundary conditions on the interfaces between regions. We know from (3.5) and (3.6) that the tangential components of the electric and magnetic fields are discontinuous across a dielectric interface by an amount equal to the magnetic and electric surface currents on that interface, respectively. However, in our integral equation formulation there are no real surface currents on dielectric interfaces, they are fictitious. This implies that at the interface between dielectric regions $R_l$ and $R_m$,

$$\mathbf{J}_l(\mathbf{r}) = -\mathbf{J}_m(\mathbf{r}) \tag{3.207}$$

$$\mathbf{M}_l(\mathbf{r}) = -\mathbf{M}_m(\mathbf{r}) \tag{3.208}$$

and therefore, (3.194) contains linearly dependent unknowns. This can be addressed by assigning an opposing orientation to the basis functions on opposite sides of an interface, so that they will have the same coefficient. As these basis functions now represent a single unknown, the corresponding columns of (3.194) can be combined. Note that by doing this, (3.194) becomes overdetermined as it has more rows than columns, and so we must address that problem as well.

#### 3.6.3.1    EFIE-CFIE-PMCHWT Approach

The overdetermined system can be handled in several ways, though the most common approach is the so-called Poggio-Miller-Chang-Harrington-Wu-Tsai (PMCHWT) formulation [16, 17]. In this method, the EFIEs and MFIEs on each side of dielectric interfaces are added together by combining the corresponding rows of the matrix. Doing so, (3.194) becomes square, and can be solved. We can combine this with the treatment for open and closed conductors (EFIE and CFIE, respectively) to yield a unified approach for composite objects that we refer to as the EFIE-CFIE-PMCHWT formulation [9]. This approach can be summarized as follows:

1. On dielectric interfaces we test the EFIE and MFIE, and eliminate linearly dependent unknowns using PMCHWT.

2. On interfaces between a dielectric and a closed conductor, we test the CFIE.

3. On interfaces between a dielectric and an open conductor, we test the EFIE only.

*Treatment at Junctions*

Points where three or more interfaces meet are referred to as *junctions*. Enforcement of the EFIE-CFIE-PMCHWT formulation at junctions requires special consideration for specific problem types. We will consider junctions in more detail in later chapters.

### 3.6.4 Physical Optics Equivalent

Though integral equation methods have existed for a long time, and can in principle treat a wide variety of problems, the size of the problems that can be solved through direct means is always limited by the currently available computing technology. Though the requirements for one- and two-dimensional problems are somewhat reasonable, three-dimensional problems present a far more formidable challenge, and historically, Moment Method solutions have been viable only for 3D problems of small electrical size (5 to 10 wavelengths, or less). Larger 3D problems, such as computing radiation patterns of large antennas (such as radio telescopes), or the radar cross section of space vehicles and aircraft, can range upwards of 10 to 100 wavelengths per linear dimension, or even more depending on the waveband of interest. Because these problems are of practical interest, it is often better to have a less accurate answer than no answer at all. Fortunately, in the absence of direct solutions, approximations can be made that make the problem tractable, while yielding fair to modest accuracy in most cases. We will now briefly consider one such approximation that is still used widely for this purpose.

Consider again the equivalent problem in Figure 3.2c, where the surface currents are specified by (3.149) and (3.150). Let us now assume that the object is electrically large, so that at any point on $S$, the surface can be considered to be locally planar and infinite in extent. If the object is perfectly conducting, $\mathbf{M}_1(\mathbf{r}) = 0$ and the scattered magnetic field is equal in amplitude and phase to the incident field. The electric surface current is then

$$\mathbf{J}_1(\mathbf{r}) = 2\hat{\mathbf{n}}_1 \times \mathbf{H}_1^i(\mathbf{r}) \qquad (3.209)$$

which is commonly referred to as the Physical Optics (PO) approximation for conductors, and is illustrated in Figure 3.4. PO also assumes that current only exists on surfaces having a direct line of sight to the source, and in shadowed areas the current is zero. In subsequent chapters, we will compare results using the PO approximation (3.209) to those from the Moment Method. Though it is reasonably accurate at near-specular angles, PO performs poorly elsewhere. As a result, PO is often supplemented by adding to it secondary effects such as edge diffraction via the Physical Theory of Diffraction [18, 19] and multiple bounce interactions via Shooting and Bouncing Rays [20]. In Appendix A, we will look at scattering using PO in additional detail.

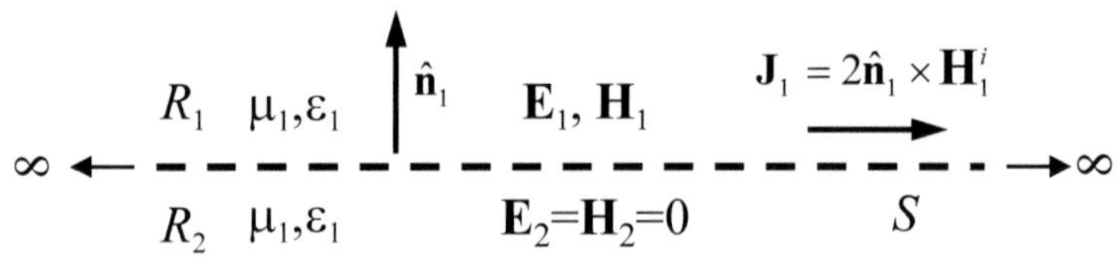

$$R_1 \quad \mu_1, \varepsilon_1 \quad \hat{\mathbf{n}}_1 \quad \mathbf{E}_1, \mathbf{H}_1 \qquad \mathbf{J}_1 = 2\hat{\mathbf{n}}_1 \times \mathbf{H}_1^i$$

$$\infty \longleftarrow - - - - - - - - - - - - - - - \cdot \longrightarrow \infty$$

$$R_2 \quad \mu_1, \varepsilon_1 \qquad \mathbf{E}_2 = \mathbf{H}_2 = 0 \qquad S$$

**FIGURE 3.4:** Physical Optics Approximation

# References

[1] C. A. Balanis, *Advanced Engineering Electromagnetics*. John Wiley and Sons, 1989.

[2] W. C. Chew, *Waves and Fields in Inhomogeneous Media*. IEEE Press, 1995.

[3] P. M. Morse and H. Feshbach, *Methods of Theoretical Physics*. McGraw-Hill, 1953.

[4] M. Abramowitz and I. Stegun, *Handbook of Mathematical Functions*. National Bureau of Standards, 1966.

[5] M. Tong and W. Chew, "Super-hyper singularity treatment for solving 3D electric field integral equations," *Microw. Opt. Technol. Lett*, vol. 49, pp. 1383–1388, June 2007.

[6] M. Tong and W. Chew, "On the near-interaction elements in integral equation solvers for electromagnetic scattering by three-dimensional thin objects," *IEEE Trans. Antennas Propagat.*, vol. 57, pp. 2500–2506, August 2009.

[7] N. Morita, N. Kumagai, and J. R. Mautz, *Integral Equation Methods for Electromagnetics*. Artech House, 1990.

[8] J. M. Rius, E. Úbeda, and J. Parrón, "On the testing of the magnetic field integral equation with RWG basis functions in the method of moments," *IEEE Trans. Antennas Propagat.*, vol. 49, pp. 1550–1553, November 2001.

[9] P. Ylä-Oijala, M. Taskinen, and J. Sarvas, "Surface integral equation method for general composite metallic and dielectric structures with junctions," *Progress in Electromagnetics Research*, vol. 52, pp. 81–108, 2005.

[10] L. N. Medgyesi-Mitschang and J. M. Putnam, "Electromagnetic scattering from axially inhomogeneous bodies of revolution," *IEEE Trans. Antennas Propagat.*, vol. 32, pp. 796–806, March 1984.

[11] L. N. Medgyesi-Mitschang and J. M. Putnam, "Combined field integral equation formulation for inhomogeneous two- and three-dimensional bodies: The junction problem," *IEEE Trans. Antennas Propagat.*, vol. 39, pp. 667–672, May 1991.

[12] R. Mautz and R. Harrington, "H-field, E-field and combined solutions for bodies of revolution," Tech. Rep. Report RADC-TR-77-109, Rome Air Development Center, Griffiss AFB, N.Y., March 1977.

[13] W. C. Chew, J. M. Jin, E. Michielssen, and J. Song, *Fast and Efficient Algorithms in Computational Electromagnetics*. Artech House, 2001.

[14] A. F. Peterson, S. L. Ray, and R. Mittra, *Computational Methods for Electromagnetics*. IEEE Press, 1998.

[15] R. A. Shora and A. D. Yaghjian, "Dual-surface integral equations in electromagnetic scattering," *IEEE Trans. Antennas Propagat.*, vol. 53, pp. 1706–1709, May 2005.

[16] K. Umashankar, A. Taflove, and S. M. Rao, "Electromagnetic scattering by arbitrary shaped three-dimensional homogeneous lossy dielectric bodies," *IEEE Trans. Antennas Propagat.*, vol. 34, pp. 758–766, June 1986.

[17] S. M. Rao, C. C. Cha, R. L. Cravey, and D. L. Wilkes, "Electromagnetic scattering from arbitrary shaped conducting bodies coated with lossy materials of arbitrary thickness," *IEEE Trans. Antennas Propagat.*, vol. 39, pp. 627–631, May 1991.

[18] K. M. Mitzner, "Incremental length diffraction coefficients," Tech. Rep. AFAL-TR-73-296, Northrop Corporation, Aircraft Division, April 1974.

[19] A. Michaeli, "Equivalent edge currents for arbitrary aspects of observation," *IEEE Trans. Antennas Propagat.*, vol. 32, pp. 252–257, March 1984.

[20] H. Ling, S. W. Lee, and R. Chou, "Shooting and bouncing rays: calculating the RCS of an arbitrarily shaped cavity," *IEEE Trans. Antennas Propagat.*, vol. 37, pp. 194–205, February 1989.

# Chapter 4

## Solution of Matrix Equations

Discretization and testing of integral equations via the Method of Moments results in a linear system which must be solved numerically, and so an understanding of how to do this is necessary. Though engineers will likely not have to implement their own linear equation solver, it is useful to understand how such solvers work, what pitfalls exist, as well as what software libraries exist for helping in this task. In this chapter we will briefly describe common methods for solving linear equations such as Gaussian Elimination and LU Decomposition, as well as iterative solvers.

## 4.1 Direct Methods

In this section we discuss Gaussian Elimination (GE) and LU Decomposition, methods that allow for a direct solution of linear systems through a direct matrix factorization. These algorithms require a compute time of the order $O(N^3)$, where $N$ is the number of unknowns. These methods work very well when $N$ is relatively small, however for larger problems, the factorization time may grow prohibitively large, or we may simply run out of memory to store the matrix. Even though the power of the computer continues to grow year to year, this will always be a challenge because larger problems are always on the mind of the engineer.

### 4.1.1 Gaussian Elimination

Gaussian Elimination is a simple method of reducing a matrix to row echelon form through the use of elementary row operations. Once this has been done, the unknown vector is obtained through simple back-substitution. To illustrate this operation, let us form the augmented matrix from the $N \times N$ matrix equation $\mathbf{Ax} = \mathbf{b}$

$$\left[\begin{array}{ccccc|c}
A_{11} & A_{12} & A_{13} & \cdots & A_{1N} & b_1 \\
A_{21} & A_{22} & A_{23} & \cdots & A_{2N} & b_2 \\
A_{31} & A_{32} & A_{33} & \cdots & A_{3N} & b_3 \\
\vdots & \vdots & \vdots & \ddots & \vdots & \vdots \\
A_{N1} & A_{N2} & A_{N3} & \cdots & A_{NN} & b_N
\end{array}\right] \tag{4.1}$$

We start with the first row and divide every entry by $A_{11}$, leaving a 1 on the diagonal. For every row below this, we then subtract a multiple of this row so that a zero remains in the first column. The result is

$$\left[\begin{array}{ccccc|c}
1 & A'_{12} & A'_{13} & \cdots & A'_{1N} & b'_1 \\
0 & A'_{22} & A'_{23} & \cdots & A'_{2N} & b'_2 \\
0 & A'_{32} & A'_{33} & \cdots & A'_{3N} & b'_3 \\
\vdots & \vdots & \vdots & \ddots & \vdots & \vdots \\
0 & A'_{N2} & A'_{N3} & \cdots & A'_{NN} & b'_N
\end{array}\right] \tag{4.2}$$

We repeat this same operation for the second row and all rows below that until only ones remain on the diagonal and the lower triangle is zero. The original matrix equation now reads

$$\left[\begin{array}{ccccc|c}
1 & A'_{12} & A'_{13} & \cdots & A'_{1N} & b'_1 \\
0 & 1 & A'_{23} & \cdots & A'_{2N} & b'_2 \\
0 & 0 & 1 & \cdots & A'_{3N} & b'_3 \\
\vdots & \vdots & \vdots & \ddots & \vdots & \vdots \\
0 & 0 & 0 & \cdots & 1 & b'_N
\end{array}\right] \tag{4.3}$$

The solution to this system is obtained through the following back-substitution operations

$$x_N = b'_N \tag{4.4}$$

$$x_i = b'_i - \sum_{k=i+1}^{N} A'_{ik} x_k \qquad i < N \tag{4.5}$$

The elimination requires a total of $N^3/3$ operations, and $N^2/2$ for the back-substitution. Note that the row operations leading to (4.3) modify the right-hand side vector as well. This is undesirable when dealing with multiple right-hand sides, as they must all be precomputed and available when performing the elimination. This limitation is resolved by using LU decomposition, which is discussed in Section 4.1.2.

#### 4.1.1.1 Pivoting

When performing the row operations for Gaussian elimination, the values of the diagonal entries (the *pivots*) were not addressed. A problem arises when one of these entries is zero, and if the relative magnitude of the pivot is small, round-off errors may become a problem in the subsequent multiplications and subtractions. Because the augmented matrix system is not altered by exchanging any two rows, the current row can be exchanged with any of the ones below it that contain the "strongest" diagonal entry (one of greatest magnitude). This is referred to as *partial pivoting*. An exchange of rows *and* columns in the matrix is called *full pivoting*, and requires recording the permutations of the solution vector as well [1]. Many linear algebra packages employ a pivoting strategy for maximum robustness.

### 4.1.2 LU Decomposition

As previously discussed, a factorization that does not modify the right-hand side vector is more advantageous than Gaussian elimination. Therefore, let us consider the factorization of the matrix **A** into lower and upper triangular parts. We can write this as

$$\mathbf{LU} = \mathbf{A} \tag{4.6}$$

where

$$\mathbf{LU} = \begin{bmatrix} L_{11} & 0 & 0 & \cdots & 0 \\ L_{21} & L_{22} & 0 & \cdots & 0 \\ L_{31} & L_{32} & L_{33} & \cdots & 0 \\ \vdots & \vdots & \vdots & \ddots & \vdots \\ L_{N1} & L_{N2} & L_{N3} & \cdots & L_{NN} \end{bmatrix} \begin{bmatrix} U_{11} & U_{12} & U_{13} & \cdots & U_{1N} \\ 0 & U_{22} & U_{23} & \cdots & U_{2N} \\ 0 & 0 & U_{33} & \cdots & U_{3N} \\ \vdots & \vdots & \vdots & \ddots & \vdots \\ 0 & 0 & 0 & \cdots & U_{NN} \end{bmatrix} \tag{4.7}$$

Using this decomposition, we can solve the following equation

$$\mathbf{Ax} = (\mathbf{LU})\mathbf{x} = \mathbf{L}(\mathbf{Ux}) = \mathbf{b} \tag{4.8}$$

by first solving the equation

$$\mathbf{Ly} = \mathbf{b} \tag{4.9}$$

where **y** is obtained through the forward substitution

$$y_1 = \frac{b_1}{L_{11}} \tag{4.10}$$

$$y_i = \frac{1}{L_{ii}} \left[ b_i - \sum_{k=1}^{i-1} L_{ik} y_k \right] \qquad i > 1 \tag{4.11}$$

and then by solving

$$\mathbf{Ux} = \mathbf{y} \qquad (4.12)$$

where $\mathbf{x}$ is obtained through the back-substitution

$$x_N = \frac{y_N}{U_{NN}} \qquad (4.13)$$

$$x_i = \frac{1}{U_{ii}} \left[ y_i - \sum_{k=i+1}^{N} U_{ik} x_k \right] \qquad i < N \qquad (4.14)$$

The remaining task is to determine the elements of $\mathbf{L}$ and $\mathbf{U}$. If we carry out the matrix multiplication $\mathbf{LU}$, we will have $N^2$ equations for the $N^2 + N$ unknowns comprising $L_{ij}$ and $U_{ij}$. Because the diagonal is represented twice, we are free to specify $N$ of the unknowns ourselves, so we choose

$$L_{ii} = 1 \quad i = 1, 2, \ldots, N \qquad (4.15)$$

hence

$$\mathbf{LU} = \begin{bmatrix} 1 & 0 & 0 & \ldots & 0 \\ L_{21} & 1 & 0 & \ldots & 0 \\ L_{31} & L_{32} & 1 & \ldots & 0 \\ \vdots & \vdots & \vdots & \ddots & \vdots \\ L_{N1} & L_{N2} & L_{N3} & \ldots & 1 \end{bmatrix} \begin{bmatrix} U_{11} & U_{12} & U_{13} & \ldots & U_{1N} \\ 0 & U_{22} & U_{23} & \ldots & U_{2N} \\ 0 & 0 & U_{33} & \ldots & U_{3N} \\ \vdots & \vdots & \vdots & \ddots & \vdots \\ 0 & 0 & 0 & \ldots & U_{NN} \end{bmatrix} \qquad (4.16)$$

and we can immediately set $U_{11} = A_{11}$. Multiplying $\mathbf{L}$ by the first column of $\mathbf{U}$ we obtain

$$L_{21} = \frac{A_{21}}{U_{11}}$$

$$L_{31} = \frac{A_{31}}{U_{11}}$$

$$\ldots \qquad (4.17)$$

and doing the same with the second column of $\mathbf{U}$ yields $U_{12} = A_{12}$ and

$$U_{22} = A_{22} - L_{21} U_{12}$$

$$L_{32} = \frac{1}{U_{22}} \left[ A_{32} - L_{31} U_{12} \right]$$

$$L_{42} = \frac{1}{U_{22}} \left[ A_{42} - L_{41} U_{12} \right]$$

$$\ldots \qquad (4.18)$$

which leads to a generalized method known as *Crout's Algorithm*. For each column $j = 1, 2, \ldots, N$, we first solve for the elements of $\mathbf{U}$ on and above the diagonal, i.e.,

$$U_{ij} = A_{ij} - \sum_{k=1}^{i-1} L_{ik}U_{kj} \qquad i = 1, 2, \ldots, j \qquad (4.19)$$

and then for elements of $\mathbf{L}$ below the diagonal

$$L_{ij} = \frac{1}{U_{jj}} \left[ A_{ij} - \sum_{k=1}^{j-1} L_{ik}U_{kj} \right] \qquad i = j + 1, j + 2, \ldots, N \qquad (4.20)$$

Following this algorithm, the elements needed at each step are already computed when they are needed. The matrix $\mathbf{A}$ can therefore be factored *in place*, requiring no additional storage for $\mathbf{L}$ and $\mathbf{U}$. A practical LU decomposition algorithm will also incorporate partial pivoting to ensure numerical stability at each step of the process. The operation count for the LU decomposition and back-substitution are of the same complexity as Gaussian elimination. Once a factorization is completed, it can then be reused for an arbitrary number of right-hand sides. The factored matrix can also be written to disk and read back in later if needed, negating the need to compute and factor the matrix a second time.

### 4.1.3 Condition Number

Given the matrix system $\mathbf{Ax} = \mathbf{b}$, it is desirable to know the impact on the solution $\mathbf{x}$ given any inaccuracies in the estimate of $\mathbf{b}$. If after computing the singular value decomposition (SVD) of $\mathbf{A}$ we observe a large difference between the largest and smallest singular values, we know that small changes in $\mathbf{b}$ may result in large changes in $\mathbf{x}$. We now define the *condition number* of a matrix, which is

$$\kappa(\mathbf{A}) = \|\mathbf{A}^{-1}\| \, \|\mathbf{A}\| \qquad (4.21)$$

where $\| \cdot \|$ is a matrix norm. If we choose the 2-norm, then the condition number is given by

$$\kappa(\mathbf{A}) = \frac{\lambda_{max}(\mathbf{A})}{\lambda_{min}(\mathbf{A})} \qquad (4.22)$$

where $\lambda_{min}(\mathbf{A})$ and $\lambda_{max}(\mathbf{A})$ are the minimum and maximum eigenvalues of $\mathbf{A}$, respectively [2]. The condition number is important because many elements of the MOM system matrix and right-hand side vectors are obtained through numerical integration. Small errors in the integration can be amplified by a poorly conditioned system matrix. The condition number is also important to iterative methods such as those in the next section, as it has a direct influence on their rate of convergence.

## 4.2   Iterative Methods

We now discuss iterative solver techniques, which have become increasingly popular in recent years. In such algorithms, the bulk of the compute time is spent computing the matrix-vector products between the system matrix and one or more vectors at each iteration. Though the full matrix must still be stored, the overall compute time is of the order $O(MN^2)$, where $N^2$ is the operations for the matrix-vector product and $M$ is the number of iterations. More recently, techniques such as the Fast Multipole Method (FMM) [3] and the Adaptive Integral Method (AIM) [4] have been developed, which allow for fast computation of interactions between basis function groups. When combined with an iterative solver, they eliminate the need to store many of the small-valued matrix elements and greatly accelerate the matrix-vector product.

Iterative methods do not modify the original matrix, and instead start with an approximation (or guess) for the solution vector and attempt to minimize a residual vector at each iteration. At the core of each lies the update operation

$$\mathbf{x}_{k+1} = \mathbf{x}_k + \alpha_k \mathbf{p}_k \qquad (4.23)$$

where the new estimate of the solution $\mathbf{x}_{k+1}$ is obtained from the previous solution $\mathbf{x}_k$ and a search direction $\mathbf{p}_k$, where $\alpha_k$ is a constant. The residual at each step is

$$\mathbf{r}_{k+1} = \mathbf{A}\mathbf{x}_{k+1} - \mathbf{b} \qquad (4.24)$$

The computational burden of these methods lies in one or more matrix-vector products per iteration. In this section we will briefly discuss several commonly used iterative methods and summarize templates that the reader can use to write routines of their own. Because the iterative solver algorithms comprise simple operations such as simple vector and matrix-vector products, they are easy and quick to implement in software, even if the user is not familiar with their inner workings. Some of these methods involve matrix-vector products involving the transpose or conjugate transpose of the system matrix, whereas some of the others do not. The latter are referred to as transpose-free. Transpose-free variants simplify the programming of some algorithms such as the FMM, which is discussed in Chapter 9.

### 4.2.1   Conjugate Gradient

The Conjugate Gradient (CG) algorithm is a method for solving a matrix system where $\mathbf{A}$ is symmetric and positive definite. It is related to the Method of Steepest Descent, and minimizes a quadratic function by generating a sequence of conjugate-direction search vectors that are $\mathbf{A}$-orthogonal at each step

---

**Algorithm 1** Preconditioned Conjugate Gradient (CGNR) Method

---

Make an initial guess for $\mathbf{x}_0$
$\mathbf{r}_0 = \mathbf{b} - \mathbf{A}\mathbf{x}_0$
$\tilde{\mathbf{r}}_0 = \mathbf{A}^\dagger\mathbf{r}_0$
Solve $\mathbf{M}\mathbf{z}_0 = \tilde{\mathbf{r}}_0$
$\mathbf{p}_0 = \mathbf{z}_0$
**for** $i = 1, 2, ...$, until convergence **do**
    $\mathbf{w}_{i-1} = \mathbf{A}\mathbf{p}_{i-1}$
    $\alpha_{i-1} = (\mathbf{z}_{i-1}^* \cdot \tilde{\mathbf{r}}_{i-1})/\|\mathbf{w}_{i-1}\|_2^2$
    $\mathbf{x}_i = \mathbf{x}_{i-1} + \alpha_{i-1}\mathbf{p}_{i-1}$
    $\mathbf{r}_i = \mathbf{r}_{i-1} - \alpha_{i-1}\mathbf{w}_{i-1}$
    Convergence check
    $\tilde{\mathbf{r}}_i = \mathbf{A}^\dagger\mathbf{r}_i$
    Solve $\mathbf{M}\mathbf{z}_i = \tilde{\mathbf{r}}_i$
    $\beta_{i-1} = (\mathbf{z}_i^* \cdot \tilde{\mathbf{r}}_i)/(\mathbf{z}_{i-1}^* \cdot \tilde{\mathbf{r}}_{i-1})$
    $\mathbf{p}_i = \mathbf{z}_i + \beta_{i-1}\mathbf{p}_{i-1}$
**end for**

---

[5, 6]. Because the system matrix obtained for integral equation problems via the Moment Method is neither symmetric or positive definite, this algorithm cannot be applied directly. Instead, we can apply it to the following system called the "normal equations"

$$\mathbf{A}^\dagger\mathbf{A}\mathbf{x} = \mathbf{A}^\dagger\mathbf{b} \qquad (4.25)$$

where $\mathbf{A}^\dagger$ denotes the conjugate transpose, and $\mathbf{A}^\dagger\mathbf{A}$ is symmetric and positive definite. Pseudocode [6] for the preconditioned CG method for the normal equations is provided in Algorithm 1, where $\mathbf{M}$ is a preconditioner matrix. This method requires two matrix-vector products per iteration, one with $\mathbf{A}$ and one with $\mathbf{A}^\dagger$.

The CG method has been used extensively for electromagnetic field problems, and the relationship between the matrix eigenvalues and its convergence studied [7, 8, 9, 10]. For a matrix system with $N$ unknowns, CG theoretically converges to the exact solution in at most $N$ iterations assuming there are no round-off errors, and the residual error decreases at each step. For many practical radiation and scattering problems, CG performs very well. However, when the system matrix is poorly conditioned, its convergence rate can be very slow and may stagnate without additional decrease in the residual norm.

---

**Algorithm 2** Preconditioned Biconjugate Gradient (BiCG) Method

---

Make an initial guess for $\mathbf{x}_0$
$\mathbf{r}_0 = \mathbf{b} - \mathbf{A}\mathbf{x}_0$
$\tilde{\mathbf{r}}_0 = \mathbf{r}_0$
**for** $i = 1, 2, ...,$ until convergence **do**
    Solve $\mathbf{M}\mathbf{z}_{i-1} = \mathbf{r}_{i-1}$
    Solve $\mathbf{M}^T\tilde{\mathbf{z}}_{i-1} = \tilde{\mathbf{r}}_{i-1}$
    $\rho_{i-1} = \mathbf{z}_{i-1} \cdot \tilde{\mathbf{r}}_{i-1}$
    **if** $\rho_{i-1} = 0$ **then**
        method fails
    **end if**
    **if** $i = 1$ **then**
        $\mathbf{p}_i = \mathbf{z}_{i-1}$
        $\tilde{\mathbf{p}}_i = \tilde{\mathbf{z}}_{i-1}$
    **else**
        $\beta_{i-i} = \rho_{i-1}/\rho_{i-2}$
        $\mathbf{p}_i = \mathbf{z}_{i-1} + \beta_{i-1}\mathbf{p}_{i-1}$
        $\tilde{\mathbf{p}}_i = \tilde{\mathbf{z}}_{i-1} + \beta_{i-1}\tilde{\mathbf{p}}_{i-1}$
    **end if**
    $\mathbf{q}_i = \mathbf{A}\mathbf{p}_i$
    $\tilde{\mathbf{q}}_i = \mathbf{A}^T\tilde{\mathbf{p}}_i$
    $\alpha_i = \rho_{i-1}/(\tilde{\mathbf{p}}_i \cdot \mathbf{q}_i)$
    $\mathbf{x}_i = \mathbf{x}_{i-1} + \alpha_i\mathbf{p}_i$
    $\mathbf{r}_i = \mathbf{r}_{i-1} - \alpha_i\mathbf{q}_i$
    $\tilde{\mathbf{r}}_i = \tilde{\mathbf{r}}_{i-1} - \alpha_i\tilde{\mathbf{q}}_i$
    Convergence check
**end for**

---

### 4.2.2   Biconjugate Gradient

The Biconjugate Gradient (BiCG) method was developed by Lanczos [11] and is applicable to general, nonsymmetric systems. BiCG approaches the problem by generating a pair of bi-orthogonal residual sequences using $\mathbf{A}$ and $\mathbf{A}^T$. Pseudocode [12] for the preconditioned BiCG method is provided in Algorithm 2. This method requires two matrix-vector products per iteration, one with $\mathbf{A}$ and one with $\mathbf{A}^T$. Though the residual error of the BiCG does not necessarily decrease at each step and its convergence may be erratic, it does exhibit good performance for many problems.

**Algorithm 3** Preconditioned Conjugate Gradient Squared (CGS) Method

Make an initial guess for $\mathbf{x}_0$
$\mathbf{r}_0 = \mathbf{b} - \mathbf{A}\mathbf{x}_0$
$\tilde{\mathbf{r}} = \mathbf{r}_0$
**for** $i = 1, 2, ...$, until convergence **do**
  $\rho_{i-1} = \tilde{\mathbf{r}} \cdot \mathbf{r}_{i-1}$
  **if** $\rho_{i-1} = 0$ **then**
    method fails
  **end if**
  **if** $i = 1$ **then**
    $\mathbf{u}_1 = \mathbf{r}_0$
    $\mathbf{p}_1 = \mathbf{u}_1$
  **else**
    $\beta_{i-i} = \rho_{i-1}/\rho_{i-2}$
    $\mathbf{u}_i = \mathbf{r}_{i-1} + \beta_{i-1}\mathbf{q}_{i-1}$
    $\mathbf{p}_i = \mathbf{u}_i + \beta_{i-1}(\mathbf{q}_{i-1} + \beta_{i-1}\mathbf{p}_{i-1})$
  **end if**
  Solve $\mathbf{M}\hat{\mathbf{p}} = \mathbf{p}_i$
  $\hat{\mathbf{v}} = \mathbf{A}\hat{\mathbf{p}}$
  $\alpha_i = \rho_{i-1}/(\tilde{\mathbf{r}} \cdot \hat{\mathbf{v}})$
  $\mathbf{q}_i = \mathbf{u}_i - \alpha_i\hat{\mathbf{v}}$
  Solve $\mathbf{M}\hat{\mathbf{u}} = \mathbf{u}_i + \mathbf{q}_i$
  $\mathbf{x}_i = \mathbf{x}_{i-1} + \alpha_i\hat{\mathbf{u}}$
  $\hat{\mathbf{q}} = \mathbf{A}\hat{\mathbf{u}}$
  $\mathbf{r}_i = \mathbf{r}_{i-1} - \alpha_i\hat{\mathbf{q}}_i$
  Convergence check
**end for**

### 4.2.3 Conjugate Gradient Squared

The Conjugate Gradient Squared (CGS) algorithm [13] is a method that avoids using the transpose of $\mathbf{A}$ and attempts to obtain a faster rate of convergence than CG and BiCG. The method does converge faster in many cases, though because of its more aggressive formulation it is also more sensitive to residual errors and may diverge quickly if the system is ill-conditioned. Pseudocode [12] for the preconditioned CGS method is provided in Algorithm 3. This method requires two matrix-vector products with $\mathbf{A}$ per iteration.

---

**Algorithm 4** Preconditioned Biconjugate Gradient Stabilized (BiCG-Stab) Method

---

Make an initial guess for $\mathbf{x}_0$
$\mathbf{r}_0 = \mathbf{b} - \mathbf{A}\mathbf{x}_0$
$\tilde{\mathbf{r}} = \mathbf{r}_0$
**for** $i = 1, 2, ...$, until convergence **do**
    $\rho_{i-1} = \tilde{\mathbf{r}} \cdot \mathbf{r}_{i-1}$
    **if** $\rho_{i-1} = 0$ **then**
        method fails
    **end if**
    **if** $i = 1$ **then**
        $\mathbf{p}_1 = \mathbf{r}_0$
    **else**
        $\beta_{i-i} = (\rho_{i-1}/\rho_{i-2})(\alpha_{i-1}/\omega_{i-1})$
        $\mathbf{p}_i = \mathbf{r}_{i-1} + \beta_{i-1}(\mathbf{p}_{i-1} - \omega_{i-1}\mathbf{v}_{i-1})$
    **end if**
    Solve $\mathbf{M}\hat{\mathbf{p}} = \mathbf{p}_i$
    $\mathbf{v}_i = \mathbf{A}\hat{\mathbf{p}}$
    $\alpha_i = \rho_{i-1}/(\tilde{\mathbf{r}} \cdot \mathbf{v}_i)$
    $\mathbf{s} = \mathbf{r}_{i-1} - \alpha_i\mathbf{v}_i$
    Check norm of $s$, if small enough set $\mathbf{x}_i = \mathbf{x}_{i-1} + \alpha_i\hat{\mathbf{p}}$, stop
    Solve $\mathbf{M}\hat{\mathbf{s}} = \mathbf{s}$
    $\mathbf{t} = \mathbf{A}\hat{\mathbf{s}}$
    $\omega_i = (\mathbf{t} \cdot \mathbf{s})/(\mathbf{t} \cdot \mathbf{t})$
    $\mathbf{x}_i = \mathbf{x}_{i-1} + \alpha_i\hat{\mathbf{p}} + \omega_i\hat{\mathbf{s}}$
    $\mathbf{r}_i = \mathbf{s} - \omega_i\mathbf{t}$
    Convergence check
    For continuation, it is necessary that $\omega_i \neq 0$
**end for**

---

### 4.2.4 Biconjugate Gradient Stabilized

The Biconjugate Gradient Stabilized (BiCG-Stab) algorithm is similar to CGS but attempts to avoid its irregular convergence patterns. Pseudocode for the preconditioned BiCG-Stab method is provided in Algorithm 4 [12]. This method requires two matrix-vector products with $\mathbf{A}$ per iteration. Note that there are two end-condition tests in this method.

## 4.2.5 GMRES

In 1986 Saad and Schultz proposed [14] an algorithm known as the Generalized Minimum Residual method (GMRES) method, which can be used for the iterative solution of general, nonsymmetric matrices. In the Conjugate Gradient method, the residuals form an orthogonal basis for the space span $\{\mathbf{r}_0, \mathbf{A}\mathbf{r}_0, \mathbf{A}^2\mathbf{r}_0, \ldots\}$. In GMRES, this basis is formed explicitly via the following algorithm:

> $\mathbf{w}_i = \mathbf{A}\mathbf{v}_i$
> **for** $k = 1, 2, \ldots, i$ **do**
> $\quad \mathbf{w}_i = \mathbf{w}_i - (\mathbf{w}_i \cdot \mathbf{v}_k)\mathbf{v}_k$
> **end for**
> $\mathbf{v}_{i+1} = \mathbf{w}_i / \|\mathbf{w}_i\|$

which is a modified Gram-Schmidt orthogonalization. Applied to the Krylov sequence $\{\mathbf{A}^k\mathbf{r}_0\}$, this is called the Arnoldi method. The inner product coefficients $\mathbf{w}_i \cdot \mathbf{v}_k$ and $\|\mathbf{w}_i\|$ are stored in an upper Hessenberg matrix. The GMRES iterates are then constructed as

$$\mathbf{x}_i = \mathbf{x}_0 + y_1\mathbf{v}_1 + y_2\mathbf{v}_2 + \cdots y_i\mathbf{v}_i \tag{4.26}$$

where the coefficients $y_k$ are chosen to minimize the residual norm $\|\mathbf{b} - \mathbf{A}\mathbf{x}_i\|$. GMRES has the property that this residual norm can be computed without the iterate having been formed. Thus, the expensive action of forming the iterate can be postponed until the residual norm has grown small enough. The drawback to GMRES is that the work and storage required per iteration increases linearly with the iteration count. Unless convergence happens quickly, the cost can quickly become prohibitive. One common way of handling this is a reset of the iteration: after $m$ iterations, the accumulated data are cleared and the intermediate results used as the initial data for the next $m$ iterations. This procedure is then repeated until convergence is achieved. The difficulty now lies in choosing an appropriate value for $m$. If $m$ is too small, the algorithm may be slow to converge, or fail to converge entirely. If $m$ is too large, the required computations and storage may be excessive.

GMRES is employed often to solve electromagnetic problems due to its good convergence properties and robust performance. It is also attractive for use in the FMM as it is transpose-free, and we use GMRES for many of the example problems in Chapter 9. Pseudocode [12] for the preconditioned, restarted GMRES$(m)$ method is provided in Algorithm 5.

---

**Algorithm 5** Preconditioned GMRES($m$) Method

---

Make an initial guess for $\mathbf{x}_0$
**for** $j = 1, 2, ...,$ **do**
   Solve $\mathbf{Mr} = \mathbf{b} - \mathbf{Ax}_0$
   $\mathbf{v}_1 = \mathbf{r}/||\mathbf{r}||_2$
   $\mathbf{s} := ||\mathbf{r}||_2 \mathbf{e}_1$
   **for** $i = 1, ..., m$ **do**
      Solve $\mathbf{Mw} = \mathbf{Av}_i$
      **for** $k = 1, ..., i$ **do**
         $h_{k,i} = \mathbf{w} \cdot \mathbf{v}_k$
         $\mathbf{w} = \mathbf{w} - h_{k,i}\mathbf{v}_k$
      **end for**
      $h_{i+1,i} = ||\mathbf{w}||_2$
      $\mathbf{v}_{i+1} = \mathbf{w}/h_{i+1,i}$
      Apply $J_1, \ldots, J_i - 1$ on $(h_{1,i}, \ldots, h_{i+1,i})$
      Construct $J_i$ acting on the $i$th and $(i+1)$st component
      of $h_{.,i}$ such that $(i+1)$st component of $J_i h_{.,i}$ is 0
      $\mathbf{s} := J_i \mathbf{s}$
      **if** $\mathbf{s}(i+1)$ small enough **then**
         UPDATE($\tilde{\mathbf{x}}, i$)
         stop
      **end if**
   **end for**
   UPDATE($\tilde{\mathbf{x}}, m$)
**end for**

**procedure** UPDATE($\tilde{\mathbf{x}}, i$)
   Solve $\mathbf{Hy} = \tilde{\mathbf{s}}$
   Where the upper $i \times i$ triangular part of $\mathbf{H}$ has $h_{i,j}$ as
   its elements, and $\tilde{\mathbf{s}}$ represents the first $i$ components of $\mathbf{s}$.
   $\tilde{\mathbf{x}} = \mathbf{x}_0 + y_1\mathbf{v}_1 + y_2\mathbf{v}_2 + \cdots y_i\mathbf{v}_i$
   $\mathbf{s}(i+1) = ||\mathbf{b} - \mathbf{A}\tilde{\mathbf{x}}||_2$
   **if** $\tilde{\mathbf{x}}$ is accurate enough **then**
      stop
   **else**
      $\mathbf{x}_0 = \tilde{\mathbf{x}}$
   **end if**
**end procedure**

---

### 4.2.6 Stopping Criteria

When using an iterative method, we need to know when to terminate the iteration. Since we do not know the solution vector $\mathbf{x}$, we cannot compute the error vector

$$\mathbf{e}_i = \mathbf{x} - \mathbf{x}_i \qquad (4.27)$$

We instead compute the residual norm $N_i$ at iteration $i$, which is

$$N_i = \frac{\|\mathbf{r}_n\|}{\|\mathbf{b}\|} = \frac{\|\mathbf{A}\mathbf{x}_i - \mathbf{b}\|}{\|\mathbf{b}\|} \qquad (4.28)$$

We terminate the iteration once the residual norm is below a certain value, such as $10^{-3}$. Note that the residual norm is only an indirect measure the amount of error, because [2]

$$\frac{\|\mathbf{e}_n\|}{\|\mathbf{e}_0\|} \le \kappa(\mathbf{A}) \frac{\|\mathbf{r}_n\|}{\|\mathbf{r}_0\|} \qquad (4.29)$$

and if $\mathbf{A}$ is badly conditioned, $\mathbf{x}_i$ may not be a very good estimate of the solution for a given $N_i$. In practical implementations, the iteration is stopped when the residual norm goes below a predefined value, or a maximum number of iterations is reached.

### 4.2.7 Preconditioning

Previous sections have emphasized the relationship between the condition number of a matrix and the convergence of iterative solvers. In many cases the solution may be unobtainable because the iteration stagnates or diverges. In others, the solution may converge but only after many iterations and a long compute time. In these cases, especially in those where the solution does not converge, a method of improving the convergence behavior of the solver is desirable. To investigate such a method, let us modify the original linear system to create the new system

$$\mathbf{M}^{-1}\mathbf{A}\mathbf{x} = \mathbf{M}^{-1}\mathbf{b} \qquad (4.30)$$

where $\mathbf{M}$ is a *preconditioner matrix*. If $\mathbf{M}^{-1}$ resembles $\mathbf{A}^{-1}$ in some way, then the solution of the above matrix system remains the same, however the eigenvalues of $\mathbf{M}^{-1}\mathbf{A}$ may be more attractive and allow for better performance in an iterative solver. The preconditioners summarized in Algorithms 1-4 solve the preconditioned system of (4.30), and require the solution of one or more auxiliary linear systems of the form $\mathbf{M}\mathbf{z} = \mathbf{r}$ at each iteration.

The problem we now face is finding the value of $\mathbf{M}^{-1}$. Obviously, if $\mathbf{M}^{-1}$ is exactly equal to $\mathbf{A}^{-1}$ we have not improved anything, as this requires a complete factorization of $\mathbf{A}$. The key is determining a value of $\mathbf{M}^{-1}$ that has a reasonable setup time and memory demand, and that does not impose an unacceptable overhead when solving the auxiliary problem at each iteration. In some cases, the application of a preconditioner is absolutely necessary as the original system does not converge at all. In other cases, the hope is that the improved rate of convergence of the new system will result in a significantly reduced number of iterations and therefore an overall savings in compute time. This is particularly attractive when the calculations involve multiple right-hand sides, as the preconditioner setup time can be amortized quickly. Many effective preconditioning schemes are discussed at length in [6], and we will discuss several of these in the context of the Fast Multipole Method in Chapter 9.

## 4.3   Software for Linear Systems

There are many off-the-shelf software libraries available for working with and solving linear systems of equations. Fortunately, many of these are well written, time-tested, robust and freely available. This means that the engineer does not have to worry about writing their own routines, and can instead just install and link their software with an existing library. In this section we will briefly discuss some of the software most widely used for matrix algebra problems.

### 4.3.1   BLAS

The BLAS (Basic Linear Algebra Subprograms) comprises a set of FORTRAN subroutines that act as building blocks for more complex vector and matrix operations [15]. BLAS comprises three levels of functions: BLAS level 1 for scalar, vector and vector-vector operations, BLAS level 2 for matrix-vector operations, and BLAS level 3 for matrix-matrix operations. Separate routines are supplied for real and complex arithmetic, and in single and double precision. The BLAS source codes are freely available from the Netlib repository at *http://www.netlib.org/blas*, and may be compiled and included in commercial applications at no cost. Machine optimized versions of BLAS are also available from various vendors such as Intel, Apple, IBM, Sun, and Cray, and are included as part of an operating system or sold separately as a runtime library. There are also various open source implementations of BLAS having varying levels of optimization, with some supporting parallelization on the host CPU, as well as on graphics processing units (GPUs).

## 4.3.2 LAPACK

LAPACK is a FORTRAN library for solving dense linear systems, least-squares solutions of linear systems of equations, eigenvalue problems, and singular value problems [16]. It is available from the Netlib repository at *http://www.netlib.org/lapack*, and may be used freely in commercial applications. LAPACK routines use BLAS functions for low-level work, and a BLAS implementation must accompany an installation of LAPACK. As with BLAS, separate versions of most LAPACK routines are available for real and complex arithmetic, and in single and double precision. There are also advanced implementations of LAPACK parallelized for host CPUs, such as ATLAS [17, 18] and PLASMA [19], and implementations on GPUs, such as MAGMA [19] and CULA [20].

## 4.3.3 MATLAB®

MATLAB is a popular numerical computing environment and programming language commercially available from The Mathworks, Inc. It was originally developed to aid in solving matrix problems and many of its matrix functions operate via calls to the BLAS and LAPACK libraries. The MATLAB scripting language is easy to learn and fairly complex simulations can be written in it. Though its scripts are interpreted and less efficient than compiled code, MATLAB is easy to learn, accurate, and can be used to solve many of the examples found in this book.

---

# References

[1] W. H. Press, B. P. Flannery, S. A. Teukolsky, and W. T. Vetterling, *Numerical Recipes in C : The Art of Scientific Computing*. Cambridge University Press, 1992.

[2] A. F. Peterson, S. L. Ray, and R. Mittra, *Computational Methods for Electromagnetics*. IEEE Press, 1998.

[3] L. Greengard and V. Rokhlin, "A fast algorithm for particle simulations," *J. Comput. Phys.*, vol. 73, pp. 325–348, 1987.

[4] E. Bleszynski, M. Bleszynski, and T. Jaroszewicz, "AIM: Adaptive integral method for solving large-scale electromagnetic scattering and radiation problems," *Radio Sci.*, vol. 31, pp. 1225–1251, 1996.

[5] M. Hestenes and E. Steifel, "Methods of conjugate gradients for solving linear systems," *J. Res. Nat. Bur. Stand.*, vol. 49, pp. 409–435, 1952.

[6] Y. Saad, *Iterative Methods for Sparse Linear Systems.* PWS, 1st ed., 1996.

[7] T. K. Sarkar, K. R. Siarkiewicz, and R. F. Stratton, "Survey of numerical methods for solution of large systems of linear equations for electromagnetic field problems," *IEEE Trans. Antennas Propagat.*, vol. 29, pp. 847–856, November 1981.

[8] T. K. Sarkar, "The conjugate gradient method as applied to electromagnetic field problems," *IEEE Antennas Propagat. Soc. Newsletter*, pp. 5–14, August 1986.

[9] A. F. Peterson and R. Mittra, "Convergence of the conjugate gradient method when applied to matrix equations representing electromagnetic scattering problems," *IEEE Trans. Antennas Propagat.*, vol. 34, pp. 1447–1454, December 1986.

[10] A. F. Peterson, C. F. Smith, and R. Mittra, "Eigenvalues of the moment-method matrix and their effect on the convergence of the conjugate gradient algorithm," *IEEE Trans. Antennas Propagat.*, vol. 36, pp. 1177–1179, August 1988.

[11] C. Lanczos, "Solution of systems of linear equations by minimized iterations," *J. Res. Nat. Bur. Standards*, vol. 49, pp. 33–53, 1952.

[12] R. Barrett, M. Berry, T. F. Chan, J. Demmel, J. Donato, J. Dongarra, V. Eijkhout, R. Pozo, C. Romine, and H. V. der Vorst, *Templates for the Solution of Linear Systems: Building Blocks for Iterative Methods.* SIAM, second ed., 1994.

[13] P. Sonneveld, "CGS, a fast Lanczos-type solver for nonsymmetric linear systems," *SIAM J. Sci. Statist. Comput.*, vol. 10, pp. 36–52, 1989.

[14] Y. Saad and M. H. Schultz, "GMRES: a generalized minimal residual algorithm for solving nonsymmetric linear systems," *SIAM J. Sci. Stat. Comput.*, vol. 7, pp. 856–869, July 1986.

[15] L. S. Blackford, J. Demmel, J. Dongarra, I. Duff, S. Hammarling, G. Henry, M. Heroux, L. Kaufman, A. Lumsdaine, A. Petitet, R. Pozo, K. Remington, and R. C. Whaley, "An updated set of Basic Linear Algebra Subprograms (BLAS)," *ACM Transactions on Mathematical Software*, vol. 28, pp. 135–151, June 2002.

[16] E. Anderson, Z. Bai, C. Bischof, S. Blackford, J. Demmel, J. Dongarra, J. Du Croz, A. Greenbaum, S. Hammarling, A. McKenney, and D. Sorensen, *LAPACK Users' Guide*. Philadelphia, PA: Society for Industrial and Applied Mathematics, third ed., 1999.

[17] R. C. Whaley and A. Petitet, "Minimizing development and maintenance costs in supporting persistently optimized BLAS," *Software: Practice and Experience*, vol. 35, pp. 101–121, February 2005.

[18] R. C. Whaley, A. Petitet, and J. J. Dongarra, "Automated empirical optimization of software and the ATLAS project," *Parallel Computing*, vol. 27, no. 1–2, pp. 3–35, 2001.

[19] E. Agullo, J. Demmel, J. Dongarra, B. Hadri, J. Kurzak, J. Langou, H. Ltaief, P. Luszczek, and S. Tomov, "Numerical linear algebra on emerging architectures: The PLASMA and MAGMA projects," *Journal of Physics: Conference Series*, vol. 180, 2009.

[20] J. R. Humphrey, D. K. Price, K. E. Spagnoli, A. L. Paolini, and E. J. Kelmelis, "CULA: Hybrid GPU accelerated linear algebra routines," *SPIE Defense and Security Symposium (DSS)*, April 2010.

# Chapter 5

## *Thin Wires*

In this chapter we will use the Method of Moments to analyze the radiation and scattering by thin, conducting wires. This is an area of great practical interest, as many realistic antennas can be modeled by wires whose radius is much smaller than its length and the wavelength of operation. We will first derive a thin wire approximation to the magnetic vector potential, and consider the well-known Hallén and Pocklington integral equations for thin, straight wires. Next, we will discuss modeling of the feed system at an antenna's terminals, apply the EFIE (3.178) to thin wires of more general shape, and compare the effectiveness of each model through a few simple examples. We will then present several examples with geometries having more general, realistic configurations.

### 5.1   Thin Wire Approximation

Consider a perfectly conducting long, $\hat{\mathbf{z}}$-oriented thin wire of length $L$ whose radius $a$ is much less than $L$ and $\lambda$. An incident electric field $\mathbf{E}^i(\mathbf{r})$ excites on this wire a surface current $\mathbf{J}(\mathbf{r})$. As the wire is very thin, we will assume that $\mathbf{J}(\mathbf{r})$ can be written in terms of a filamentary electric current $I_z(\mathbf{r})$ as

$$\mathbf{J}(\mathbf{r}) = \frac{I_z(z)}{2\pi a}\hat{\mathbf{z}} \tag{5.1}$$

where there is no dependence on wire azimuthal angle $\phi$. We also assume that the current goes to zero at the ends of a wire without any current flow onto the wire end-caps. In cylindrical coordinates, we can write the corresponding magnetic vector potential $A_z$ as

$$A_z(\rho, \phi, z) = \mu \int_{-L/2}^{L/2} \int_0^{2\pi} \frac{I_z(z')}{2\pi} \frac{e^{-jkr}}{4\pi r} \, d\phi' \, dz' \tag{5.2}$$

where

$$r = |\mathbf{r} - \mathbf{r}'| = \sqrt{(z - z')^2 + |\boldsymbol{\rho} - \boldsymbol{\rho}'|^2} \tag{5.3}$$

Using the fact that $\rho' = a$ allows us to write

$$|\rho - \rho'| = \rho^2 + a^2 - 2\rho \cdot \rho' = \rho^2 + a^2 - 2\rho a \cos(\phi' - \phi) \qquad (5.4)$$

Since the above is a function of $\phi' - \phi$, the result is cylindrically symmetric. Therefore, we can replace $\phi' - \phi$ by just $\phi'$, and write

$$A_z(\rho, z) = \mu \int_{-L/2}^{L/2} \frac{I_z(z')}{2\pi} \int_0^{2\pi} \frac{e^{-jkr}}{4\pi r} \, d\phi' \, dz' \qquad (5.5)$$

where

$$r = \sqrt{(z - z')^2 + \rho^2 + a^2 - 2\rho a \cos \phi'} \qquad (5.6)$$

and the integral

$$\int_0^{2\pi} \frac{e^{-jkr}}{4\pi r} \, d\phi' \qquad (5.7)$$

is referred to as the cylindrical wire kernel in the literature [1, 2]. If we assume $a$ to be very small, we can approximate $r$ as

$$r = \sqrt{(z - z')^2 + \rho^2} \qquad (5.8)$$

and the innermost integral is no longer a function of $\phi'$, resulting in

$$A_z(\rho, z) = \mu \int_{-L/2}^{L/2} I_z(z') \frac{e^{-jkr}}{4\pi r} \, dz' \qquad (5.9)$$

The above is often referred to as a *thin wire approximation* with reduced kernel. The original surface integral has now been effectively replaced by a line integral along the axis of the wire. In cases where the dimensions of the problem invalidate the assumptions of the reduced kernel, the cylindrical wire kernel should be evaluated by more exact means, such as that in [3]. The total radiated field is obtained via (3.73), and is

$$-j\omega\left[1 + \frac{1}{k^2}\frac{\partial^2}{\partial z^2}\right]A_z = E_z^s \qquad (5.10)$$

By enforcing the boundary condition of zero tangential electric fields on the surface of the wire, we can now write the above in terms of the incident field $E_z^i$

$$j\omega\left[1 + \frac{1}{k^2}\frac{\partial^2}{\partial z^2}\right]A_z = E_z^i \qquad (5.11)$$

When solving the thin wire equations in this chapter, we will assume the testing points to be located on the axis of the wire and the source points to be on the surface. Also note that (5.11) comprises a simplified version of (3.178) where

**M** = 0. There are two common forms by which (5.11) is commonly written. The first retains the differential operator outside the integral, and can be written as

$$j\omega\mu\left[1 + \frac{1}{k^2}\frac{\partial^2}{\partial z^2}\right]\int_{-L/2}^{L/2} I_z(z')\frac{e^{-jkr}}{4\pi r}\,dz' = E_z^i(z) \tag{5.12}$$

This is called *Hallén's Integral Equation* [4]. We can also move the differential operator under the integral sign, yielding

$$j\omega\mu\int_{-L/2}^{L/2} I_z(z')\left[1 + \frac{1}{k^2}\frac{\partial^2}{\partial z^2}\right]\frac{e^{-jkr}}{4\pi r}\,dz' = E_z^i(z) \tag{5.13}$$

which is called *Pocklington's Integral Equation* [5]. Pocklington's equation is not as well behaved as Hallén's, as the differential operator is acting on the Green's function. As we will see, results obtained using it typically exhibit slower convergence and less accuracy than those obtained from Hallén's.

---

## 5.2 Thin Wire Excitations

In antenna problems, the unknowns of interest are most often the input impedance at a particular feedpoint and the resulting radiation pattern, directivity, and gain. The easiest way to compute these is to consider the antenna in its transmitting mode, which requires a reasonable model of the feed system at the input terminals. In practical situations, the antenna might be fed by an open-wire transmission line, or by a coaxial line through a ground plane. These various feed systems impact the antenna impedance characteristics in different ways. With this in mind, we need a way to model the field introduced by the feed system without having to model the system itself. In this section we consider two common feed methods used in thin wire problems, the delta-gap source and the magnetic frill. The delta-gap source treats the feed as if the field due to the feedline exists only in the gap between the antenna terminals, with a value of zero outside (no fringing effects). This method typically produces less accurate results for input impedance, though it still performs well in computing radiation patterns. The magnetic frill models the feed as a coaxial line that terminates in a monopole over a ground plane. Use of the frill results in more accurate input impedance values at the expense of increased computations in computing the right-hand side vector elements.

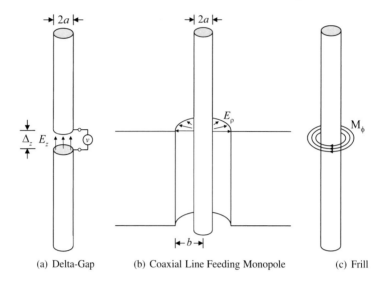

(a) Delta-Gap        (b) Coaxial Line Feeding Monopole        (c) Frill

**FIGURE 5.1:** Thin Wire Excitation Models

## 5.2.1 Delta-Gap Source

The delta-gap source model assumes that the impressed electric field in the thin gap between the antenna terminals can be expressed as

$$\mathbf{E}^i = \frac{V_o}{\Delta_z} \hat{\mathbf{z}} \tag{5.14}$$

where $\Delta_z$ is the width of the gap. This is illustrated in Figure 5.1a. In our numerical simulation, we will assume that this field exists on a single wire segment and is zero on all others. The resulting excitation vector will have nonzero elements only for basis functions having support on that segment.

## 5.2.2 Magnetic Frill

In Figure 5.1b, we depict a coaxial line feeding a monopole antenna over an infinite ground plane. If we assume the field distribution in the aperture to be purely TEM, we can use the method of images and replace the ground plane and aperture with the magnetic frill shown in Figure 5.1c. With an aperture electric field given by

$$\mathbf{E}(\rho) = \frac{1}{2\rho \log(b/a)} \hat{\boldsymbol{\rho}} \tag{5.15}$$

the equivalent magnetic current density is

$$\mathbf{M}(\rho) = -2\,\hat{\mathbf{n}} \times \mathbf{E}(\rho) = \frac{-1}{\rho \log(b/a)} \hat{\boldsymbol{\phi}} \qquad a \le \rho \le b \tag{5.16}$$

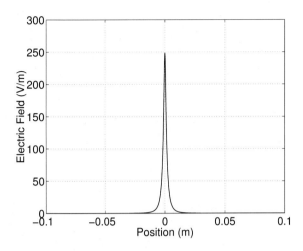

**FIGURE 5.2:** Field Intensity Due To Frill

This current generates an electric field along the wire. For a frill centered at the origin, the field intensity on the axis of the wire ($\rho = 0$) is [6]

$$E_z^i(z) = \frac{1}{2\log(b/a)} \left( \frac{e^{-jkR_1}}{R_1} - \frac{e^{-jkR_2}}{R_2} \right) \tag{5.17}$$

where

$$R_1 = \sqrt{z^2 + a^2} \tag{5.18}$$

$$R_2 = \sqrt{z^2 + b^2} \tag{5.19}$$

The $\hat{\mathbf{z}}$-directed field in (5.17) is the incident field used for the right-hand side vector. Note that by using this model, we are effectively modeling a dipole with the feed system of a monopole. A true monopole will have an input impedance only half that of the dipole as it only radiates into the half space above the ground. A plot of the axis electric field intensity due to a frill with $a = 1$ mm, $b = 5$ mm and $\lambda = 1$ m is shown in Figure 5.2. The fast drop-off of the field near $z = 0$ suggests that special care should be given to the numerical integrations used in computing the right-hand side vector elements to yield maximum accuracy.

### 5.2.3 Plane Wave

Given a plane wave incident along the direction vector $\mathbf{r}^i$, the tangential electric field intensity on a filamentary wire is given by

$$E_{tan}(\mathbf{r}) = \hat{\mathbf{t}}(\mathbf{r}) \cdot \mathbf{E}(\mathbf{r}) \tag{5.20}$$

where $\hat{\mathbf{t}}(\mathbf{r})$ is the vector tangent to the wire at $\mathbf{r}$. For a $\hat{\mathbf{z}}$-oriented wire illuminated by a $\hat{\boldsymbol{\theta}}$-oriented plane wave of unit amplitude, the above is

$$E_{tan}(\mathbf{r}) = \hat{\mathbf{z}} \cdot \hat{\boldsymbol{\theta}} \, e^{-j\mathbf{kr}\cdot\mathbf{r}^i} = \sin\theta^i e^{jkz\cos\theta^i} \qquad (5.21)$$

Because the thin-wire approximation assumes no azimuthally excited currents, an $\hat{\boldsymbol{\phi}}$-polarized incident wave will induce no current on the wire.

## 5.3  Hallén's Equation

We will now consider the solution to Hallén's Equation (5.12). If we consider it in the following form

$$j\omega\left[1 + \frac{1}{k^2}\frac{\partial^2}{\partial z^2}\right]A_z(z) = E_z^i(z) \qquad (5.22)$$

we see that it is an inhomogeneous scalar Helmholtz Equation for $A_z(z)$, which can be solved by the Green's function method. The general solution to the homogeneous equation

$$\left[1 + \frac{1}{k^2}\frac{\partial^2}{\partial z^2}\right]A_z(z) = 0 \qquad (5.23)$$

is

$$A_z(z) = C_1 e^{jkz} + C_2 e^{-jkz} \qquad (5.24)$$

To obtain a particular solution we must obtain the Green's function $G(z)$ which satisfies the equation

$$j\omega\left[1 + \frac{1}{k^2}\frac{\partial^2}{\partial z^2}\right]G(z) = \delta(z) \qquad (5.25)$$

Once $G(z)$ is known, the solution for $A_z(z)$ can be obtained by

$$A_z(z) = C_1 e^{jkz} + C_2 e^{-jkz} + \int_{-L/2}^{L/2} G(z,z')\, E_z^i(z')\, dz' \qquad (5.26)$$

To obtain the Green's function, let us use the trial function

$$G(z) = C\sin(k|z|) \qquad (5.27)$$

which is continuous at $z = 0$ with discontinuous derivative at $z = 0$ as is required for the Green's function [7]. To determine the constant C, we integrate (5.25) from $-\epsilon$ to $\epsilon$, yielding

$$j\omega C \int_{-\epsilon}^{\epsilon} \sin(k|z|)\, dz - \frac{jw}{k} C[\cos(-kz)]|_{-\epsilon}^{0} + \frac{jw}{k} C[\cos(kz)]|_{0}^{\epsilon} = 1 \quad (5.28)$$

If we allow $\epsilon$ to approach zero, the first term goes to zero and after evaluating the remaining terms we get

$$C = -\frac{jk}{2\omega} = -\frac{j\mu}{2\eta} \quad (5.29)$$

hence

$$G(z) = -\frac{j\mu}{2\eta} \sin(k|z|) \quad (5.30)$$

and the solution for $A_z(z)$ is

$$A_z(z) = C_1 e^{jkz} + C_2 e^{-jkz} - \frac{j\mu}{2\eta} \int_{-L/2}^{L/2} \sin(k|z - z'|)\, E_z^i(z')\, dz' \quad (5.31)$$

Substituting the original expression for $A_z$ on the left-hand side of the above yields

$$\int_{-L/2}^{L/2} I_z(z') \frac{e^{-jkr}}{4\pi r}\, dz' = C_1 e^{jkz} + C_2 e^{-jkz}$$
$$- \frac{j}{2\eta} \int_{-L/2}^{L/2} \sin(k|z - z'|)\, E_z^i(z')\, dz' \quad (5.32)$$

By similar reasoning, a second solution for the Green's function is found to be

$$G_z(z) = \frac{\mu}{2\eta} e^{-jk|z|} \quad (5.33)$$

which leads to a second expression for $A_z$

$$\int_{-L/2}^{L/2} I_z(z') \frac{e^{-jkr}}{4\pi r}\, dz' = C_1 e^{jkz} + C_2 e^{-jkz} + \frac{1}{2\eta} \int_{-L/2}^{L/2} e^{-jk|z-z'|}\, E_z^i(z')\, dz'$$
$$(5.34)$$

Note also that the two homogeneous terms in the above can also be written as

$$C_1 e^{jkz} + C_2 e^{-jkz} = D_1 \cos(kz) + D_2 \sin(kz) \quad (5.35)$$

where $D_1$ and $D_2$ are complex. The form we use will depend on the symmetry of the problem.

## 5.3.1    Symmetric Problems

We will first solve Hallén's Equation for symmetric problems such as the induced current and input impedance of a center-fed dipole antenna. We first expand the left-hand side of (5.32) using $N$ weighted basis functions, yielding

$$\sum_{n=1}^{N} a_n \int_{f_n} f_n(z') \frac{e^{-jkr}}{4\pi r} \, dz' \tag{5.36}$$

Applying the Method of Moments and testing (5.32) using $N$ testing functions $f_m$ yields matrix elements $Z_{mn}$ given by

$$Z_{mn} = \int_{f_m} f_m(z) \int_{f_n} f_n(z') \frac{e^{-jkr}}{4\pi r} \, dz' \, dz \tag{5.37}$$

The choice of right-hand side depends on the symmetry of the problem being considered. Since we are feeding the antenna at its center, we expect that the induced current will be symmetric. Therefore, we will retain the right-hand side of (5.32) but rewrite the homogeneous terms following (5.35), yielding a right-hand side given by

$$D_1 \cos(kz) + D_2 \sin(kz) - \frac{j}{2\eta} \int_{-L/2}^{L/2} \sin(k|z - z'|) \, E_z^i(z') \, dz' \tag{5.38}$$

As we expect the solution to be symmetric, we set $D_2$ in the above to zero. Applying the MOM to the right-hand side yields

$$D_1 \int_{f_m} f_m(z) \cos(kz) \, dz - \frac{j}{2\eta} \int_{f_m} f_m(z) \int_{-L/2}^{L/2} \sin(k|z - z'|) \, E_z^i(z') \, dz' \, dz \tag{5.39}$$

The above expressions comprise a linear system of the form

$$\mathbf{Za} = D_1 \mathbf{s} + \mathbf{b} \tag{5.40}$$

To obtain the solution vector $\mathbf{a}$, we must determine the constant $D_1$, which can be done by enforcing the boundary conditions at the ends of the wire, which are

$$I_z(-L/2) = I_z(L/2) = 0 \tag{5.41}$$

We accomplish this in our discretized version by forcing the basis function coefficient at each end to be zero [8]. This can be expressed vectorially as $\mathbf{u}^T \mathbf{a} = 0$, where $\mathbf{u}^T = [1, 0, \ldots, 0, 1]$. Solving (5.40) for $\mathbf{a}$ we obtain

$$\mathbf{a} = D_1 \mathbf{Z}^{-1} \mathbf{s} + \mathbf{Z}^{-1} \mathbf{b} \tag{5.42}$$

and multiplying both sides by $\mathbf{u}^T$ we obtain

$$\mathbf{u}^T\mathbf{a} = D_1\mathbf{u}^T\mathbf{Z}^{-1}\mathbf{s} + \mathbf{u}^T\mathbf{Z}^{-1}\mathbf{b} = 0 \qquad (5.43)$$

and solving for $D_1$ yields

$$D_1 = -\frac{\mathbf{u}^T\mathbf{Z}^{-1}\mathbf{b}}{\mathbf{u}^T\mathbf{Z}^{-1}\mathbf{s}} \qquad (5.44)$$

The complete right-hand side vector can now be built from $\mathbf{s}$ and $\mathbf{b}$ using $D_1$.

### 5.3.1.1 Solution Using Pulse Functions and Point Matching

To solve the symmetric Hallén's Equation with pulse basis functions and point matching, we subdivide the wire into $N$ equally spaced segments of length $L/N$. The matrix elements of (5.37) are then

$$Z_{mn} = \int_{z_n-\Delta_z/2}^{z_n+\Delta_z/2} \frac{e^{-jkR}}{4\pi R}\, dz' \qquad (5.45)$$

where the matching is done at the center of the testing segment ($z = z_m$), and $R = \sqrt{(z_m - z')^2 + a^2}$. The non-self terms are computed from (5.45) using an $M$-point numerical quadrature.

*Self Terms*

When the source and testing segments overlap, we use a small-argument approximation to the Green's function to write

$$Z_{mm} = \int_{-\Delta_z/2}^{\Delta_z/2} \frac{e^{-jkR}}{4\pi R}\, dz' \approx \int_{-\Delta_z/2}^{\Delta_z/2} \frac{1 - jkR}{4\pi R}\, dz' \qquad (5.46)$$

which evaluates to [9] (200.01)

$$Z_{mm} = \frac{1}{4\pi} \log\left[\frac{\sqrt{1 + 4a^2/\Delta_z^2} + 1}{\sqrt{1 + 4a^2/\Delta_z^2} - 1}\right] - \frac{jk\Delta_z}{4\pi} \qquad (5.47)$$

We assume that the approximation in (5.47) is valid when the segment is small compared to the wavelength.

*Excitation*

The components of the right-hand side vector elements are

$$s_m = \cos(kz_m) \qquad (5.48)$$

and

$$b_m = -\frac{j}{2\eta} \int_{-L/2}^{L/2} \sin\left(k|z_m - z'|\right) E_z^i(z') \, dz' \tag{5.49}$$

where we have not yet specified the incident field $E_z^i(z')$. If we use a delta-gap source located at the center of the wire, $E_z^i(z') = \delta(z')$, and (5.49) simplifies to

$$b_m = -\frac{j}{2\eta} \sin(k|z_m|) \tag{5.50}$$

If we use the magnetic frill of (5.17), the convolution of (5.49) must be performed numerically for each $z_m$. Because of the sharp drop-off in the frill amplitude, special care should be given to this integral to ensure its accuracy.

## 5.3.2   Asymmetric Problems

When the feedpoint is not at the center of the antenna or the wire is excited by an incident wave, we can no longer assume the solution to be symmetric. In this case, the elements of the system matrix remain the same as in (5.37), however we will use the more general right-hand side of (5.34). Testing the right-hand side by testing function $f_m$ yields

$$C_1 \int_{f_m} f_m(z) e^{jkz} \, dz + C_2 \int_{f_m} f_m(z) e^{-jkz} \, dz$$

$$+ \frac{1}{2\eta} \int_{f_m} f_m(z) \int_{-L/2}^{L/2} e^{-jk|z_m - z'|} E_z^i(z') \, dz' \, dz \tag{5.51}$$

The resulting matrix equation is of the form

$$\mathbf{Za} = C_1 \mathbf{s}_1 + C_2 \mathbf{s}_2 + \mathbf{b} \tag{5.52}$$

which can be rewritten as

$$\mathbf{Za} = [\mathbf{s}_1, \mathbf{s}_2] \, [C_1, C_2]^T + \mathbf{b} \tag{5.53}$$

Solving for **a**, we obtain

$$\mathbf{a} = \mathbf{Z}^{-1}[\mathbf{s}_1, \mathbf{s}_2] \, [C_1, C_2]^T + \mathbf{Z}^{-1}\mathbf{b} \tag{5.54}$$

We must solve for the constants $C_1$ and $C_2$ to obtain **a**, which can be done by a method similar to the symmetric case [8]. Let us define the matrix $[\mathbf{U}] = [\mathbf{u}_1^T, \mathbf{u}_2^T]$, where the vectors $\mathbf{u}_1^T = [1, 0, \ldots, 0, 0]$ and $\mathbf{u}_2^T = [0, 0, \ldots, 0, 1]$ select the elements on the end of the wire and enforce the condition

$$\mathbf{U}^T \mathbf{a} = 0 \tag{5.55}$$

Applying this to the above yields

$$\mathbf{U}^T \mathbf{a} = \mathbf{U}^T \mathbf{Z}^{-1}[\mathbf{s}_1, \mathbf{s}_2] \, [C_1, C_2]^T + \mathbf{U}^T \mathbf{Z}^{-1} \mathbf{b} = 0 \qquad (5.56)$$

Solving for $[C_1, C_2]^T$ yields

$$[C_1, C_2]^T = \left[ \mathbf{U}^T \mathbf{Z}^{-1}[\mathbf{s}_1, \mathbf{s}_2] \right]^{-1} \mathbf{U}^T \mathbf{Z}^{-1} \mathbf{b} \qquad (5.57)$$

### 5.3.2.1 Solution Using Pulse Functions and Point Matching

The solution of the asymmetric Hallén's Equation by point matching is similar to the symmetric case, as the system matrix elements remain the same. The only difference lies in the right-hand side vector elements.

*Excitation*

Using point matching, the components of the right-hand side vector elements in (5.52) are

$$s_{1,m} = e^{jkz_m} \qquad (5.58)$$

$$s_{2,m} = e^{-jkz_m} \qquad (5.59)$$

and

$$b_m = -\frac{j}{2\eta} \int_{-L/2}^{L/2} e^{-jk|z_m - z'|} \, E_z^i(z') \, dz' \qquad (5.60)$$

where we have not yet specified the incident field $E_z^i(z')$. If we place a delta-gap source within a wire segment centered at $z_c$, $E_z^i(z') = \delta(z_c)$, and (5.60) becomes

$$b_m = -\frac{j}{2\eta} e^{-jk|z_m - z_c|} \qquad (5.61)$$

If instead the incident field comprises a magnetic frill or other incident field, (5.60) must be computed numerically for each $z_m$.

---

## 5.4 Pocklington's Equation

Pocklington's Equation (5.13) is relatively straightforward to solve using the Method of Moments, as the differential operator acts only on the Green's function in the innermost integral. Expanding the current into a sum of $N$

weighted basis functions, and using $N$ testing functions, we obtain a linear
system with matrix elements given by

$$Z_{mn} = \frac{j\omega\mu}{4\pi} \int_{f_m} f_m(z) \int_{f_n} f_n(z') \left[1 + \frac{1}{k^2} \frac{\partial^2}{\partial z^2}\right] \frac{e^{-jkr}}{r} \, dz' \, dz \qquad (5.62)$$

and excitation vector elements given by

$$b_m = \int_{f_m} f_m(z) \, E_z^i(z) \, dz \qquad (5.63)$$

## 5.4.1   Solution Using Pulse Functions and Point Matching

We again subdivide the wire into $N$ equally spaced segments of length
$L/N$. Using pulse basis functions point matching at $z_m$, the first part of (5.62)
is

$$\frac{j\omega\mu}{4\pi} \int_{z_n - \Delta_z/2}^{z_n + \Delta_z/2} \frac{e^{-jkR}}{R} \, dz' \qquad (5.64)$$

where

$$R = \sqrt{(z_m - z')^2 + a^2} \qquad (5.65)$$

We note that (5.64) has the same form as (5.45), and can be computed the same
way. The second part of (5.62) can be written as

$$\frac{j}{4\pi\omega\epsilon} \int_{f_m} f_m(z) \int_{f_n} f_n(z') \frac{\partial^2}{\partial z^2} \frac{e^{-jkr}}{r} \, dz' \, dz \qquad (5.66)$$

Evaluating the first partial derivative in the integrand yields

$$\frac{\partial}{\partial z} \frac{e^{-jkr}}{r} = -(z - z') \frac{1 + jkr}{r^3} e^{-jkr} \qquad (5.67)$$

and inserting the above into (5.66) yields

$$\frac{j}{4\pi\omega\epsilon} \int_{f_m} f_m(z) \int_{f_n} f_n(z') \frac{\partial}{\partial z} \left[ -(z - z') \frac{1 + jkr}{r^3} e^{-jkr} \right] dz' \qquad (5.68)$$

Using pulse basis functions point matching at $z_m$, the above evaluates to

$$\frac{j}{4\pi\omega\epsilon} \left[ (z_m - z') \frac{1 + jkR}{R^3} e^{-jkR} \right] \Bigg|_{z'=z_n - \Delta_z/2}^{z'=z_n + \Delta_z/2} \qquad (5.69)$$

which allows us to write the matrix elements as

$$Z_{mn} = \frac{j\omega\mu}{4\pi} \int_{z_n - \Delta_z/2}^{z_n + \Delta_z/2} \frac{e^{-jkR}}{R} \, dz'$$

$$+ \frac{j}{4\pi\omega\epsilon} \left[ (z_m - z') \frac{1 + jkR}{R^3} e^{-jkR} \right] \Bigg|_{z'=z_n - \Delta_z/2}^{z'=z_n + \Delta_z/2} \qquad (5.70)$$

The second term is analytic and can be used as-is in computing each matrix element. Because of the strongly singular $1/R^3$ term, we will find that our results converge less quickly than those obtained using Hallén's equation for the same problem. The corresponding right-hand side vector elements are

$$b_m = E_z^i(z_m) \tag{5.71}$$

## 5.5 Thin Wires of Arbitrary Shape

The formulations considered thus far treated wires that are perfectly straight. Because realistic antennas have curves, bends, and wire-to-wire junctions, we need to develop a thin wire treatment that takes on a more general form. We will do so in this section.

### 5.5.1 Method of Moments Discretization

Consider a curved thin wire, illustrated in Figure 5.3a. We subdivide the curve into $N$ segments with $N+1$ endpoints as shown in Figure 5.3b. Though the segments need not be of equal length, they should be small enough to reflect the curvature of the wire, and should support the required number of basis functions per wavelength. The tangent vector on the curve is now a piecewise continuous function, as is illustrated with the arrows. As there is only electric current, we use the EFIE (3.178). The filamentary current $\mathbf{J}(\mathbf{r})$ is expanded using electric basis functions $\mathbf{f}(\mathbf{r})$ whose vector components are tangent to the

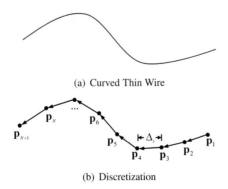

(a) Curved Thin Wire

(b) Discretization

**FIGURE 5.3:** Curved Thin Wire Discretization

segment(s) on which it resides. Testing the EFIE with these same functions yields the matrix elements (3.185) for the thin wire case, which are

$$Z_{mn}^{EJ} = j\omega\mu \int_{\mathbf{f}_m} \mathbf{f}_m(\mathbf{r}) \cdot \int_{\mathbf{f}_n} \mathbf{f}_n(\mathbf{r}') \, G(\mathbf{r}, \mathbf{r}') \, d\mathbf{r}' \, d\mathbf{r}$$
$$- \frac{j}{\omega\epsilon} \int_{\mathbf{f}_m} \nabla \cdot \mathbf{f}_m(\mathbf{r}) \int_{\mathbf{f}_n} \nabla' \cdot \mathbf{f}_n(\mathbf{r}') \, G(\mathbf{r}, \mathbf{r}') \, d\mathbf{r}' \, d\mathbf{r} \qquad (5.72)$$

where we have re-distributed the differential operators following (3.204). The right-hand side vector elements (3.191) are

$$V_m^E = \int_{\mathbf{f}_m} \mathbf{f}_m(\mathbf{r}) \cdot \mathbf{E}^i(\mathbf{r}) \, d\mathbf{r} \qquad (5.73)$$

We will refer to the approach in this section as a thin-wire EFIE. Similar approaches are found in the literature.

## 5.5.2   Solution Using Triangle Basis and Testing Functions

We will use triangle functions to solve the thin-wire EFIE. As the current must go to zero on the ends of the wire, we assign triangle functions everywhere except at the wire endpoints. and because the basis functions have vector components, they must be oriented properly. Consider a triangle function which has support on segments $s+$ and $s-$ whose endpoints comprise $(\mathbf{p}_{k-1}, \mathbf{p}_k, \mathbf{p}_{k+1})$, as shown in Figure 5.4. The vector components are $\mathbf{t}^+$ on $s+$, and $\mathbf{t}^-$ on $s-$. Although the vector orientations are shown going right to left, it is equally valid to orient them from left to right; the only requirement is that the orientations satisfy Kirchoff's law at $p_k$.

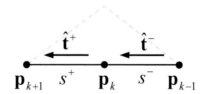

**FIGURE 5.4:** Vector Triangle Function Orientation

### 5.5.2.1 Non-Self Terms

For segments that do not overlap, the matrix elements (5.72) are computed using an $M$-point numerical quadrature formula, yielding

$$Z_{mn}^{EJ} = \frac{1}{4\pi} \sum_{p=1}^{M} \sum_{q=1}^{M} w_p(\mathbf{r}_p) w_q(\mathbf{r}_q') \left[ j\omega\mu \, \mathbf{f}_m(\mathbf{r}_p) \cdot \mathbf{f}_n(\mathbf{r}_q') \pm \frac{j}{\omega\epsilon\Delta_p\Delta_q} \right] \frac{e^{-jkR_{pq}}}{R_{pq}}$$

(5.74)

where $\mathbf{r}_p$ and $\mathbf{r}_q$ are quadrature points along the wire axis, and

$$R_{pq} = |\mathbf{r}_p - \mathbf{r}_q'|$$

(5.75)

The sums in (5.74) are performed over all segments where the source and testing functions reside. For a triangle function that flows into or out of its anchor point on a segment of length $\Delta_l$, its divergence is $1/\Delta_l$ or $-1/\Delta_l$ on that segment, respectively. The sign of the second term then depends on the orientation of the triangle functions on the source and testing segments. Note that it is more efficient to compute the matrix elements by looping over pairs of segments instead of pairs of basis and testing functions.

### 5.5.2.2 Self Terms

For overlapping segments, we will evaluate the innermost integral analytically and the outermost integral numerically. To calculate the first term on the right in (5.72), consider the following innermost integral with the triangle function of positive slope $x'/\Delta_l$

$$S_1(x) = \int_0^{\Delta_l} \frac{x'}{\Delta_l} \frac{e^{-jkr}}{r} \, dx'$$

(5.76)

where $r = \sqrt{(x-x')^2 + a^2}$. Using the small-argument approximation to the Green's function this becomes

$$\int_0^{\Delta_l} \frac{x'}{\Delta_l} \frac{e^{-jkr}}{r} \, dx' \approx \int_0^{\Delta_l} \frac{x'}{\Delta_l} \frac{1-jkr}{r} \, dx'$$

(5.77)

which is

$$\int_0^{\Delta_l} \frac{x'}{\Delta_l} \frac{1-jkr}{r} \, dx' = \frac{1}{\Delta_l} \int_0^{\Delta_l} \frac{x'}{r} \, dx' - \frac{jk\Delta_l}{2}$$

(5.78)

The first term on the right then evaluates to [10]

$$\frac{1}{\Delta_l}\sqrt{a^2 + (x-\Delta_l)^2} - \frac{1}{\Delta_l}\sqrt{a^2 + x^2} + \frac{x}{\Delta_l} \log\left[ \frac{x + \sqrt{a^2 + x^2}}{x - \Delta_l + \sqrt{a^2 + (x-\Delta_l)^2}} \right]$$

resulting in

$$S_1(x) = \frac{1}{\Delta_l}\sqrt{a^2 + (x - \Delta_l)^2} - \frac{1}{\Delta_l}\sqrt{a^2 + x^2}$$
$$+ \frac{x}{\Delta_l}\log\left[\frac{x + \sqrt{a^2 + x^2}}{x - \Delta_l + \sqrt{a^2 + (x - \Delta_l)^2}}\right] - \frac{jk\Delta_l}{2} \qquad (5.79)$$

The integral involving the triangle of positive slope is sufficient to obtain all the self terms because of symmetry. We will compute the second term on the right in (5.72) the same way as the first term. The innermost integral is

$$S_2(x) = \pm\frac{1}{\Delta_l^2}\int_0^{\Delta_l}\frac{1 - jkr}{r}\,dx' \qquad (5.80)$$

where we have again used the small-argument approximation to the Green's function, and the sign of the integral depends on whether the derivatives of the source and testing functions are of opposing sign. The above evaluates to [10]

$$S_2(x) = \pm\frac{1}{\Delta_l^2}\left[\log\left[\frac{x + \sqrt{a^2 + x^2}}{x - \Delta_l + \sqrt{a^2 + (x - \Delta_l)^2}}\right] - jk\Delta_l\right] \qquad (5.81)$$

The self-term contributions to (5.72) can then be obtained via numerical quadrature as

$$\frac{1}{4\pi}\sum_{p=1}^{M}w_p(x_p)\left[j\omega\mu f_m(x_p)\,S_1(x_p) - \frac{j}{\omega\epsilon}S_2(x_p)\right] \qquad (5.82)$$

### 5.5.3    Solution Using Sinusoidal Basis and Testing Functions

The solution using sinusoidal basis and testing functions is practically the same as with triangle functions. The difference lies in the self terms, which we compute using numerical outer and analytic inner integrations as before.

#### 5.5.3.1    Self Terms

To calculate the first term on the right in (5.72), consider the following innermost integral with the sinusoid function of positive slope

$$S_1(x_p) = \frac{1}{\sin(k\Delta_l)}\int_0^{\Delta_l}\sin(kx')\frac{e^{-jkr}}{r}\,dx' \qquad (5.83)$$

Again using the small-argument approximation to the Green's Function this becomes

$$\int_0^{\Delta_l}\sin(kx')\frac{e^{-jkr}}{r}\,dx' \approx \int_0^{\Delta_l}\sin(kx')\left[\frac{1 - jkr}{r}\right]dx' \qquad (5.84)$$

which is

$$\int_0^{\Delta_l} \sin(kx') \left[ \frac{1 - jkr}{r} \right] dx' = \int_0^{\Delta_l} \frac{\sin(kx')}{r} dx' + j \left[ \cos(k\Delta_l) - 1 \right] \quad (5.85)$$

The first term on the right is not tractable in its present form, so we will approximate the sine by its third-order polynomial approximation. This yields

$$\int_0^{\Delta_l} \frac{\sin(kx')}{r} dx' \approx \int_0^{\Delta_l} \frac{kx' - (kx')^3/6}{\sqrt{(x' - x_p)^2 + a^2}} dx' \quad (5.86)$$

Evaluating the above results in [10]

$$S_1(x_p) = C \frac{kx_p}{12} \left( \left[ -3a^2 k^2 + 2x_p^2 k^2 - 12 \right] \log \left[ x_p - x' + \sqrt{a^2 + (x_p - x')^2} \right] \right.$$

$$\left. - \frac{k}{36} \left[ -4a^2 k^2 + k^2 \left( 11x_p^2 + 5x' x_p - 2(x')^2 \right) - 36 \right] \cdot \right.$$

$$\left. \left. \sqrt{a^2 + (x_p - x')^2} \right) \right|_0^{\Delta_l} + jC \left[ \cos(k\Delta_l) - 1 \right] \quad (5.87)$$

where $C = 1/\sin(k\Delta_l)$. This result is lengthy but can be evaluated numerically in a straightforward manner. We compute the second term on the right of (5.72) the same way. Under the small-argument approximation to the Green's function, the innermost integral in this case is

$$S_2(x_p) = \frac{k}{\sin(k\Delta_l)} \int_0^{\Delta_l} \frac{\cos(kx')}{r} dx' - jk \quad (5.88)$$

We approximate the cosine by the second-order approximation, yielding the integral

$$\int_0^{\Delta_l} \frac{\cos(kx')}{r} dx' \approx \int_0^{\Delta_l} \frac{1 - (kx')^2/2}{\sqrt{(x' - x_p)^2 + a^2}} dx' \quad (5.89)$$

The above evaluates to [10]

$$S_2(x_p) = -\frac{C}{4} \left( k^2 \left[ 3x_p + x' \right] \sqrt{a^2 + (x_p - x')^2} + (a^2 k^2 - 2x_p^2 k^2 + 4) \cdot \right.$$

$$\left. \log \left[ x_p - x' + \sqrt{a^2 + (x_p - x')^2} \right] \right) \Bigg|_0^{\Delta_l} - jk \quad (5.90)$$

The contributions to the self terms can then be computed via numerical quadrature as

$$\frac{1}{4\pi} \sum_{p=1}^M w_p(x_p) \left[ j\omega\mu \, s_m(x_p) S_1(x_p) - \frac{j}{\omega\epsilon} s'_m(x_p) S_2(x_p) \right] \quad (5.91)$$

where $s_m(x)$ is the sinusoidal testing function.

### 5.5.4   Lumped and Distributed Impedances

In some cases it may be necessary to modify the impedance characteristics of an antenna by loading it at one or more points. Feedpoint matching is one of the most common adjustments, and is done by placing an impedance in parallel at the feedpoint, or by inserting coils or capacitors at one or more points along the wire. To model these lumped impedances in the thin wire EFIE, we must take into account the boundary conditions on the segments where the loads exist. Given a complex impedance $Z_l$ on a segment of length $\Delta_l$, the boundary condition on this segment is

$$E_{tan}^i + E_{tan}^s = E_{load} = \frac{V_l}{\Delta_l} = \frac{Z_l I_s}{\Delta_l} \tag{5.92}$$

The right-hand side of the EFIE now becomes

$$\frac{Z_l I_l}{\Delta_l} - E_{tan}^i \tag{5.93}$$

where the current $I_l$ on the segment is supported by one or more basis functions. If we use triangle functions in the thin wire EFIE, the testing operation with the basis function $f_m$ creates a "self term" on the right-hand side of the form

$$b_m' = \frac{Z_l}{2} \tag{5.94}$$

which can then be moved to the left-hand side and subtracted from the diagonal element $Z_{mm}^{EJ}$. Any basis function having support on this segment will have its self term modified in this manner.

## 5.6 Examples

In this section we consider several thin wire antenna problems. We will first compare the relative performance of the Hallén, Pocklington, and thin-wire EFIE models in computing the input impedance of a straight wire. We will then consider problems involving a half-wavelength dipole, circular loop, folded dipole, and a multi-element Yagi. For each problem, we will use the thin-wire EFIE with triangular basis functions.

### 5.6.1 Comparison of Thin Wire Models

For this comparison, we will compute the input impedance and induced current distribution on center-fed dipole antennas. To judge the effectiveness of each method, we will compare the rates of convergence versus the number of wire segments used in the discretization. We use a delta-gap voltage source and an odd number of wire segments so the voltage source is exactly at the center of the antenna. For numerical integration, a 5-point Gaussian quadrature is used on each segment. Each dipole has a radius of $10^{-3}\lambda$, and in computing the induced current distributions, the amplitude of the applied voltage is unity.

#### 5.6.1.1 Input Impedance

We first compare the convergence of the input impedance versus the number of wire segments used. We first consider a dipole of length $\lambda/2$, with the resistance and reactance plotted in Figures 5.5a and 5.5b, respectively. The results from Hallén's and Pocklington's equation slowly converge toward a value approaching that of the thin wire EFIE models, which converge quickly. This is not surprising, since triangle and sinusoidal functions are better at modeling the current distribution than pulses, and the boundary conditions are enforced more accurately than with point matching. Similar comparisons for the resistance and reactance of a $3\lambda/2$ dipole are made in Figures 5.6a and 5.6b, respectively. The results from the thin wire EFIE models again converge quickly, whereas those of Pocklington's Equation are again poor at low segment counts, demonstrating the limitations in pulse basis functions and point matching, as well as the $1/R^3$ singularity in the integrand. This singularity would be even further compounded were the radius of the wire made smaller.

We next set the number of segments per wavelength to 25, and vary the length of the dipole from 0.1 to $3\lambda$. The results are shown in Figure 5.7. The thin-wire EFIE results agree very well at almost every data point, suggesting that triangles and sinusoids have similar performance characteristics for thin wire problems. The impedances obtained from Hallén's and Pocklington's Equation do not agree very well at this level of discretization, as expected.

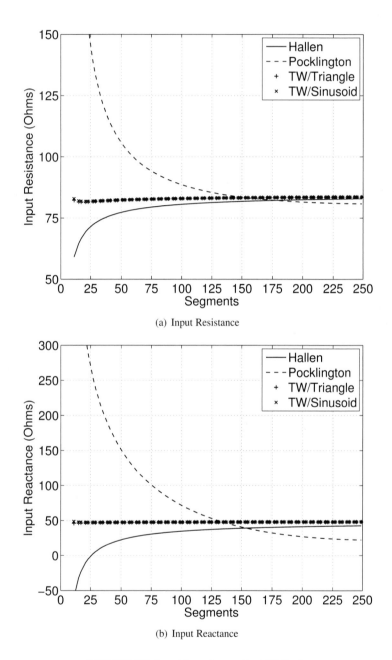

(a) Input Resistance

(b) Input Reactance

**FIGURE 5.5:** Input Impedance of a $\lambda/2$ Dipole

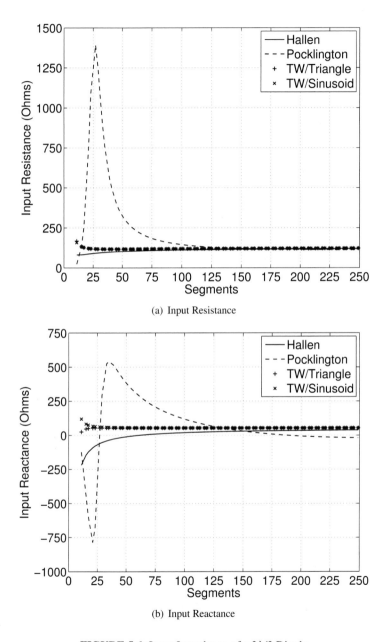

(a) Input Resistance

(b) Input Reactance

**FIGURE 5.6:** Input Impedance of a $3\lambda/2$ Dipole

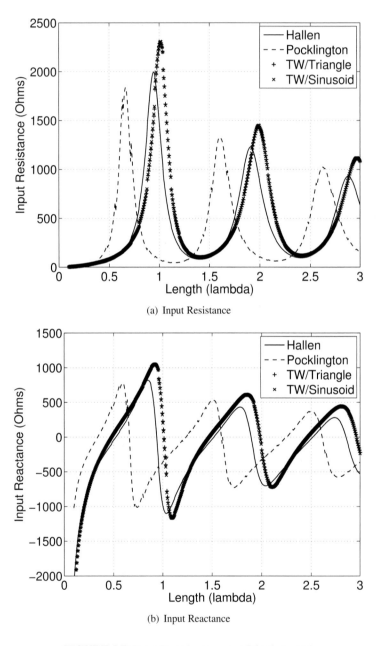

(a) Input Resistance

(b) Input Reactance

**FIGURE 5.7:** Input Impedance versus Dipole Length

### 5.6.1.2 Induced Current Distribution

We next compare the currents induced on a center-fed dipole as computed by each model. For the comparison, we will plot the current at the center of each wire segment. Note that the solution via Hallén's equation will yield exactly zero current in the first and last segments, as this was enforced explicitly. In Figures 5.8a and 5.8b are shown the currents induced on a $\lambda/2$ dipole computed using 31 and 81 wire segments, respectively. The thin-wire EFIE models show almost no change between the two figures indicating excellent convergence, whereas the currents obtained from Hallén's and Pocklington's equation vary significantly. Results for a $2\lambda$ dipole are shown in Figures 5.9a and 5.9b for 81 and 181 wire segments, respectively. Hallén's equation does somewhat better at the lower segment count than does Pocklington's equation, though they are both fairly good at the higher count. The convergence of the thin-wire EFIE models is again excellent.

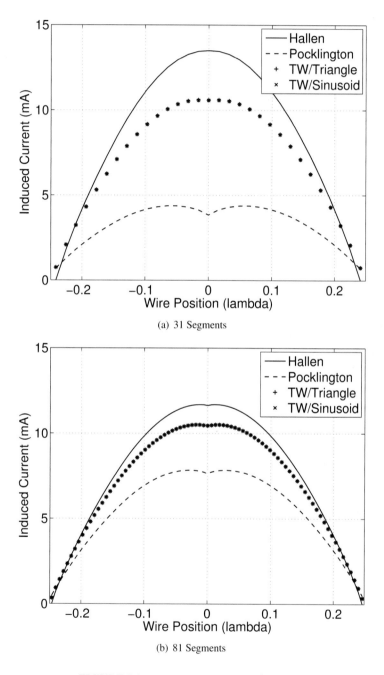

(a) 31 Segments

(b) 81 Segments

**FIGURE 5.8:** Induced Current on a $\lambda/2$ Dipole

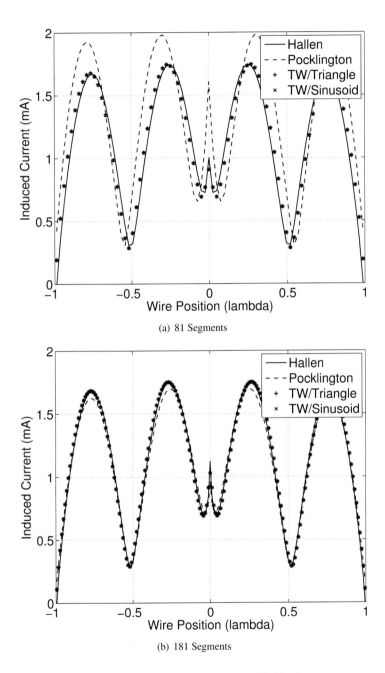

(a) 81 Segments

(b) 181 Segments

**FIGURE 5.9:** Induced Current on a $2\lambda$ Dipole

## 5.6.2   Half-Wavelength Dipole

It is known that a dipole in free space has its first resonance at a length just under $\lambda/2$. Using the thin-wire EFIE and an iterative procedure, we find that a resonant dipole of radius $10^{-3}\lambda$ has a length of approximately $0.48\lambda$ and a purely resistive input impedance of 74 $\Omega$. Such an antenna could be fed directly with a 50 $\Omega$ or 75 $\Omega$ coaxial cable (RG-58 or RG-59, respectively) and a balun, with a reasonable SWR.

Let us move the feedpoint to the origin and align the antenna with the $z$ axis. Using (3.108), we then compute the near field in a $20\lambda \times 20\lambda$ meter region of the $xz$ plane. The real part of the electric field (in V/m) is plotted in Figure 5.10. We note that the near field region is small in extent, and the field decays fast along the $z$ axis. Next, we compare the computed field pattern to an ideal dipole from antenna theory. Recall that the radiated electric field of an infinitesimal, $\hat{z}$-oriented electric current has $\hat{r}$ and $\hat{\theta}$ components given by

$$E_r(r, \theta) = -\frac{k^2 \eta I \Delta_l}{2\pi}\left[\frac{1}{(jkr)^2} + \frac{1}{(jkr)^3}\right]\cos\theta\, e^{-jkr} \qquad (5.95)$$

and

$$E_\theta(r, \theta) = -\frac{k^2 \eta I \Delta_l}{2\pi}\left[\frac{1}{jkr} + \frac{1}{(jkr)^2} + \frac{1}{(jkr)^3}\right]\sin\theta\, e^{-jkr} \qquad (5.96)$$

where $I$ is the current amplitude and $\Delta_l$ is its length, where $\Delta_l \ll \lambda$. If we know $I(z)$, we can then integrate (5.95) and (5.96) to determine the total field. For the $\lambda/2$ dipole, we will use an ideal sinusoidal distribution model for current, where

$$I(z) = I_0 \sin\left[k(L/2 - z)\right] \qquad (5.97)$$

and $I_0$ is the amplitude of the current at the feedpoint. We now compute the real part of the total field in the same region as before, with results (in V/m) shown in Figure 5.11. The results compare very well, though they are not identical. We next compare the computed currents to those of (5.97) for the same feedpoint amplitude in Figure 5.12, where we see that the computed current has a slightly different amplitude envelope.

Finally, we compute the radiated field in the $xz$ plane at a fixed distance of $10\lambda$ from the antenna for $0 \le \theta \le 2\pi$. To this, we compare the normalized far electric field of an ideal $\lambda/2$ dipole, which is [11]

$$E_\theta(\theta) = \frac{\cos\left(\frac{\pi}{2}\cos\theta\right)}{\sin\theta} \qquad (5.98)$$

The normalized amplitudes are compared in Figure 5.13, where the agreement is seen to be quite good.

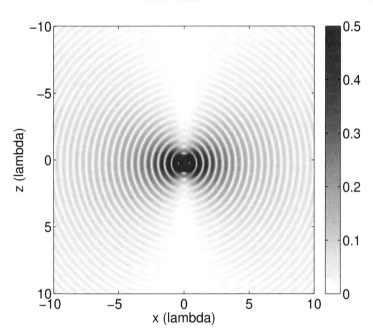

**FIGURE 5.10:** Radiated Near Field of a $\lambda/2$ Dipole ($xz$ plane, MOM)

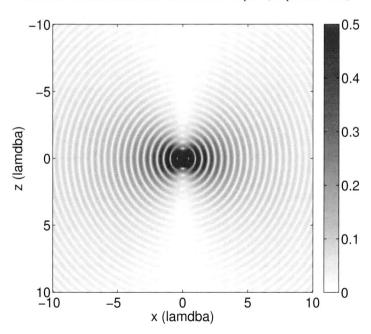

**FIGURE 5.11:** Radiated Near Field of a $\lambda/2$ Dipole ($xz$ plane, Equation)

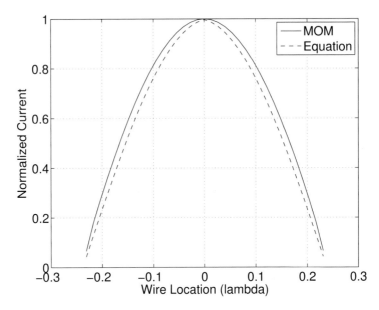

**FIGURE 5.12:** Comparison of Current Amplitudes on the $\lambda/2$ Dipole

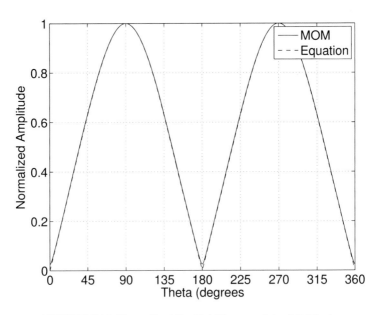

**FIGURE 5.13:** Normalized Far Field Pattern of the $\lambda/2$ Dipole

### 5.6.3 Circular Loop Antenna

We next consider the input impedance of a circular loop antenna in the $xy$ plane. Setting the radius of the wire at $10^{-4}\lambda$, we vary the circumference of the loop up to $2.5\lambda$. At each step, we subdivide the loop into approximately 10 segments per wavelength, using no fewer than 36 segments at the smallest electrical length. This ensures that the model retains the shape of a circle at the lowest frequency. The feedpoint of the antenna is placed at the point where the $+x$ axis intersects the loop. The real and imaginary parts of the input impedance are shown in Figure 5.14. For lengths under $0.5\lambda$, the loop has a high inductance with virtually no radiation resistance. At approximately $\lambda/2$, the loop becomes an open circuit. We see first and second resonances at approximately $1.03\lambda$ and $2.05\lambda$, where the input impedance is about 140 $\Omega$ and 182 $\Omega$, respectively.

The real part of the radiated electric field in the $xy$ plane is shown (in V/m) in Figure 5.15, and the corresponding normalized field amplitude at a distance of $10\lambda$ is shown versus $\phi$ in Figure 5.16. Four nulls are visible, and a slight asymmetry in the pattern is seen. Similar plots of the fields in the $xz$ plane are shown in Figures 5.17 and 5.18, where the pattern is seen to be significantly different. To better visualize the overall radiation pattern, we compute the normalized radiated field for all $\theta$ and $\phi$ on the unit sphere at a distance of $10\lambda$, which is rendered in Figure 5.19.

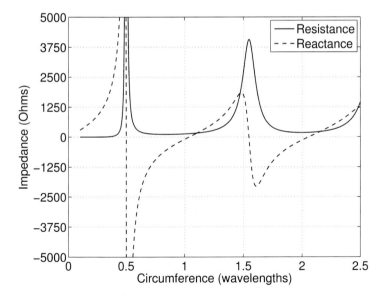

**FIGURE 5.14:** Circular Loop Input Impedance

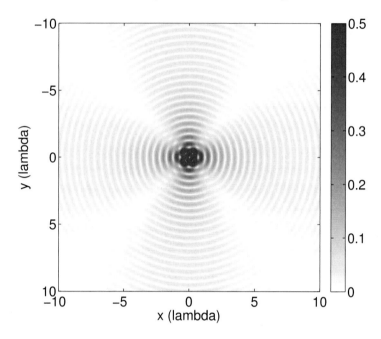

**FIGURE 5.15:** Radiated Near Field of the Circular Loop ($xy$ plane)

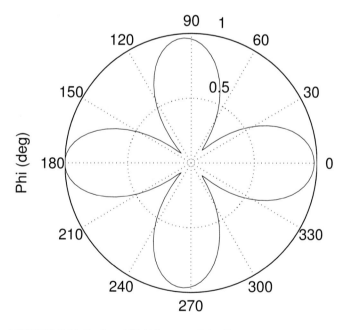

**FIGURE 5.16:** Radiated Field Pattern of the Circular Loop ($xy$ plane)

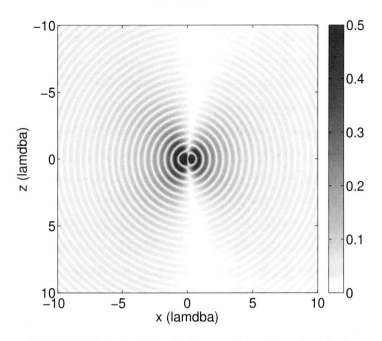

**FIGURE 5.17:** Radiated Near Field of the Circular Loop ($xz$ plane)

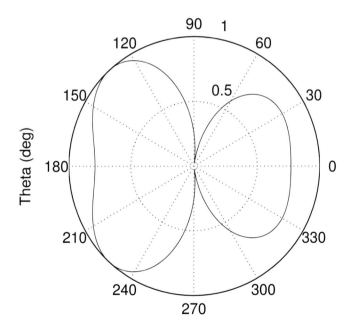

**FIGURE 5.18:** Radiated Field Pattern of the Circular Loop ($xz$ plane)

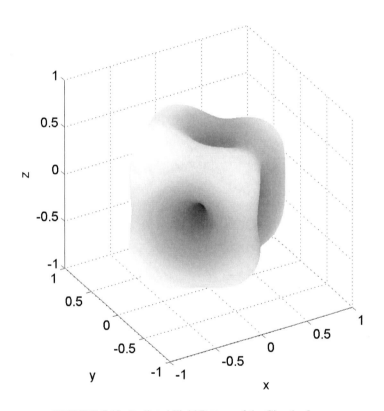

**FIGURE 5.19:** Radiated Field Pattern of the Circular Loop

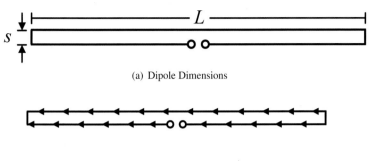

(a) Dipole Dimensions

(b) Current Distribution at $\lambda/2$

**FIGURE 5.20:** Folded Dipole Antenna

### 5.6.4 Folded Dipole Antenna

The folded dipole is an antenna commonly used in HF and VHF systems. It has this name as it is a dipole that is essentially "folded over on itself," comprising a square loop that is very short in one dimension. This is illustrated in Figure 5.20a, where the loop has a length of $L$ and width of $s$, and a feedpoint is placed in the center of one of the long sides. At DC, this antenna is a short circuit, but with increasing length the current distribution changes and assumes the characteristics of Figure 5.20b when $L$ approaches $\lambda/2$. In this case, the currents on the secondary side of the loop take on the same magnitude and phase as those on the primary side. Because $s$ is very small, the far radiated field remains much the same, however the current is divided evenly between the two halves. Therefore, the input impedance of the folded dipole will be four times that of an ordinary dipole at its first resonance, or approximately 288 $\Omega$. Folded dipoles are often used because of this property, as they can be matched to open-wire feed lines having a higher intrinsic impedance than coaxial cable.

In Figure 5.21 we plot the input impedance of a folded dipole for $L \leq 1\lambda$, where $s = L/20$ and the radius of the wire is $10^{-4}\lambda$. We use 30 segments for each long conductor and 5 segments for the short ends. We see that when the dipole is electrically small it is similar to the circular loop, and has virtually no radiation resistance. It becomes an open circuit at the antenna terminals at approximately $0.3\lambda$. The first resonance occurs at a length of about $0.46\lambda$ where the input resistance is 284 $\Omega$, close to the expected value. We next center a $0.46\lambda$ folded dipole at the origin, and orient it in the $xz$ plane. We then compute the radiated near field in the $xz$ plane, where the real part of the electric field (in V/m) is plotted in Figure 5.22.

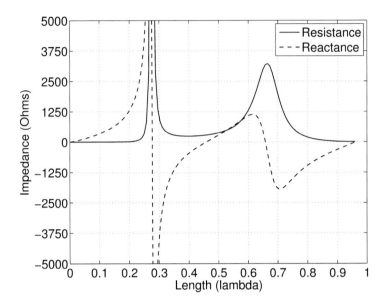

**FIGURE 5.21:** Folded Dipole Input Impedance

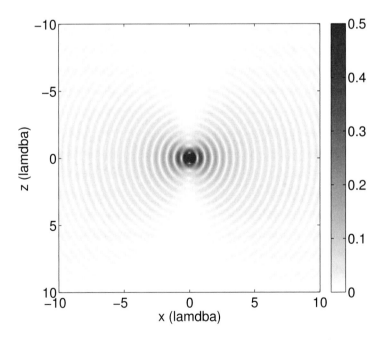

**FIGURE 5.22:** Radiated Near Field of the Folded Dipole ($xz$ plane)

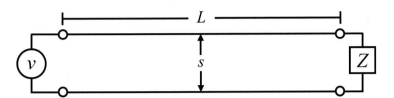

**FIGURE 5.23:** Two-Wire Transmission Line

### 5.6.5 Two-Wire Transmission Line

As transmission lines are implicit in virtually all antenna problems, let us next examine the behavior of a common two-wire transmission line illustrated in Figure 5.23. Two-wire line (also called "ladder line") is typically constructed of two multistranded copper wires, held a precise distance apart by a vinyl sheath or flat plastic ribbon. In two-wire line, this ribbon has rectangular holes cut into it at regular intervals to lighten the line, giving it the appearance of a ladder. For the purposes of simulation, we will ignore the presence of a sheath or ribbon, and treat the two-wire line in an identical manner to the folded dipole of Section 5.6.4, except that the feedpoint and load are placed at the short ends. We will analyze a two-wire line 2.5 meters long, constructed from wires 1 mm in radius having a separation distance of 3 cm. We use 30 segments for the main wires and 5 segments for each end.

Our first task is to determine the characteristic impedance of the line. The theoretical value for a two-wire line is obtained from transmission line theory [8] and is

$$Z_o = \sqrt{L/C} \tag{5.99}$$

where L is the inductance per unit length

$$L = \frac{\mu_0}{\pi} \cosh^{-1} \left( \frac{d}{2a} \right) \tag{5.100}$$

and C is the capacitance per unit length

$$C = (\pi \epsilon_0) / \cosh^{-1} \left( \frac{d}{2a} \right) \tag{5.101}$$

where $d$ is the distance between center of the wires, and $a$ is the radius of each wire. Using (5.99 - 5.101), we find that the impedance of our line is approximately $Z_o = 407$ Ω. The true impedance is determined numerically as

$$Z_o = \sqrt{Z_{oc} Z_{sc}} \tag{5.102}$$

where $Z_{oc}$ and $Z_{sc}$ are the input impedances with an opened and shorted load end, respectively. We must be careful, however, that we do not take the measurement at a frequency where the length of the line would result in an extremely small or large impedance at the input end, as numerical inaccuracy would result. We therefore make the measurement at a frequency of 15 MHz, where the length of the line is approximately $\lambda/8$. Using the thin-wire EFIE, we find that the impedance is $Z_o = 405\ \Omega$, very close to the theoretical value.

We next compare the computed input impedance to that of an ideal lossless transmission line, which is [8]

$$Z_{in} = Z_o \frac{Z_l + jZ_o \tan(k_0 l)}{Z_o + jZ_l \tan(k_0 l)} \tag{5.103}$$

where $Z_l$ is the load impedance, and $Z_{in}$ the impedance measured at a distance $l$ from the load. For this comparison, we use a load with a pure resistance of $200\ \Omega$. The results are plotted in Figure 5.24 for frequencies between 100 kHz and 100 MHz. The comparison is quite good. At lower frequencies, the input impedance tends toward the DC value, as expected.

Two-wire lines are useful only at frequencies where they do not radiate much energy, which is typically satisfied when $d \ll \lambda$. Let us retain the 200 $\Omega$ load, and place the line in the $xy$ plane with the feedpoint on the $-x$ axis. The real part of the radiated near electric field (in V/m) at 15 MHz is shown in the $xz$ plane in Figure 5.25. Note that even though the load is not matched and there are standing waves on the line, the fields remain concentrated in the region directly surrounding the line, with virtually no field radiated. Increasing the frequency to 500 MHz, we see in Figure 5.26 that the near field is no longer confined. Increasing the frequency further still to 1 GHz, we see in Figure 5.27 that the line is clearly acting as an antenna, making it unsuitable as a transmission line at this frequency.

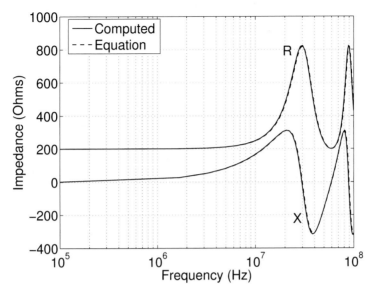

**FIGURE 5.24:** Two-Wire Line Input Impedance

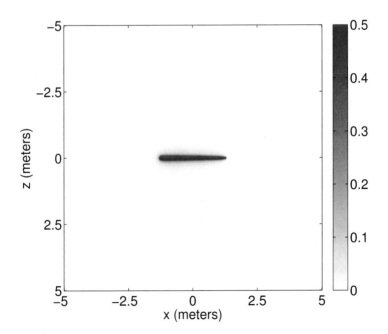

**FIGURE 5.25:** Radiated Near Field of the Two-Wire Line at 15 MHz ($xz$ plane)

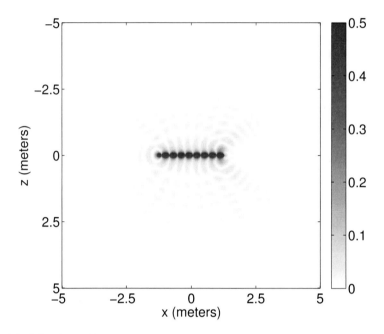

**FIGURE 5.26:** Radiated Near Field of the Two-Wire Line at 500 MHz ($xz$ plane)

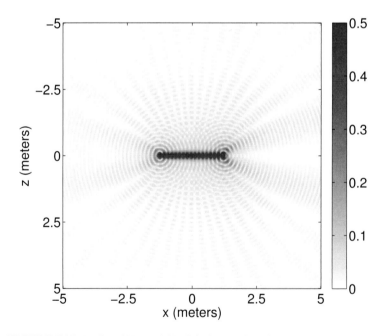

**FIGURE 5.27:** Radiated Near Field of the Two-Wire Line at 1 GHz ($xz$ plane)

### 5.6.6    Yagi Antenna for 146 MHz

We now examine a more practical antenna: a Yagi antenna for the 2-meter amateur radio band (144–148 MHz). The dimensions and locations for the antenna elements are summarized in Table 5.1 [12], and the diameter for all elements is 6.35 mm (1/4 inch). Our thin-wire EFIE model comprises separate straight wires in the $xy$ plan, as shown in Figure 5.28. The feedpoint is at the center of the *driven element*, and is noted with a dot in Figure 5.28. Though a real-world Yagi antenna would use a rigid piece of material such as metal or plastic tubing called a *boom* to affix the antenna elements, the boom does not typically have much current flow on it and we omit it from our analysis.

Most Yagi antennas have a single element placed behind the driven element which is called a *reflector*. This element reduces the amount of power radiated along the $-x$ axis in Figure 5.28 (the "back" side). The elements to the right of the driven element are called *directors*, and they act to focus or *direct* most of the power along the $+x$ axis (the "front" side). Though these elements are not connected to the feedline, they interact with the other elements as they are in the near field, and are often referred to as *parasitic* elements.

The input resistance and reactance of the Yagi antenna over the band are shown in Figures 5.29a and 5.29b, respectively. The antenna has a feedpoint resistance ranging from 18 to 33 $\Omega$ over the frequency band, and a capacitive reactance, which is typical of Yagi antennas. This antenna can be matched to a 50 $\Omega$ coaxial feed using a lumped inductive element at the feedpoint (a coil) or a tuning stub such as the hairpin match [13] and a 2:1 balun. We will match the antenna at about 146.5 MHz, where its reactance is approximately $-j18.8$ $\Omega$. To achieve the match we follow Section 5.29, and place an inductance of about 20 nH at the feedpoint, which will tune out the capacitive reactance. The reactance after matching is shown in Figure 5.29c, and the standing wave ratio (SWR) on a 50 $\Omega$ line with 2:1 balun in Figure 5.29d. The front-to-back ratio and the forward gain (in dB relative to a dipole, or "dBd") are shown in Figures 5.29e and 5.29f, respectively.

We next look at the radiated field pattern in the $xy$ plane. The total radiated electric field in a $40 \times 40$ meter region around the antenna is shown in Figure 5.30. The far radiated field pattern is shown versus $\phi$ in Figure 5.31. The total radiated field in the $xz$ plane is shown in Figure 5.32. The corresponding $H$-plane far radiated field pattern is shown versus $\theta$ in Figure 5.33. The normalized total far field pattern on the unit sphere for all $\theta$ and $\phi$ is shown in Figure 5.34.

**TABLE 5.1:** 2 Meter Yagi Antenna Parameters

| Element | REF | DE | D1 | D2 | D3 | D4 | D5 | D6 | D7 |
|---|---|---|---|---|---|---|---|---|---|
| Length (mm) | 1038 | 955 | 956 | 932 | 916 | 906 | 897 | 891 | 887 |
| $x$ location (mm) | 0 | 312 | 447 | 699 | 1050 | 1482 | 1986 | 2553 | 3168 |

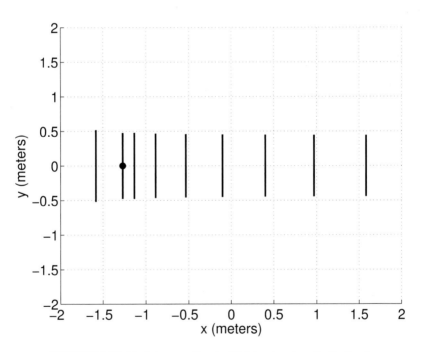

**FIGURE 5.28:** Yagi Antenna For the 2 Meter Amateur Radio Band

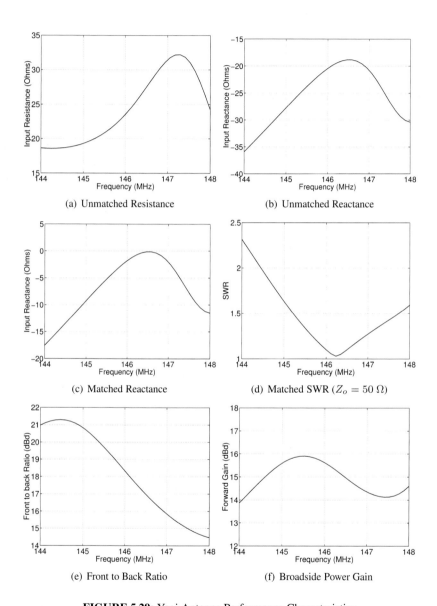

(a) Unmatched Resistance

(b) Unmatched Reactance

(c) Matched Reactance

(d) Matched SWR ($Z_o = 50\ \Omega$)

(e) Front to Back Ratio

(f) Broadside Power Gain

**FIGURE 5.29:** Yagi Antenna Performance Characteristics

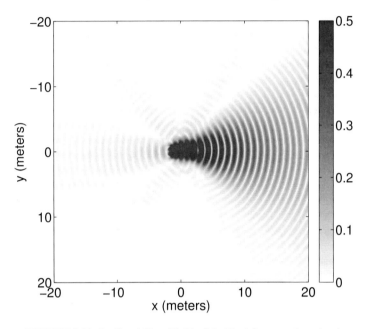

**FIGURE 5.30:** Radiated Near Field of the Yagi Antenna ($xy$ plane)

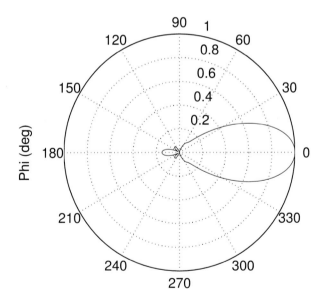

**FIGURE 5.31:** Far Field Pattern of the Yagi Antenna ($xy$ plane)

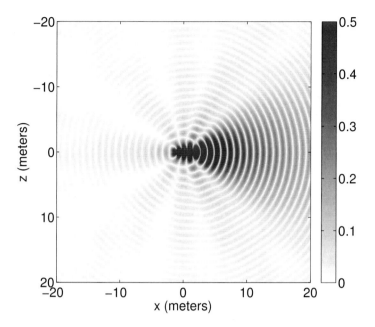

**FIGURE 5.32:** Radiated Near Field of the Yagi Antenna ($xz$ plane)

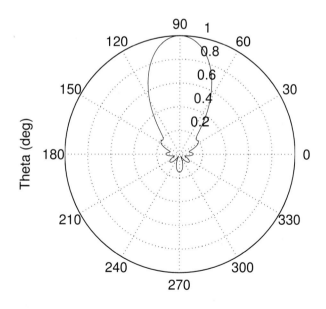

**FIGURE 5.33:** Far Field Pattern of the Yagi Antenna ($xz$ plane)

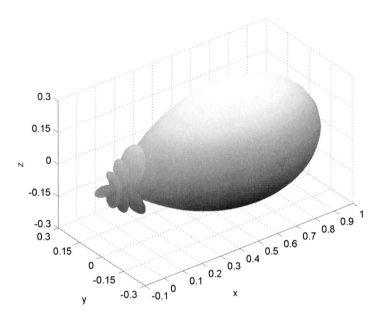

**FIGURE 5.34:** 3D Far Field Pattern of the Yagi Antenna

---

## References

[1] R. W. P. King, "The linear antenna – Eighty years of progress," *Proc. IEEE*, vol. 55, pp. 2–26, January 1967.

[2] W. Wang, "The exact kernel for cylindrical antenna," *IEEE Trans. Antennas Propagat.*, vol. 39, pp. 434–435, April 1991.

[3] D. R. Wilton and N. J. Champagne, "Evaluation and integration of the thin wire kernel," *IEEE Trans. Antennas Propagat.*, vol. 54, pp. 1200–1206, April 2006.

[4] E. Hallen, "Theoretical investigations into the transmitting and receiving qualities of antennae," *Nova Acta (Uppsala)*, vol. 11, pp. 1–44, 1938.

[5] H. C. Pocklington, "Electrical oscillations in wires," *Proc. Camb. Phil. Soc.*, vol. 9, 1897.

[6] W. Stutzman and G. Thiele, *Antenna Theory and Design*. John Wiley and Sons, 1981.

[7] C. A. Balanis, *Advanced Engineering Electromagnetics*. John Wiley and Sons, 1989.

[8] S. J. Orfanidis, "Electromagnetic waves and antennas." Electronic version: http://www.ece.rutgers.edu/ orfanidi/ewa/.

[9] H. B. Dwight, *Tables of Integrals and Other Mathematical Data*. The Macmillan Company, 1961.

[10] *The Integrator*. Wolfram Research, http://integrals.wolfram.com.

[11] C. A. Balanis, *Antenna Theory: Analysis and Design*. John Wiley and Sons, second ed., 1997.

[12] *The ARRL Antenna Book*. The Amateur Radio Relay League, 17th ed., 1994.

[13] W. Orr, *Radio Handbook*. Sams, 1992.

# Chapter 6

## Two-Dimensional Problems

In this chapter we will investigate scattering problems in the context of two-dimensional structures. The discussion and examples here will serve to illustrate the approach outlined in Section 3.6.3.1, as well as practical implementation of the matrix elements (3.185–3.190) and the choice of basis and testing functions. We will first consider conducting objects, and discuss the implementation of the $\mathcal{L}$ and $\hat{\mathbf{n}} \times \mathcal{K}$ operators in the EFIE and nMFIE. We will then consider dielectric objects and the $\mathcal{K}$ operator in the EFIE/MFIE, and the $\hat{\mathbf{n}} \times \mathcal{L}$ operator in the nMFIE. Because general two-dimensional problems can be split into TM- and TE-polarized sub-problems, we will consider each of these individually.

## 6.1 Conducting Objects

First we will consider conducting objects and the solution of the EFIE and nMFIE in TM and TE polarization. As the $\mathcal{L}$ operator can be used on thin, open structures, we will first apply the EFIE to a thin, conducting strip of finite length, and develop the needed expressions for the self terms. We will then generalize the EFIE to a cylindrical contour of arbitrary shape. Because the $\hat{\mathbf{n}} \times \mathcal{K}$ operator is valid only on closed surfaces, we will consider the nMFIE on closed cylindrical geometries only.

### 6.1.1 EFIE: TM Polarization

Consider a thin, perfectly conducting strip illuminated by a TM-polarized plane wave as illustrated in Figure 6.1. The electric field has unit amplitude and is given by

$$\mathbf{E}^i(\boldsymbol{\rho}) = e^{jk\hat{\boldsymbol{\rho}}_i \cdot \boldsymbol{\rho}}\, \hat{\mathbf{z}} = e^{jk(x\cos\phi^i + y\sin\phi^i)}\, \hat{\mathbf{z}} \tag{6.1}$$

where $\hat{\boldsymbol{\rho}}^i = \cos\phi^i \hat{\mathbf{x}} + \sin\phi^i \hat{\mathbf{y}}$. In the absence of magnetic current, the EFIE (3.178) is

$$j\omega\mu\big[(\mathcal{L}\mathbf{J}_l)(\mathbf{r})\big]_{tan} = \big[\mathbf{E}_l^i(\mathbf{r})\big]_{tan} \tag{6.2}$$

125

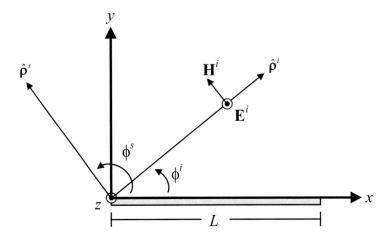

**FIGURE 6.1:** Thin Strip: TM Polarization

On the thin strip, this can be written as

$$j\omega\mu\left[1 + \frac{1}{k^2}\nabla\nabla\cdot\right]\int_0^L \mathbf{J}(x')\, G(\boldsymbol{\rho}, \boldsymbol{\rho}')\, dx' = e^{jkx\cos\phi^i} \qquad (6.3)$$

where the integration is only in $x'$ as the strip is assumed to be very thin. We note that because the TM-polarized field has only a $\hat{\mathbf{z}}$ component, so will the electric current, therefore $\nabla' \cdot \mathbf{J}(x') = 0$. Using this fact along with (3.49) and (3.79), the above becomes

$$\frac{\omega\mu}{4}\int_0^L \mathbf{J}(x')\, H_0^{(2)}\left(k|x - x'|\right)\, dx' = e^{jkx\cos\phi^i}\hat{\mathbf{z}} \qquad (6.4)$$

The scattered far electric field (3.142) can be computed as

$$\mathbf{E}^s(\boldsymbol{\rho}) = -\omega\mu\sqrt{\frac{j}{8\pi k}}\,\frac{e^{-jk\rho}}{\sqrt{\rho}}\int_0^L \mathbf{J}(x')\, e^{jkx'\cos\phi^s}\, dx' \qquad (6.5)$$

The corresponding Physical Optics current on the strip for the TM case is

$$\mathbf{J}(\boldsymbol{\rho}) = \frac{2}{\eta}\,\hat{\mathbf{y}} \times \left[-\hat{\boldsymbol{\rho}}^i \times \mathbf{E}^i(\boldsymbol{\rho})\right] = \frac{2}{\eta}\sin\phi^i e^{jkx\cos\phi^i}\hat{\mathbf{z}} \qquad (6.6)$$

#### 6.1.1.1   Solution Using Pulse Functions

We will first solve this problem using pulse basis functions and point matching for testing. Dividing the strip into $N$ segments of width $\Delta_x = L/N$

yields $N$ basis functions. Point matching at the center of the testing segment $x_m$ yields the matrix elements

$$Z_{mn} = \frac{\omega\mu}{4} \int_{x_n}^{x_{n+1}} H_0^{(2)}\left(k|x_m - x'|\right) dx' \tag{6.7}$$

and right-hand side vector elements

$$V_m = e^{jkx_m \cos\phi^i} \tag{6.8}$$

Note that the system matrix has a symmetric Toeplitz structure due to the choice of basis functions and the linear geometry of the strip.

*Non-Self Terms*

For non-overlapping segments separated by sufficient distance, we apply a centroidal approximation to (6.7), which yields

$$Z_{mn} = \frac{\omega\mu\Delta_x}{4} H_0^{(2)}\left(k|x_m - x_n|\right) \tag{6.9}$$

where $x_m$ and $x_n$ are at the center of the field and source points, respectively.

*Self Terms*

For overlapping segments, we use the small argument approximation to the Hankel function

$$H_0^{(2)}(kx) \approx 1 - j\frac{2}{\pi}\log\left(\frac{\gamma kx}{2}\right) \qquad x \to 0 \tag{6.10}$$

where $\gamma = 1.781$. Inserting this into (6.7) yields

$$Z_{mm} = \frac{\omega\mu}{4} \int_0^{\Delta_x} \left[1 - j\frac{2}{\pi}\log\left(\frac{\gamma k|x - x'|}{2}\right)\right] dx' \tag{6.11}$$

The logarithmic part of this integral has a singularity, so we must integrate it carefully. To do so will use [1], which contains solutions for potential integrals of the form $\log x$ and $x \log x$. Performing the integration yields

$$Z_{mm} = \frac{\omega\mu}{4}\left[\Delta_x - j\frac{2}{\pi}\right]\left[\Delta_x \log\left(\gamma k\frac{\Delta_x - x}{2e}\right) + x\log\left(\frac{x}{\Delta_x - x}\right)\right] \tag{6.12}$$

and substitution of the field location $x = \Delta_x/2$ yields

$$Z_{mm} = \frac{\omega\mu\Delta_x}{4}\left[1 - j\frac{2}{\pi}\log\left(\frac{\gamma k\Delta_x}{4e}\right)\right] \tag{6.13}$$

*Near Terms*

When the segments are adjacent but not overlapping, the integrand may still exhibit strongly singular behavior that can lead to inaccurate matrix elements when the centroidal approximation is used. For ($|m - n| = 1$), we will again use (6.12), this time with $x = 3\Delta_x/2$. The result is

$$Z_{mn} = \frac{\omega\mu\Delta_x}{4}\left[1 - \frac{j}{\pi}\right]\left[3\log\left(\frac{3\gamma k\Delta_x}{4}\right) - \log\left(\frac{\gamma k\Delta_x}{4}\right) - 2\right] \quad (6.14)$$

Note that due to the linear geometry of the strip, it is possible to use (6.13) to obtain all the matrix elements in this case. However, evaluating near-term expressions can be time consuming for larger problems, particularly in the three-dimensional case (Chapter 8). Therefore, the use of near-term expressions is usually limited to interaction distances where they are needed to obtain maximum accuracy.

*Solution*

Consider a strip of width $3\lambda$ centered at the origin, divided into 45 segments (15 segments/$\lambda$). In Figure 6.2a, we plot the induced surface current amplitude at broadside incidence ($\phi^i = \pi/2$). The current is singular near the edges, and the magnitude falls off quickly with distance from each edge. To this we compare the PO current of (6.6), which matches the EFIE fairly well near the center of the strip, but does poorly near the edges. In Figure 6.2b we plot the 2D RCS for incident angles ranging from 0 to 180 degrees. PO compares well to the EFIE at angles near broadside and within the first main lobe, but the comparison is poor beyond that, as edge diffraction effects are omitted by PO.

Using (3.120), we next compute the scattered near field in a $20\lambda \times 20\lambda$ region around the strip. The real component of the electric field $E_z^s$ is shown (in V/m) for $\phi^i = \pi/4$ and $\phi^i = \pi/2$ in Figures 6.3a and 6.3b, respectively.

### 6.1.1.2   Solution Using Triangle Functions

We next solve this problem using triangle basis and testing functions. Because the current has only a $\hat{\mathbf{z}}$-component, we will omit the vector notation and work with the scalar components, which we refer to as $f_m^z$ and $f_n^z$, respectively. The strip is again divided into $N$ segments of width $\Delta_x = L/N$. Because the current is nonzero at the ends of the strip, we use $N - 2$ whole triangle functions and a half-triangle function at each end of the strip, which yields $N$ unknowns. The matrix elements are

$$Z_{mn} = \frac{\omega\mu}{4}\int_{f_m^z} f_m^z(x)\int_{f_n^z} f_n^z(x') H_0^{(2)}(k|x - x'|)\,dx'\,dx \quad (6.15)$$

To evaluate (6.15) we will use an $M$-point quadrature rule over the source and field triangles, which yields

$$Z_{mn} = \frac{\omega\mu}{4} \sum_{p=1}^{M} w(x_p) \, f_m^z(x_p) \sum_{q=1}^{M} w(x_q) \, f_n^z(x_q) \, H_0^{(2)}\left(k|x_p - x_q|\right) \quad (6.16)$$

where $x_q$ and $x_p$ are the source and field locations, respectively. The corresponding right-hand side vector elements are

$$V_m = \sum_{p=1}^{M} w(x_p) f_m^z(x_p) e^{jkx_p \cos\phi^i} \quad (6.17)$$

*Self Terms*

When any of the segments overlap, a singularity occurs in the Hankel function. To address this, we will evaluate the innermost integral in (6.15) analytically and the outermost integral numerically. Using the small-argument approximation to the Hankel function from (6.10), the innermost integral becomes

$$I(x_p) = \int_0^{\Delta_x} f_n(x')\left[1 - j\frac{2}{\pi}\log\left(\frac{\gamma k|x_p - x'|}{2}\right)\right] dx' \quad (6.18)$$

The observation point $x_p$ lies within the limits of integration, and the function $f_n(x')$ represents the half of the triangle with positive or negative slope. This integral can be decomposed into a sum of integrals involving $x\log x$ and $\log x$. For the triangle with positive slope of the form $f_n(x') = x'/\Delta_x$, the integral evaluates to

$$I(x_p) = \frac{\Delta_x}{2} - j\frac{2}{\pi}\left[\frac{x_p^2}{2\Delta_x}\log\left(\frac{x_p}{\Delta_x - x_p}\right) + \frac{\Delta_x}{2}\log\left(b[\Delta_x - x_p]\right) - \frac{x_p}{2} - \frac{\Delta_x}{4}\right] \quad (6.19)$$

where $b = \gamma k/2$. Due to symmetry, this result is sufficient for computing the singular terms for itself as well as the triangle having negative slope.

*Solution*

We again examine the $3\lambda$ strip using 45 segments. For integration, a 5-point Gaussian quadrature is used on each segment. In Figure 6.2a is shown the computed surface current amplitude at broadside incidence. The current is seen to take on a higher value at the edges than it did with pulse functions, however the overall difference is slight. Figure 6.2b depicts the 2D RCS, which is nearly identical to that obtained using pulse functions. This suggests that pulse functions may be sufficient to accurately characterize the RCS for this particular problem.

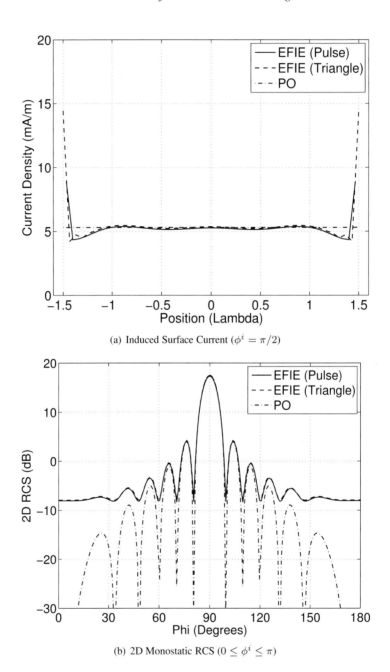

(a) Induced Surface Current ($\phi^i = \pi/2$)

(b) 2D Monostatic RCS ($0 \le \phi^i \le \pi$)

**FIGURE 6.2:** $3\lambda$ Strip: Currents and RCS, TM Polarization

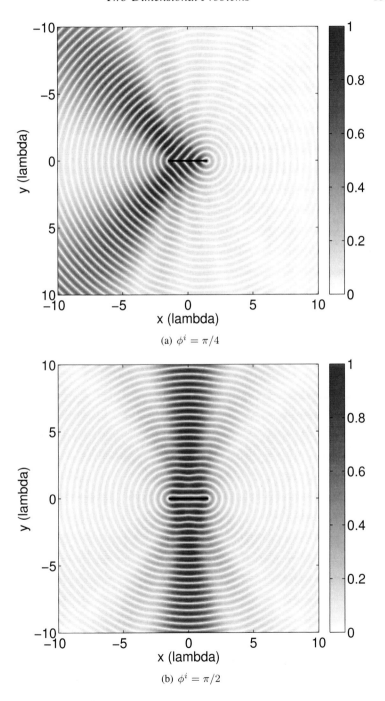

(a) $\phi^i = \pi/4$

(b) $\phi^i = \pi/2$

**FIGURE 6.3:** $3\lambda$ Strip: Scattered TM Near Field (V/m)

## 6.1.2  Generalized EFIE: TM Polarization

Following (6.3) and (6.4), the EFIE (3.178) for a conducting, two-dimensional contour $C$ of arbitrary shape can be written as

$$\frac{\omega\mu}{4} \int_C \mathbf{J}(\boldsymbol{\rho}') \, H_0^{(2)}\left(k|\boldsymbol{\rho}-\boldsymbol{\rho}'|\right) d\rho' = \mathbf{E}^i(\boldsymbol{\rho}) \qquad (6.20)$$

### 6.1.2.1  MOM Discretization

The contour $C$ is divided into a number of linear segments, and the current $\mathbf{J}(\boldsymbol{\rho})$ is expanded using $\hat{z}$-oriented electric basis functions $\mathbf{f}^z$. Testing with these same functions yields the matrix elements (3.185) for the TM case

$$Z_{mn}^{EJ} = \mathrm{L}^{zz}(\mathbf{f}_m^z, \mathbf{f}_n^z) \qquad (6.21)$$

where

$$\mathrm{L}^{zz}(\mathbf{f}_m^z, \mathbf{f}_n^z) = \frac{\omega\mu}{4} \int_{\mathbf{f}_m^z} \mathbf{f}_m^z(\boldsymbol{\rho}) \cdot \int_{\mathbf{f}_n^z} \mathbf{f}_n^z(\boldsymbol{\rho}') \, H_0^{(2)}\left(k|\boldsymbol{\rho}-\boldsymbol{\rho}'|\right) d\rho' \, d\rho \qquad (6.22)$$

The right-hand side vector elements (3.191) are

$$V_m^{E,TM} = \int_{\mathbf{f}_m^z} \mathbf{f}_m^z(\boldsymbol{\rho}) \cdot \mathbf{E}_z^i(\boldsymbol{\rho}) \, d\rho \qquad (6.23)$$

### 6.1.2.2  Solution Using Triangle Functions

Using triangle basis and testing functions, (6.22) can be evaluated on non-overlapping segments using an $M$-point numerical quadrature, yielding

$$\mathrm{L}^{zz}(\mathbf{f}_m^z, \mathbf{f}_n^z) = \frac{\omega\mu}{4} \sum_{p=1}^{M} w(\boldsymbol{\rho}_p) f_m^z(\boldsymbol{\rho}_p) \sum_{q=1}^{M} w(\boldsymbol{\rho}_q) f_n^z(\boldsymbol{\rho}_q) \, H_0^{(2)}\left(k|\boldsymbol{\rho}_p-\boldsymbol{\rho}_q|\right) \qquad (6.24)$$

where we have again omitted the vector notation for convenience. Any self-terms can be computed via (6.19). In generating the matrix elements, the right-hand side of (6.24) should be evaluated inside a loop over segment pairs and the resulting values placed into the corresponding matrix locations. As each segment supports at most two triangles, (6.24) will contribute to a maximum of four matrix elements. The right-hand side vector elements (6.23) are

$$V_m^{E,TM} = \sum_{p=1}^{M} w(\boldsymbol{\rho}_p) f_m^z(\boldsymbol{\rho}_p) E_z^i(\boldsymbol{\rho}_p) \qquad (6.25)$$

For a plane wave having an electric field of unit amplitude, (6.25) becomes

$$V_m^{E,TM} = \sum_{p=1}^{M} w(x_p) f_m^z(x_p) e^{jk(x_p \cos\phi_i + y_p \sin\phi_i)} \qquad (6.26)$$

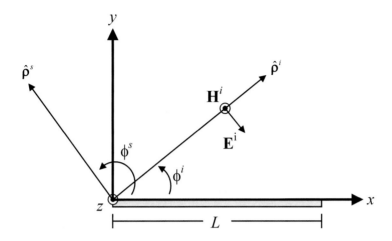

**FIGURE 6.4:** Thin Strip: TE Polarization

### 6.1.3 EFIE: TE Polarization

We next consider the thin strip for TE polarization, as illustrated in Figure 6.4, where an incident electric field of unit amplitude is now given by

$$\mathbf{E}^i(\boldsymbol{\rho}) = (\sin\phi^i\hat{\mathbf{x}} - \cos\phi^i\hat{\mathbf{y}})e^{jk(x\cos\phi^i + y\sin\phi^i)} \tag{6.27}$$

Since the incident field has $\hat{\mathbf{x}}$ and $\hat{\mathbf{y}}$ components, the induced electric currents will as well. However, since $\hat{\mathbf{y}}$-directed currents (normal to the strip) cannot exist in this case, the current has only an $\hat{\mathbf{x}}$ component. Therefore, (3.178) can be written as

$$\frac{\omega\mu}{4}\int_0^L J_x(x')\left[1 + \frac{1}{k^2}\frac{\partial^2}{\partial x^2}\right]H_0^{(2)}(k\rho)\,dx' = \sin\phi^i e^{jkx\cos\phi^i} \tag{6.28}$$

where $\rho = \sqrt{(x-x')^2 + (y-y')^2}$. Let us now evaluate the expression

$$\left[1 + \frac{1}{k^2}\frac{\partial^2}{\partial x^2}\right]H_0^{(2)}(k\rho) \tag{6.29}$$

To begin, we note that [2]

$$\frac{d}{du}H_n^{(2)}(u) = -H_{n+1}^{(2)}(u) + \frac{n}{u}H_n^{(2)}(u) \tag{6.30}$$

and

$$\frac{\partial}{\partial x}H_n^{(2)}(k\rho) = \frac{d}{du}\left[H_n^{(2)}(u)\right]\frac{\partial u}{\partial x} \tag{6.31}$$

where $u = k\rho$. Therefore,

$$\frac{\partial u}{\partial x} = \frac{\partial (k\rho)}{\partial x} = \frac{k(x - x')}{\sqrt{(x - x')^2 + (y - y')^2}} \tag{6.32}$$

and so

$$\frac{\partial}{\partial x} H_0^{(2)}(k\rho) = -H_1^{(2)}(k\rho) \frac{k(x - x')}{\sqrt{(x - x')^2 + (y - y')^2}} \tag{6.33}$$

Differentiating a second time yields

$$\frac{\partial^2}{\partial x^2} H_0^{(2)}(k\rho) = -\frac{k(x - x')}{\sqrt{(x - x')^2 + (y - y')^2}} \left[ \frac{\partial}{\partial x} H_1^{(2)}(k\rho) \right]$$
$$- H_1^{(2)}(k\rho) \left[ \frac{\partial}{\partial x} \frac{k(x - x')}{\sqrt{(x - x')^2 + (y - y')^2}} \right] \tag{6.34}$$

which is

$$\frac{\partial^2}{\partial x^2} H_0^{(2)}(k\rho) = -\frac{k^2(x - x')^2}{(x - x')^2 + (y - y')^2} \left[ -H_2^{(2)}(k\rho) + \frac{1}{k\rho} H_1^{(2)}(k\rho) \right]$$
$$- H_1^{(2)}(k\rho) \left[ \frac{k}{(x - x')^2 + (y - y')^2} - \frac{k^2(x - x')^2}{[(x - x')^2 + (y - y')^2]^{3/2}} \right] \tag{6.35}$$

The second and fourth term cancel, leaving

$$\frac{\partial^2}{\partial x^2} H_0^{(2)}(k\rho) = \frac{k^2(x - x')^2}{(x - x')^2 + (y - y')^2} H_2^{(2)}(k\rho)$$
$$- \frac{k}{(x - x')^2 + (y - y')^2} H_1^{(2)}(k\rho) \tag{6.36}$$

Using the recurrence relationship [2]

$$\frac{2}{k\rho} H_1^{(2)}(k\rho) = H_0^{(2)}(k\rho) + H_2^{(2)}(k\rho) \tag{6.37}$$

we can then write

$$\left[ 1 + \frac{1}{k^2} \frac{\partial^2}{\partial x^2} \right] H_0^{(2)}(k\rho) = \frac{1}{2} \left[ \frac{2(x - x')^2}{(x - x')^2 + (y - y')^2} - 1 \right] H_2^{(2)}(k\rho) + \frac{1}{2} H_0^{(2)}(k\rho) \tag{6.38}$$

We note from the geometry of Figure 6.4 that

$$\cos\theta = \frac{(x - x')}{\sqrt{(x - x')^2 + (y - y')^2}} \tag{6.39}$$

where the angle $\theta$ is measured in the right-hand sense from the positive x-axis. Using this, the previous expression can be written as

$$\left[1 + \frac{1}{k^2}\frac{\partial^2}{\partial x^2}\right]H_0^{(2)}(k\rho) = \frac{1}{2}\left[\cos(2\theta)H_2^{(2)}(k\rho) + H_0^{(2)}(k\rho)\right] \qquad (6.40)$$

and we can now write (6.28) as

$$\frac{\omega\mu}{8}\int_0^L J_x(x')\left[\cos(2\theta)H_2^{(2)}(k\rho) + H_0^{(2)}(k\rho)\right]dx' = \sin\phi^i e^{jkx\cos\phi^i} \qquad (6.41)$$

Note that for observations on the strip, $\rho = |x - x'|$, and for any point $x'$ where $x' \neq x$, $\theta$ is either $0$ or $\pi$, thus $\cos(2\theta) = 1$. In these cases, (6.37) can be used to write (6.41) as

$$\frac{\omega\mu}{4}\int_0^L J_x(x')\frac{H_1^{(2)}(k\rho)}{k\rho}dx' = \sin\phi^i e^{jkx\cos\phi^i} \qquad (6.42)$$

Note that for any matrix elements where $x = x'$, we will need to use (6.41) along with (6.39).

The corresponding PO current on the strip for the TE case is

$$\mathbf{J}(\rho) = 2\,\hat{\mathbf{y}}\times\mathbf{H}^i(\rho) = 2\,\hat{\mathbf{y}}\times\frac{\hat{\mathbf{z}}}{\eta} = \frac{2}{\eta}e^{jkx\cos\phi^i}\hat{\mathbf{x}} \qquad (6.43)$$

which at broadside incidence is equivalent in magnitude to the PO current for the TM case.

### 6.1.3.1 Pulse Function Solution

We will approach this problem numerically as we did for the TM case. We subdivide the strip into $N$ segments with $N$ pulse basis functions, and point match at the center of each segment. For segments that are not "close" ($|m - n| > 2$), we will compute the matrix elements using a centroidal approximation of (6.42), which is

$$Z_{mn} = \frac{\omega\mu\Delta_x}{4}\frac{H_1^{(2)}\left(k|x_m - x_n|\right)}{k|x_m - x_n|} \qquad |m - n| > 2 \qquad (6.44)$$

The right-hand side elements are

$$b_m = \sin\phi^i e^{jkx_m\cos\phi^i} \qquad (6.45)$$

*Near Terms*

When segments are adjacent or separated by one segment, we will treat these as near terms and perform the integration analytically. The integral for these matrix elements is

$$Z_{mn} = \frac{\omega\mu}{4} \int_{-\Delta_x/2}^{\Delta_x/2} \frac{H_1^{(2)}(k|x_m - x_n| - x')}{k(|x_m - x_n| - x')} dx' \qquad |m - n| = 1, 2 \quad (6.46)$$

Using the small-argument approximation to $H_1^{(2)}(k\rho)$ [2]

$$H_1^{(2)}(k\rho) \approx \frac{k\rho}{2} + \frac{2j}{\pi k\rho} \qquad r \to 0 \qquad (6.47)$$

this integral can be written as

$$Z_{mn} = \frac{\omega\mu}{8} \int_{-\Delta_x/2}^{\Delta_x/2} \left[ 1 + \frac{4j}{\pi k^2 (|x_m - x_n| - x')^2} \right] dx' \qquad (6.48)$$

which evaluates to

$$Z_{mn} = \frac{\omega\mu\Delta_x}{8} \left[ 1 + \frac{4j}{\pi k^2} \frac{1}{|x_m - x_n|^2 - \Delta_x^2/4} \right] \qquad |m - n| = 1, 2 \quad (6.49)$$

*Self Terms*

For the self terms, we can again use (6.13) for the portion of the integral involving $H_0^{(2)}(k\rho)$, however we must be careful with the Hankel function of order 2 as it has a non-integrable singularity. To overcome this problem, we will do the integration and then apply a limiting argument as the observation point approaches the strip. The integral to be evaluated is

$$I = \int_{-\Delta_x/2}^{\Delta_x/2} \cos(2\theta) H_2^{(2)}(k\rho) \, dx' \qquad (6.50)$$

First we will employ the small-argument approximation to the Hankel function $H_2^{(2)}(k\rho)$ [2]

$$H_2^{(2)}(k\rho) \approx \frac{(k\rho)^2}{8} + \frac{4j}{\pi(k\rho)^2} \qquad r \to 0 \qquad (6.51)$$

We next fix the observation point at $x = 0, y = \delta$, where $\delta$ is very small. The distance $r$ is then

$$r = \sqrt{(x')^2 + \delta^2} \qquad (6.52)$$

and we can write the integral as

$$I = \frac{k^2}{4} \int_0^{\Delta_x/2} (x'^2 + \delta^2) \left[ \frac{2x'^2}{x'^2 + \delta^2} - 1 \right] dx'$$
$$+ \frac{8j}{\pi k^2} \int_0^{\Delta_x/2} \frac{1}{x'^2 + \delta^2} \left[ \frac{2x'^2}{x'^2 + \delta^2} - 1 \right] dx' \tag{6.53}$$

where we have reintroduced the expression for $\cos(2\theta)$ from (6.39). The above is then

$$I = \frac{k^2}{4} \int_0^{\Delta_x/2} (x'^2 - \delta^2) \, dx' + \frac{8j}{\pi k^2} \int_0^{\Delta_x/2} \left[ \frac{2x'^2}{(x'^2 + \delta^2)^2} - \frac{1}{x'^2 + \delta^2} \right] dx' \tag{6.54}$$

The first integral is evaluated easily and is

$$I_1 = \frac{k^2}{4} \left[ \frac{\Delta_x^3}{24} - \frac{\delta^2 \Delta_x}{2} \right] \tag{6.55}$$

The second integral is obtained by way of [3] (120.1 and 122.2) and is

$$I_2 = \frac{8j}{\pi k^2} \left[ \frac{1}{\delta} \tan^{-1}\left(\frac{\Delta_x}{2\delta}\right) - \frac{\Delta_x/2}{\delta^2 + (\Delta_x/2)^2} - \frac{1}{\delta} \tan^{-1}\left(\frac{\Delta_x}{2\delta}\right) \right] \tag{6.56}$$

which reduces to

$$I_2 = \frac{-8j}{\pi k^2} \left[ \frac{\Delta_x/2}{\delta^2 + (\Delta_x/2)^2} \right] \tag{6.57}$$

If we now take the limit as the observation point approaches the strip ($\delta \to 0$), the results are

$$I_1 = k^2 \frac{\Delta_x^3}{96} \tag{6.58}$$

and

$$I_2 = \frac{-16j}{\pi k^2 \Delta_x} \tag{6.59}$$

and the resulting self-term is

$$Z_{mm} = \frac{\omega \mu}{8} \left[ \Delta_x + k^2 \frac{\Delta_x^3}{96} - \frac{j\Delta_x}{\pi} \left( 2 \log \left[ \frac{\gamma k \Delta_x}{4e} \right] + \frac{16}{(k\Delta_x)^2} \right) \right] \tag{6.60}$$

*Solution*

In Figure 6.5a is shown the amplitude of the computed surface current on the $3\lambda$ strip at broadside incidence, using 45 segments. The current tends toward zero near the ends of the strip as expected. In Figure 6.5b we compare the 2D RCS obtained from the EFIE versus that of PO. Again, PO does well near normal incidence, but poorly beyond that. Next, we compute the scattered near field in a $20\lambda \times 20\lambda$ region around the strip. The real component of the electric field is shown (in V/m) for $\phi^i = \pi/4$ and $\phi^i = \pi/2$ in Figures 6.6a and 6.6b, respectively.

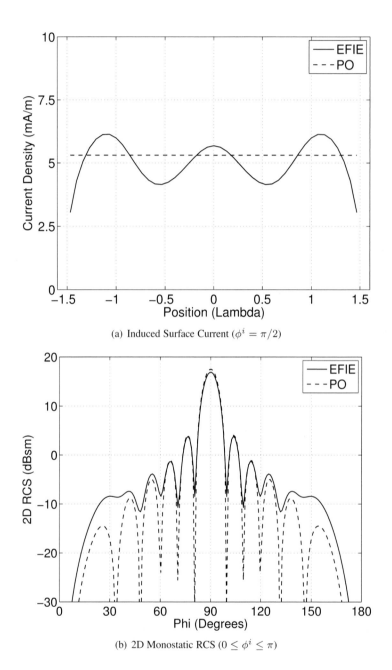

(a) Induced Surface Current ($\phi^i = \pi/2$)

(b) 2D Monostatic RCS ($0 \leq \phi^i \leq \pi$)

**FIGURE 6.5:** Strip with Pulse Functions: TE Polarization

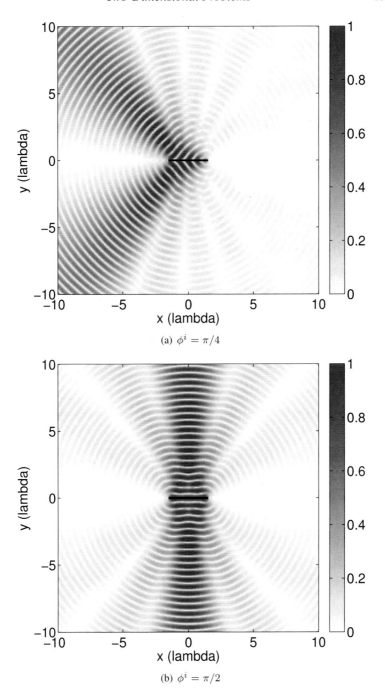

(a) $\phi^i = \pi/4$

(b) $\phi^i = \pi/2$

**FIGURE 6.6:** $3\lambda$ Strip: Scattered TE Near Field (V/m)

### 6.1.4   Generalized EFIE: TE Polarization

Extending (3.178) to a conducting, two-dimensional contour $C$ of arbitrary shape yields

$$\frac{\omega\mu}{4}\,\hat{\mathbf{t}}(\boldsymbol{\rho})\cdot\int_C\left[1+\frac{1}{k^2}\nabla\nabla\cdot\right]\mathbf{J}(\boldsymbol{\rho}')H_0^{(2)}\left(k|\boldsymbol{\rho}-\boldsymbol{\rho}'|\right)d\boldsymbol{\rho}'=\hat{\mathbf{t}}(\boldsymbol{\rho})\cdot\mathbf{E}^i(\boldsymbol{\rho}) \quad (6.61)$$

where $\hat{\mathbf{t}}(\mathbf{r})$ and $\mathbf{J}(\boldsymbol{\rho})$ are everywhere tangent to $C$.

#### 6.1.4.1   MOM Discretization

The contour $C$ is divided into a number of linear segments, and the current $\mathbf{J}(\boldsymbol{\rho})$ expanded using vector basis functions that are tangent to each segment. Testing with those same functions yields the matrix elements (3.185) for the TE case, which are

$$Z_{mn}^{EJ}=\mathrm{L}^{tt}(\mathbf{f}_m^t,\mathbf{f}_n^t) \quad (6.62)$$

where

$$\mathrm{L}^{tt}(\mathbf{f}_m^t,\mathbf{f}_n^t)=\frac{\omega\mu}{4}\int_{\mathbf{f}_m^t}\mathbf{f}_m^t(\boldsymbol{\rho})\cdot\int_{\mathbf{f}_n^t}\mathbf{f}_n^t\boldsymbol{\rho}')\,H_0^{(2)}\left(k|\boldsymbol{\rho}-\boldsymbol{\rho}'|\right)d\boldsymbol{\rho}'\,d\boldsymbol{\rho}$$
$$+\frac{1}{4\omega\epsilon}\int_{\mathbf{f}_m^t}\mathbf{f}_m^t(\boldsymbol{\rho})\cdot\int_{\mathbf{f}_n^t}\nabla\nabla\cdot\left[\mathbf{f}_n^t(\boldsymbol{\rho}')\,H_0^{(2)}\left(k|\boldsymbol{\rho}-\boldsymbol{\rho}'|\right)\right]d\boldsymbol{\rho}'\,d\boldsymbol{\rho}$$
$$(6.63)$$

The analysis of the strip in Section 6.1.3 was complicated by the differential operators acting on the Green's function, which will make the analysis of complex geometries even more difficult. Assuming that the basis and testing functions have a well-behaved divergence, we can re-distribute the differential operators following (3.204). Doing so allows us to rewrite (6.63) as

$$\mathrm{L}^{tt}(\mathbf{f}_m^t,\mathbf{f}_n^t)=\frac{\omega\mu}{4}\int_{\mathbf{f}_m^t}\mathbf{f}_m^t(\boldsymbol{\rho})\cdot\int_{\mathbf{f}_n^t}\mathbf{f}_n^t\boldsymbol{\rho}')\,H_0^{(2)}\left(k|\boldsymbol{\rho}-\boldsymbol{\rho}'|\right)d\boldsymbol{\rho}'\,d\boldsymbol{\rho}$$
$$+\frac{1}{4\omega\epsilon}\int_{\mathbf{f}_m^t}\nabla\cdot\mathbf{f}_m^t(\boldsymbol{\rho})\int_{\mathbf{f}_n^t}\nabla'\cdot\mathbf{f}_n^t(\boldsymbol{\rho}')\,H_0^{(2)}\left(k|\boldsymbol{\rho}-\boldsymbol{\rho}'|\right)d\boldsymbol{\rho}'\,d\boldsymbol{\rho}$$
$$(6.64)$$

The right-hand side vector elements (3.191) are

$$V_m^{E,TE}=\int_{\mathbf{f}_m^t}\mathbf{f}_m^t(\boldsymbol{\rho})\cdot\mathbf{E}^i(\boldsymbol{\rho})\,d\boldsymbol{\rho} \quad (6.65)$$

### 6.1.4.2 Solution Using Triangle Functions

The geometry in the TE case is similar to the thin-wire problem of Section 5.5.2, thus the orientation of the basis and testing functions follow that of Figure 5.4. We can evaluate (6.64) on non-overlapping segments using an $M$-point numerical quadrature, yielding

$$L^{tt}(\mathbf{f}_m^t, \mathbf{f}_n^t) = \sum_{p=1}^{M} \sum_{q=1}^{M} w(\boldsymbol{\rho}_p) w(\boldsymbol{\rho}_q) \left[ \frac{\omega\mu}{4} \mathbf{f}_m^t(\boldsymbol{\rho}_p) \cdot \mathbf{f}_n^t(\boldsymbol{\rho}_q) \pm \frac{1}{4\omega\epsilon\Delta_l^2} \right] H_0^{(2)}\left(k|\boldsymbol{\rho}_p - \boldsymbol{\rho}_q|\right)$$

(6.66)

where we have used the fact that the divergence of each triangle function is

$$\nabla \cdot \mathbf{f}^t(\boldsymbol{\rho}) = \pm \frac{1}{\Delta_l}$$

(6.67)

The sign of this term depends on the orientation of the basis and testing functions on each segment. The right-hand side of (6.66) should be evaluated inside a loop over segment pairs and the resulting values placed into the corresponding matrix locations. On overlapping segments, the first integral in (6.64) can be computed using (6.19), and the second integral has the form

$$\int_0^{\Delta_x} \int_0^{\Delta_x} H_0^{(2)}\left(k|x - x'|\right) \, dx' \, dx$$

(6.68)

We will evaluate the outer integral in (6.68) numerically, and the innermost integral analytically using (6.12). The right-hand side vector elements are

$$V_m^{E,TE} = \sum_{p=1}^{M} w(\boldsymbol{\rho}_p) \mathbf{f}_m^t(\boldsymbol{\rho}_p) \cdot \mathbf{E}^i(\boldsymbol{\rho}_p)$$

(6.69)

For an plane wave having an electric field of unit amplitude, (6.69) becomes

$$V_m^{E,TE} = \sum_{p=1}^{M} w(\boldsymbol{\rho}_p) \mathbf{f}_m^t(\boldsymbol{\rho}_p) \cdot (\sin \phi^i \hat{\mathbf{x}} - \cos \phi^i \hat{\mathbf{y}}) e^{jk(x_p \cos \phi^i + y_p \sin \phi^i)}$$

(6.70)

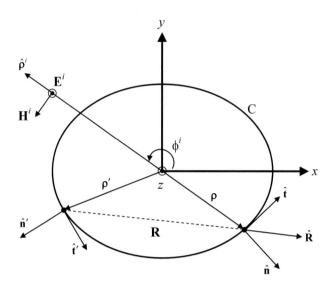

**FIGURE 6.7:** 2D Contour: TM Polarization

## 6.1.5  nMFIE: TM Polarization

Consider the closed, conducting cylindrical contour $C$ of Figure 6.7. Given the incident field of (6.1), the corresponding magnetic field is

$$\mathbf{H}^i(\boldsymbol{\rho}) = -\frac{1}{\eta}\,\hat{\boldsymbol{\rho}}^i \times \mathbf{E}^i(\boldsymbol{\rho}) = \frac{1}{\eta}\left(-\sin\phi^i\hat{\mathbf{x}} + \cos\phi^i\hat{\mathbf{y}}\right)e^{jk(x\cos\phi^i + y\sin\phi^i)}$$

(6.71)

Using (3.49) and (3.94), we can write the nMFIE (3.180) as

$$\frac{1}{2}\mathbf{J}(\boldsymbol{\rho}) - \frac{j}{4}\,\hat{\mathbf{n}}(\boldsymbol{\rho}) \times \int_{C'} \mathbf{J}(\boldsymbol{\rho}') \times \nabla H_0^{(2)}(k\rho)\,d\rho' = \hat{\mathbf{n}}(\boldsymbol{\rho}) \times \mathbf{H}^i(\boldsymbol{\rho}) \quad (6.72)$$

where we have used the notation $C'$ to indicate the integration is not valid for $\rho = \rho'$. Using the fact that

$$\nabla H_0^{(2)}(k\rho) = -k\left[\frac{x-x'}{R}\hat{\mathbf{x}} + \frac{y-y'}{R}\hat{\mathbf{y}}\right]H_1^{(2)}(k\rho) = -k\,\hat{\mathbf{R}}(\boldsymbol{\rho},\boldsymbol{\rho}')H_1^{(2)}(k\rho)$$

(6.73)

where

$$\hat{\mathbf{R}}(\boldsymbol{\rho},\boldsymbol{\rho}') = \frac{x-x'}{R}\hat{\mathbf{x}} + \frac{y-y'}{R}\hat{\mathbf{y}} = \frac{\boldsymbol{\rho} - \boldsymbol{\rho}'}{|\boldsymbol{\rho} - \boldsymbol{\rho}'|} \quad (6.74)$$

we can write (6.72) as

$$\frac{1}{2}\mathbf{J}(\boldsymbol{\rho}) + \frac{jk}{4}\hat{\mathbf{n}}(\boldsymbol{\rho}) \times \int_{C'} \mathbf{J}(\boldsymbol{\rho}') \times \hat{\mathbf{R}}(\boldsymbol{\rho},\boldsymbol{\rho}')\,H_1^{(2)}(k\rho)\,d\rho' = \hat{\mathbf{n}}(\boldsymbol{\rho}) \times \mathbf{H}^i(\boldsymbol{\rho}) \quad (6.75)$$

The TM CFIE is formed by combining the TM EFIE (6.20) and the TM nMFIE (6.75) following (3.181).

*MOM Discretization*

The contour $C$ is divided into a number of linear segments, and the current $\mathbf{J}(\boldsymbol{\rho})$ is expanded using $\hat{\mathbf{z}}$-oriented electric basis functions. Testing with these same functions yields the matrix elements (3.189) for the TM case, which are

$$Z_{mn}^{nHJ} = \mathrm{nK}^{zz}(\mathbf{f}_m^z, \mathbf{f}_n^z) \tag{6.76}$$

where

$$\mathrm{nK}^{zz}(\mathbf{f}_m^z, \mathbf{f}_n^z) = \frac{1}{2} \int_{\mathbf{f}_m^z = \mathbf{f}_n^z} \mathbf{f}_m^z(\boldsymbol{\rho}) \cdot \mathbf{f}_n^z(\boldsymbol{\rho}) \, d\boldsymbol{\rho} + \frac{jk}{4} \int_{\mathbf{f}_m^z} \mathbf{f}_m^z(\boldsymbol{\rho}) \cdot$$
$$\hat{\mathbf{n}}(\boldsymbol{\rho}) \times \int_{\mathbf{f}_n^z} \mathbf{f}_n^z(\boldsymbol{\rho}') \times \hat{\mathbf{R}}(\boldsymbol{\rho}, \boldsymbol{\rho}') \, H_1^{(2)}(k\rho) \, d\boldsymbol{\rho}' \, d\boldsymbol{\rho} \tag{6.77}$$

The first term in (6.77) is evaluated over boundary segments where $\mathbf{f}_m^z$ and $\mathbf{f}_n^z$ overlap, and the second where they do not. If we use the vector identity

$$\mathbf{A} \times \mathbf{B} \times \mathbf{C} = (\mathbf{A} \cdot \mathbf{C})\mathbf{B} - (\mathbf{A} \cdot \mathbf{B})\mathbf{C} \tag{6.78}$$

we can write

$$\hat{\mathbf{n}}(\boldsymbol{\rho}) \times \mathbf{f}_n^z(\boldsymbol{\rho}') \times \hat{\mathbf{R}}(\boldsymbol{\rho}, \boldsymbol{\rho}') = \mathbf{f}_n^z(\boldsymbol{\rho}') \left[ \mathbf{n}(\boldsymbol{\rho}) \cdot \hat{\mathbf{R}}(\boldsymbol{\rho}, \boldsymbol{\rho}') \right] \tag{6.79}$$

since $\hat{\mathbf{n}}(\boldsymbol{\rho}) \cdot \mathbf{f}_n^z(\boldsymbol{\rho}') = 0$. This allows us to write (6.77) as

$$\mathrm{nK}^{zz}(\mathbf{f}_m^z, \mathbf{f}_n^z) = \frac{1}{2} \int_{\mathbf{f}_m^z = \mathbf{f}_n^z} \mathbf{f}_m^z(\boldsymbol{\rho}) \cdot \mathbf{f}_n^z(\boldsymbol{\rho}) \, d\boldsymbol{\rho} + \frac{jk}{4} \int_{\mathbf{f}_m^z} \mathbf{f}_m^z(\boldsymbol{\rho}) \cdot$$
$$\int_{\mathbf{f}_n^z} \mathbf{f}_n^z(\boldsymbol{\rho}') \left[ \hat{\mathbf{n}}(\boldsymbol{\rho}) \cdot \hat{\mathbf{R}}(\boldsymbol{\rho}, \boldsymbol{\rho}') \right] H_1^{(2)}(k\rho) \, d\boldsymbol{\rho}' \, d\boldsymbol{\rho} \tag{6.80}$$

The right-hand side vector elements (3.193) are

$$V_m^{nH,TM} = \int_{\mathbf{f}_m^z} \mathbf{f}_m^z(\boldsymbol{\rho}) \cdot \left[ \hat{\mathbf{n}}(\boldsymbol{\rho}) \times \mathbf{H}^i(\boldsymbol{\rho}) \right] d\boldsymbol{\rho} \tag{6.81}$$

### 6.1.5.1 Solution Using Triangle Functions

Using triangle basis and testing functions, (6.80) is computed using an $M$-point numerical quadrature yielding

$$\mathrm{nK}^{zz}(\mathbf{f}_m^z, \mathbf{f}_n^z) = \frac{1}{2} \sum_{p=1}^M w(\boldsymbol{\rho}_p) f_m^z(\boldsymbol{\rho}_p) f_n^z(\boldsymbol{\rho}_p) + \frac{jk}{4} \sum_{p=1}^M w(\boldsymbol{\rho}_p) f_m^z(\boldsymbol{\rho}_p) \cdot$$
$$\sum_{q=1}^M w(\boldsymbol{\rho}_q) f_n^z(\boldsymbol{\rho}_q)(\hat{\mathbf{n}}_p \cdot \hat{\mathbf{R}}_{pq}) H_1^{(2)}\left(k|\boldsymbol{\rho}_p - \boldsymbol{\rho}_q|\right) \tag{6.82}$$

where we have again omitted the vector notation for $\mathbf{f}_m^z$ and $\mathbf{f}_n^z$. The right-hand side elements are

$$V_m^{nH,TM} = \sum_{p=1}^{M} w(\boldsymbol{\rho}_p)\, \mathbf{f}_m^z(\boldsymbol{\rho}_p) \cdot \left[\hat{\mathbf{n}}(\boldsymbol{\rho}_p) \times \mathbf{H}^i(\boldsymbol{\rho}_p)\right] \qquad (6.83)$$

For a plane wave having an electric field of unit amplitude, (6.83) becomes

$$V_m^{nH,TM} = \frac{1}{\eta} \sum_{p=1}^{M} w(\boldsymbol{\rho}_p)\, \mathbf{f}_m^z(\boldsymbol{\rho}_p) \cdot \left[\hat{\mathbf{n}}(\boldsymbol{\rho}_p) \times (-\sin\phi^i \hat{\mathbf{x}} \cos\phi^i \hat{\mathbf{y}})\right] e^{jk(x_p \cos\phi^i + y_p \sin\phi^i)}$$

$$(6.84)$$

### 6.1.6 nMFIE: TE Polarization

Consider the same contour as before, with a TE-polarized incident field as shown in Figure 6.8. Given the electric field of (6.27), the corresponding magnetic field is

$$\mathbf{H}^i(\boldsymbol{\rho}) = -\frac{1}{\eta}\,\hat{\boldsymbol{\rho}}^i \times \mathbf{E}^i(\boldsymbol{\rho}) = \frac{1}{\eta}\, e^{jk(x\cos\phi^i + y\sin\phi^i)}\,\hat{\mathbf{z}} \qquad (6.85)$$

We know from the discussion in Section 6.1.3 that the induced current $\mathbf{J}(\boldsymbol{\rho})$

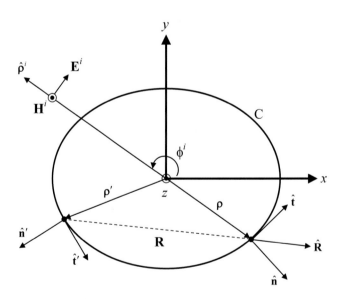

**FIGURE 6.8:** 2D Contour: TE Polarization

only has components along the vector tangent to $C$. The nMFIE for the TE case is very similar to that of the TM case (6.75), with some small differences that will appear in the MOM discretization. The TE CFIE can be created by combining the TE EFIE (6.61) and the TE nMFIE (6.75) following (3.181)

*MOM Discretization*

We again divide the boundary $C$ into $N$ segments, and the current $\mathbf{J}(\boldsymbol{\rho})$ is expanded using electric basis functions that are tangent to $C$. Testing with these same functions yields the matrix elements (3.189) for the TE case, which are

$$Z_{mn}^{nHJ} = nK^{tt}(\mathbf{f}_m^t, \mathbf{f}_n^t) \tag{6.86}$$

where

$$nK^{tt}(\mathbf{f}_m^t, \mathbf{f}_n^t) = \frac{1}{2} \int_{\mathbf{f}_m^t = \mathbf{f}_n^t} \mathbf{f}_m^t(\boldsymbol{\rho}) \cdot \mathbf{f}_n^t(\boldsymbol{\rho}) \, d\boldsymbol{\rho} + \frac{jk}{4} \int_{\mathbf{f}_m^t} \mathbf{f}_m^t(\boldsymbol{\rho}) \cdot$$

$$\hat{\mathbf{n}}(\boldsymbol{\rho}) \times \int_{\mathbf{f}_n^t} \mathbf{f}_n^t(\boldsymbol{\rho}') \times \hat{\mathbf{R}}(\boldsymbol{\rho}, \boldsymbol{\rho}') \, H_1^{(2)}(k\rho) \, d\boldsymbol{\rho}' \, d\boldsymbol{\rho} \tag{6.87}$$

where the first term is evaluated over boundary segments where $\mathbf{a}^t$ and $\mathbf{b}^t$ overlap, and the second where they do not. The right-hand side vector elements (3.193) are

$$V_m^{nH,TE} = \int_{\mathbf{f}_m^t} \mathbf{f}_m^t(\boldsymbol{\rho}) \cdot \left[ \hat{\mathbf{n}}(\boldsymbol{\rho}) \times \mathbf{H}^i(\boldsymbol{\rho}) \right] d\boldsymbol{\rho} \tag{6.88}$$

### 6.1.6.1 Solution Using Triangle Functions

The triangle basis and testing functions are oriented in the same manner as in Section 6.1.4.2. Applying an $M$-point numerical quadrature to (6.87) yields

$$nK^{tt}(\mathbf{f}_m^t, \mathbf{f}_n^t) = \frac{1}{2} \sum_{p=1}^M w(\boldsymbol{\rho}_p) \, \mathbf{f}_m^t(\boldsymbol{\rho}_p) \cdot \mathbf{f}_n^t(\boldsymbol{\rho}_p) + \frac{jk}{4} \sum_{p=1}^M w(\boldsymbol{\rho}_p) \, \mathbf{f}_m^t(\boldsymbol{\rho}_p) \cdot$$

$$\left[ \hat{\mathbf{n}}(\boldsymbol{\rho}_p) \times \sum_{q=1}^M w(\boldsymbol{\rho}_q) \, \mathbf{f}_n^t(\boldsymbol{\rho}_q) \times \hat{\mathbf{R}}_{pq} \, H_1^{(2)}(k|\boldsymbol{\rho}_p - \boldsymbol{\rho}_q|) \right] \tag{6.89}$$

The right-hand side vector elements are

$$V_m^{nH,TE} = \sum_{p=1}^M w(\boldsymbol{\rho}_p) \, \mathbf{f}_m^t(\boldsymbol{\rho}_p) \cdot \left[ \hat{\mathbf{n}}(\boldsymbol{\rho}_p) \times \mathbf{H}^i(\boldsymbol{\rho}_p) \right] \tag{6.90}$$

For a plane wave having an electric field of unit amplitude, (6.90) becomes

$$V_m^{nH,TE} = \frac{1}{\eta} \sum_{p=1}^M w(\boldsymbol{\rho}_p) \, \mathbf{f}_m^t(\boldsymbol{\rho}_p) \cdot \left[ \hat{\mathbf{n}}(\boldsymbol{\rho}_p) \times \hat{\mathbf{z}} \right] e^{jk(x_p \cos \phi^i + y_p \sin \phi^i)} \tag{6.91}$$

## 6.1.7  Examples

We will now use the generalized EFIE and nMFIE to analyze the scattering of plane waves by a two-dimensional conducting circular cylinder for TM and TE polarizations. The integral equation solution uses the MOM discretizations formulated via triangle basis and testing functions, and we use $\alpha = 0.5$ in the CFIE.

### 6.1.7.1  Conducting Cylinder: TM Polarization

For TM polarization, the exact solution for the scattered electric field of a conducting cylinder is [4]

$$E_z^s(\boldsymbol{\rho}, \phi^s) = \sum_{n=0}^{\infty} j^n c_n A_n H_n^{(2)}(k_0 \rho) \cos(n\phi^s) \qquad (6.92)$$

where $|E_z^i| = 1$ for convenience, $\phi^s$ is the bistatic scattering angle, and $c_n = 1$ for $n = 0$, and $c_n = 2$ otherwise. For the conducting cylinder, the coefficients $A_n$ are

$$A_n = -\frac{J_n(k_0 a)}{H_n^{(2)}(k_0 a)} \qquad (6.93)$$

where $a$ is the radius of the cylinder. The scattered far electric field is

$$E_z^s(\rho, \phi^s) = \sqrt{\frac{2}{\pi}} \frac{e^{-j(k_0\rho - \pi/4)}}{\sqrt{k_0\rho}} \sum_{n=0}^{\infty} (-1)^n c_n A_n \cos n\phi^s \qquad (6.94)$$

The induced electric current $J_z(\phi^a)$ is

$$J_z(\phi^a) = \frac{2}{\pi \eta_0 k_0 a} \sum_{n=0}^{\infty} \frac{(j)^n c_n \cos n\phi^a}{H_n^{(2)}(k_0 a)} \qquad (6.95)$$

where $\phi^a$ is the azimuthal angle on the surface of the cylinder. Using (6.71), the PO current is

$$J_z^{PO}(\phi^a) = \frac{2}{\eta}(\cos\phi^a \cos\phi^i + \sin\phi^a \sin\phi^i)e^{jk(x\cos\phi^i + y\sin\phi^i)} \qquad (6.96)$$

In the high-frequency limit ($k_o \to \infty$), the radar cross section in the backscattering direction ($\phi^s = 0$) approaches $\pi a$, and in the forward-scattering direction ($\phi^s = \pi$) it approaches $4k_o a$ [4].

*Results*

We first compute the radar cross section of a conducting cylinder having a radius of 1 meter for frequencies up to 800 MHz. For the MOM solution, the number of segments was determined by using 10 per $\lambda$ at 800 MHz, yielding 167 segments of equal length. For the modal solution, we terminate the summation at $N = 40$. The backscattering results are summarized in Figures 6.9a–6.9c using EFIE, nMFIE and CFIE, respectively. The EFIE does fairly well, at least within the range of frequencies considered, whereas the nMFIE shows many errors due to internal resonances. The CFIE is free of errors, as expected. The RCS in each case is seen to approach the expected high-frequency value of $\pi a$ m$^2$, which is approximately 4.97 dBsm in this case. The forward-scattering results are shown in Figures 6.9d–6.9f. The nMFIE again shows some errors but they are not quite as apparent as they were in the backscattering case. The RCS in each case tends toward the expected high-frequency value of $4k_o$ m$^2$, which at 800 MHz is approximately 18.26 dBsm.

We next compute the bistatic RCS for $0 \leq \phi^s \leq 2\pi$ and $\phi^i = 0$ at 800 MHz. The CFIE results are compared to the exact solution in Figure 6.10a, where the agreement is seen to be excellent. The results obtained using the PO current of (6.96) are shown in Figure 6.10b, where the agreement is good near the specular angles but fair to poor elsewhere.

We then compute the electric current $J_z(\phi^a)$ across all azimuthal angles for $\phi^i = 0$ at 800 MHz. The real and imaginary components of the currents obtained via the CFIE and PO are compared to the exact solution (6.95) in Figures 6.11a–6.11d, and the amplitudes compared in Figures 6.11e and 6.11f. The agreement between the modal solution and the CFIE is very good. The PO current compares fairly well in the illuminated region, however in the shadow region ($\pi/2 \leq \phi^a \leq 3\pi/2$) the PO current is zero.

Finally, we compute the scattered near field in a $10 \times 10$ meter region around the cylinder at 800 MHz for $\phi^i = 0$. The real component of the electric field (in V/m) obtained using the CFIE is shown in Figure 6.12a. The modal solution is indistinguishable to the eye and is not shown. The solution obtained using Physical Optics is shown in Figure 6.12b.

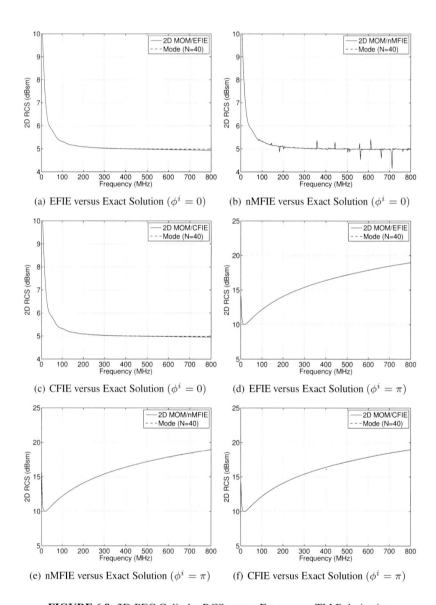

**FIGURE 6.9:** 2D PEC Cylinder RCS versus Frequency: TM Polarization

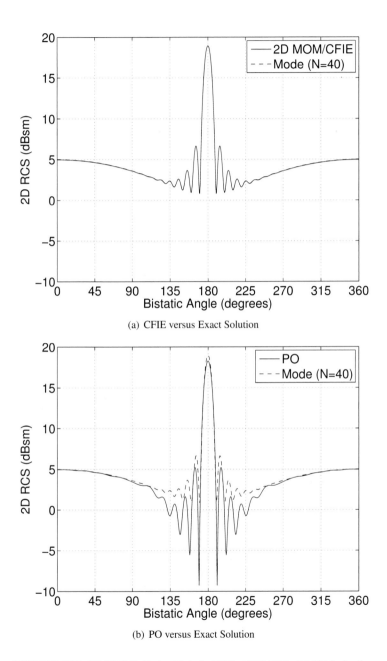

(a) CFIE versus Exact Solution

(b) PO versus Exact Solution

**FIGURE 6.10:** 2D PEC Cylinder Bistatic RCS at 800 MHz: TM Polarization

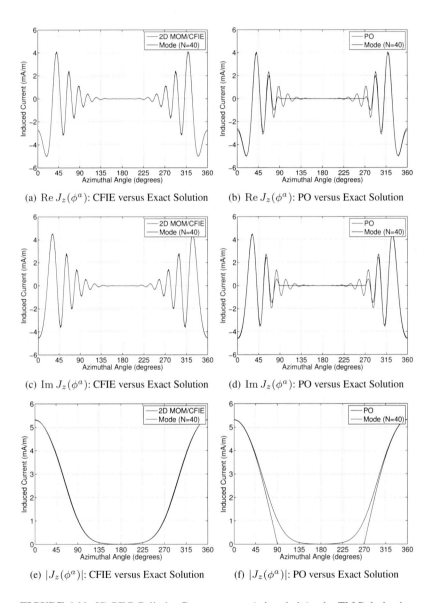

(a) Re $J_z(\phi^a)$: CFIE versus Exact Solution

(b) Re $J_z(\phi^a)$: PO versus Exact Solution

(c) Im $J_z(\phi^a)$: CFIE versus Exact Solution

(d) Im $J_z(\phi^a)$: PO versus Exact Solution

(e) $|J_z(\phi^a)|$: CFIE versus Exact Solution

(f) $|J_z(\phi^a)|$: PO versus Exact Solution

**FIGURE 6.11:** 2D PEC Cylinder Current versus Azimuthal Angle: TM Polarization

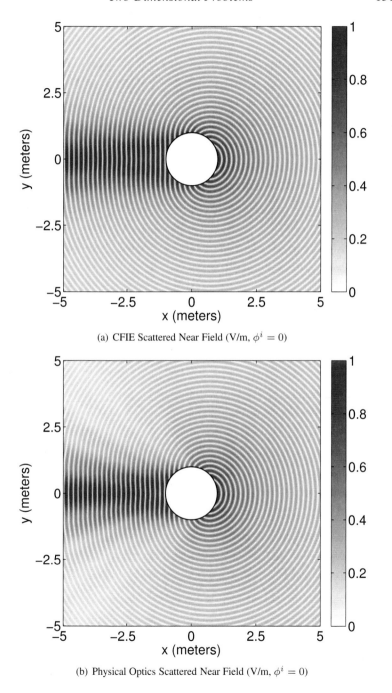

(a) CFIE Scattered Near Field (V/m, $\phi^i = 0$)

(b) Physical Optics Scattered Near Field (V/m, $\phi^i = 0$)

**FIGURE 6.12:** 2D PEC Cylinder Scattered Near Field at 800 MHz: TM Polarization

### 6.1.7.2   Conducting Cylinder: TE Polarization

For TE polarization, with a $\hat{\mathbf{z}}$-directed incident magnetic field, the exact solution for the scattered magnetic field is [4]

$$H_z^s(\rho, \phi^s) = \frac{1}{\eta_0} \sum_{n=0}^{\infty} j^n c_n B_n H_n^{(2)}(k_0 \rho) \cos(n\phi^s) \qquad (6.97)$$

where $|E_\phi^i| = 1$ for convenience. For the conducting cylinder, the coefficients $B_n$ are

$$B_n = -\frac{J_n'(k_0 a)}{H_n'^{(2)}(k_0 a)} \qquad (6.98)$$

The scattered far magnetic field is

$$H_z^s(\rho, \phi^s) = \frac{1}{\eta_0} \sqrt{\frac{2}{\pi}} \frac{e^{-j(k_0\rho - \pi/4)}}{\sqrt{k_0 \rho}} \sum_{n=0}^{\infty} (-1)^n c_n B_n \cos n\phi^s \qquad (6.99)$$

The induced azimuthal electric current is

$$\mathbf{J}(\phi^a) = \frac{2j}{\eta_0 \pi k_0 a} \sum_{n=0}^{\infty} \frac{(j)^n c_n \cos n\phi^a}{H_n'^{(2)}(k_0 a)} \hat{\boldsymbol{\phi}} \qquad (6.100)$$

where $\phi^a$ is the azimuthal angle on the surface of the cylinder. The recurrence relationship

$$Z_n'(ka) = \frac{n}{ka} Z_n(ka) - Z_{n+1}(ka) \qquad (6.101)$$

can be used to compute the derivatives, where $Z_n$ represents either $J_n$ or $H_n$ [2]. Using (6.85), the PO current is

$$\mathbf{J}^{PO}(\phi^a) = \frac{2}{\eta} (\sin \phi^a \hat{\mathbf{x}} - \cos \phi^a \hat{\mathbf{y}}) e^{jk(x \cos \phi^i + y \sin \phi^i)} \qquad (6.102)$$

*Results*

For TE polarization, we use the same 1-meter conducting cylinder and numerical parameters as in the TM case. The backscatter results are summarized in Figures 6.13a–6.13c using EFIE, nMFIE, and CFIE, respectively. The nMFIE again shows errors due to internal resonances, and the CFIE is free of errors, as expected. The RCS in all cases again approaches the high-frequency limit of approximately 4.97 dBsm. The forward-scattering results are shown in Figures 6.13d–6.13f. The nMFIE again shows some errors but they are again not as apparent as in the backscattering direction. The RCS in all cases approaches the expected high-frequency limit of approximately 18.26 dBsm at 800 MHz.

We next compute the bistatic RCS for $0 \leq \phi^s \leq 2\pi$ and $\phi^i = 0$ at 800

MHz. The CFIE results are compared to the exact solution in Figure 6.14a, and the agreement is again excellent. The results obtained using the PO current of (6.102) are shown in Figure 6.14b. The agreement is fair near the specular angles, but poor elsewhere.

We then we compute the electric current $J_z(\phi^a)$ across all azimuthal angles for $\phi^i = 0$ at 800 MHz. The real and imaginary components of the currents obtained via CFIE and PO are compared to the exact solution (6.95) in Figures 6.15a–6.15d, and the amplitudes compared in Figures 6.15e and 6.15f. The agreement between the modal solution and the CFIE is fairly good, however the PO current does not agree very well at all, as its amplitude comprises a step function in the TE case.

Finally, we compute the scattered near field in a $10 \times 10$ meter region about the cylinder at 800 MHz for $\phi^i = 0$. The real component of the electric field (in V/m) obtained using the CFIE is shown in Figure 6.16a. The modal solution is indistinguishable to the eye and is not shown. The corresponding solution obtained using Physical Optics is shown in Figure 6.16b.

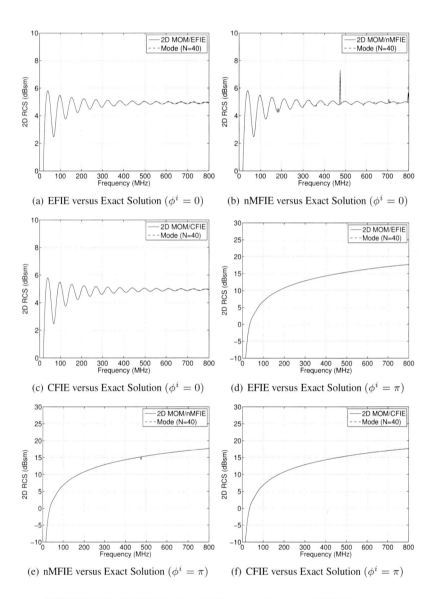

(a) EFIE versus Exact Solution ($\phi^i = 0$)

(b) nMFIE versus Exact Solution ($\phi^i = 0$)

(c) CFIE versus Exact Solution ($\phi^i = 0$)

(d) EFIE versus Exact Solution ($\phi^i = \pi$)

(e) nMFIE versus Exact Solution ($\phi^i = \pi$)

(f) CFIE versus Exact Solution ($\phi^i = \pi$)

**FIGURE 6.13:** 2D PEC Cylinder RCS versus Frequency: TE Polarization

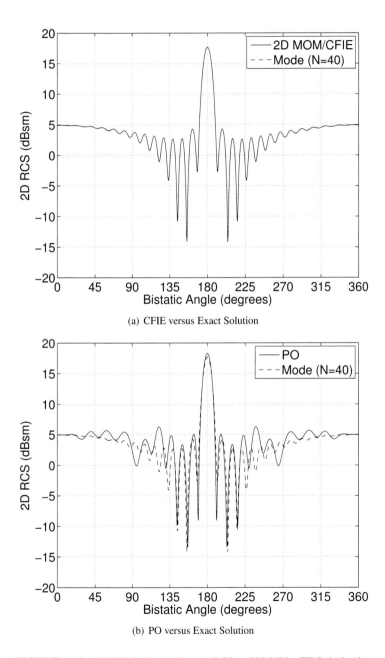

(a) CFIE versus Exact Solution

(b) PO versus Exact Solution

**FIGURE 6.14:** 2D PEC Cylinder Bistatic RCS at 800 MHz: TE Polarization

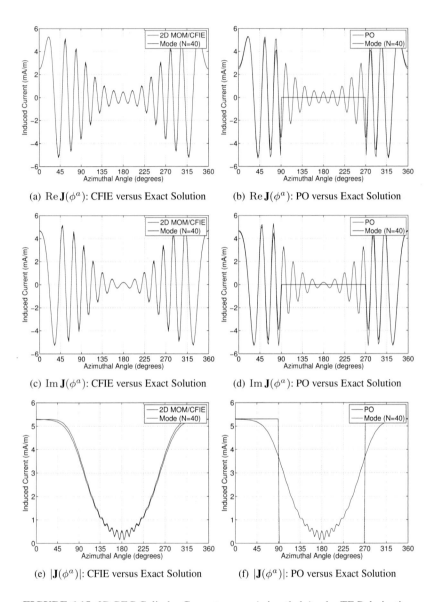

(a) $\mathrm{Re}\,\mathbf{J}(\phi^a)$: CFIE versus Exact Solution

(b) $\mathrm{Re}\,\mathbf{J}(\phi^a)$: PO versus Exact Solution

(c) $\mathrm{Im}\,\mathbf{J}(\phi^a)$: CFIE versus Exact Solution

(d) $\mathrm{Im}\,\mathbf{J}(\phi^a)$: PO versus Exact Solution

(e) $|\mathbf{J}(\phi^a)|$: CFIE versus Exact Solution

(f) $|\mathbf{J}(\phi^a)|$: PO versus Exact Solution

**FIGURE 6.15:** 2D PEC Cylinder Current versus Azimuthal Angle: TE Polarization

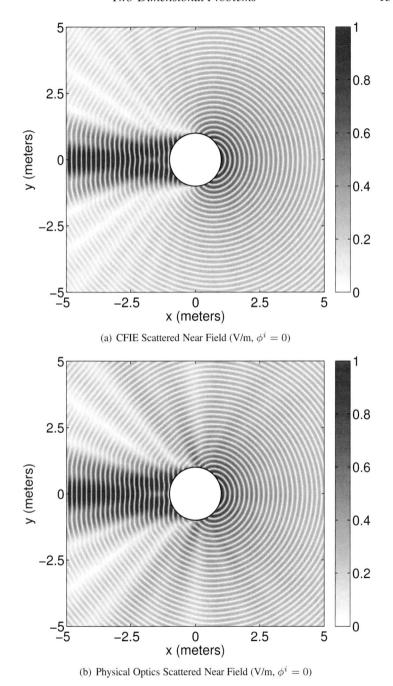

(a) CFIE Scattered Near Field (V/m, $\phi^i = 0$)

(b) Physical Optics Scattered Near Field (V/m, $\phi^i = 0$)

**FIGURE 6.16:** 2D PEC Cylinder Scattered Near Field at 800 MHz: TE Polarization

## 6.2    Dielectric and Composite Objects

In this section, we build upon the work in Section 6.1 to build a framework for treating two-dimensional dielectric and composite dielectric/conducting objects (without junctions). We will use the expressions previously derived for the EFIE and nMFIE that involved the $\mathcal{L}$ and $\hat{\mathbf{n}} \times \mathcal{K}$ operators, and develop new expressions involving the $\mathcal{K}$ and $\hat{\mathbf{n}} \times \mathcal{L}$ operators. As before, we will consider each integral equation in TM and TE polarization separately, and we will present generalized MOM discretizations exclusively in terms of triangle basis and testing functions. We will then present several examples.

### 6.2.1    Basis Function Orientation

As we begin to consider multi-region problems in more detail, the orientation of the basis functions on the various dielectric interfaces must be considered carefully. As there are $\hat{\mathbf{z}}$-directed electric currents and tangentially directed magnetic currents in the TM case, and vice-versa for the TE case, we will treat the orientations of both currents in a self-consistent manner. Consider a triangle function with support on segments $s+$ and $s-$, whose endpoints comprise $(\mathbf{p}_{k-1}, \mathbf{p}_k, \mathbf{p}_{k+1})$. The segments reside on the interface $S_{12}$ between dielectric regions $R_1$ and $R_2$, as shown in Figure 6.17. Let us define the unit vectors

$$\hat{\mathbf{t}}^{s+} = \frac{\mathbf{p}_{k+1} - \mathbf{p}_k}{|\mathbf{p}_{k+1} - \mathbf{p}_k|} \tag{6.103}$$

$$\hat{\mathbf{t}}^{s-} = \frac{\mathbf{p}_k - \mathbf{p}_{k-1}}{|\mathbf{p}_k - \mathbf{p}_{k-1}|} \tag{6.104}$$

We know from Section 3.6.3 that the currents on each side of the interface must

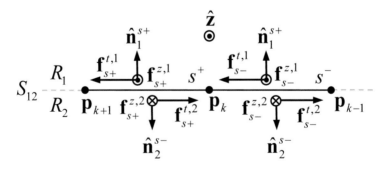

**FIGURE 6.17:** Basis Function Orientation for 2D Problems

flow in opposite directions. Therefore, let us define $R_1$ as the "positive" side of the interface. On the positive side, the vectors normal to each segment are

$$\hat{\mathbf{n}}_1^{s+} = \hat{\mathbf{t}}^{s+} \times \hat{\mathbf{z}} \tag{6.105}$$

$$\hat{\mathbf{n}}_1^{s-} = \hat{\mathbf{t}}^{s-} \times \hat{\mathbf{z}} \tag{6.106}$$

As we are still free to orient the basis functions, we begin in $R_1$, and make the following assignment for the tangential basis function

$$\hat{\mathbf{f}}_{s+}^{t,1} = \hat{\mathbf{t}}^{s+} \tag{6.107}$$

$$\hat{\mathbf{f}}_{s-}^{t,1} = \hat{\mathbf{t}}^{s-} \tag{6.108}$$

and the following assignment for the $\hat{\mathbf{z}}$-oriented basis function

$$\hat{\mathbf{f}}_{s+}^{z,1} = \hat{\mathbf{z}} \tag{6.109}$$

$$\hat{\mathbf{f}}_{s-}^{z,1} = \hat{\mathbf{z}} \tag{6.110}$$

If we then move to the opposite or "negative" side of the interface in $R_2$, we need only reverse the direction of the normal vector and the basis functions in $R_1$, as shown in Figure 6.17. We refer to this as a right-handed boundary orientation. Note that all currents have opposite orientations on each side of the interface, and for tangentially oriented currents, Kirchoff's law is satisfied at $p_k$. We will use this orientation again when discussing bodies of revolution in Chapter 7.

## 6.2.2 EFIE: TM Polarization

The EFIE (3.179) for TM polarization has an $\mathcal{L}$ operator acting on the $\hat{\mathbf{z}}$-directed electric current $\mathbf{J}(\boldsymbol{\rho})$, and a $\mathcal{K}$ operator acting on the tangential magnetic current $\mathbf{M}(\boldsymbol{\rho})$. In region $R_l$, these expressions are

$$\frac{\omega\mu_l}{4}\,\hat{\mathbf{t}}(\boldsymbol{\rho}) \cdot \int_C \left[1 + \frac{1}{k_l^2}\nabla\nabla \cdot\right] \mathbf{J}(\boldsymbol{\rho}')H_0^{(2)}\left(k_l|\boldsymbol{\rho} - \boldsymbol{\rho}'|\right) d\boldsymbol{\rho}' \tag{6.111}$$

and

$$\frac{1}{2}\hat{\mathbf{n}}_l(\boldsymbol{\rho}) \times \mathbf{M}(\boldsymbol{\rho}) - \frac{jk_l}{4}\int_{C'} \mathbf{M}(\boldsymbol{\rho}') \times \hat{\mathbf{R}}(\boldsymbol{\rho}, \boldsymbol{\rho}')\,H_1^{(2)}(k_l\rho)\,d\boldsymbol{\rho}' \tag{6.112}$$

#### 6.2.2.1  MOM Discretization

Testing (6.111) and (6.112) with the $\hat{\mathbf{z}}$-oriented electric basis functions yields the matrix elements (3.185) and (3.186) for the TM case, which are

$$Z_{mn}^{EJ(l),TM} = \mathrm{L}^{zz}(\mathbf{f}_m^z, \mathbf{f}_n^z) \tag{6.113}$$

and

$$Z_{mn}^{EM(l),TM} = \mathrm{K}^{zt}(\mathbf{f}_m^z, \mathbf{g}_n^t) \tag{6.114}$$

The function $\mathrm{L}^{zz}$ was summarized in (6.22). The function $\mathrm{K}^{zt}$ is

$$
\mathrm{K}^{zt}(\mathbf{f}_m^z, \mathbf{g}_n^t) = \frac{1}{2} \int_{\mathbf{f}_m^z = \mathbf{g}_n^t} \mathbf{f}_m^z(\boldsymbol{\rho}) \cdot \left[ \hat{\mathbf{n}}_l(\boldsymbol{\rho}) \times \mathbf{g}_n^t(\boldsymbol{\rho}) \right] d\boldsymbol{\rho}
$$
$$
- \frac{jk_l}{4} \int_{\mathbf{f}_m^z} \mathbf{f}_m^z(\boldsymbol{\rho}) \cdot \int_{\mathbf{g}_n^t} \mathbf{g}_n^t(\boldsymbol{\rho}') \times \hat{\mathbf{R}}(\boldsymbol{\rho}, \boldsymbol{\rho}')\, H_1^{(2)}(k_l\rho)\, d\boldsymbol{\rho}'\, d\boldsymbol{\rho} \tag{6.115}
$$

where the first term is evaluated over boundary segments where $\mathbf{f}_m^z$ and $\mathbf{g}_n^t$ overlap, and the second where they do not.

### 6.2.3  MFIE: TM Polarization

The MFIE (3.179) for TM-polarization has an $\mathcal{L}$ operator acting on the tangential magnetic current $\mathbf{M}(\boldsymbol{\rho})$, and a $\mathcal{K}$ operator acting on the $\hat{\mathbf{z}}$-directed electric current $\mathbf{J}(\boldsymbol{\rho})$. In region $R_l$, these expressions are

$$\frac{\omega\epsilon_l}{4}\, \hat{\mathbf{t}}(\boldsymbol{\rho}) \cdot \int_C \left[ 1 + \frac{1}{k_l^2} \nabla\nabla \cdot \right] \mathbf{M}(\boldsymbol{\rho}')H_0^{(2)}\left(k_l|\boldsymbol{\rho} - \boldsymbol{\rho}'|\right) d\boldsymbol{\rho}' \tag{6.116}$$

and

$$-\frac{1}{2}\hat{\mathbf{n}}_l(\boldsymbol{\rho}) \times \mathbf{J}(\boldsymbol{\rho}) + \frac{jk_l}{4} \int_{C'} \mathbf{J}(\boldsymbol{\rho}') \times \hat{\mathbf{R}}(\boldsymbol{\rho}, \boldsymbol{\rho}')\, H_1^{(2)}(k_l\rho)\, d\boldsymbol{\rho}' \tag{6.117}$$

#### 6.2.3.1  MOM Discretization

Testing (6.116) and (6.117) with the tangential magnetic basis functions yields the matrix elements (3.187) and (3.188), which are

$$Z_{mn}^{HM(l),TM} = \frac{\epsilon_l}{\mu_l}\mathrm{L}^{tt}(\mathbf{g}_m^t, \mathbf{g}_n^t) \tag{6.118}$$

and

$$Z_{mn}^{HJ(l),TM} = -\mathrm{K}^{tz}(\mathbf{g}_m^t, \mathbf{f}_n^z) \tag{6.119}$$

The function $\mathrm{L}^{tt}$ was summarized in (6.64). The function $\mathrm{K}^{tz}$ is functionally identical to (6.115) except that $t$ and $z$ are interchanged. Therefore, we can write this as

$$K^{tz}(\mathbf{g}_m^t, \mathbf{f}_n^z) = \frac{1}{2} \int_{\mathbf{g}_m^t = \mathbf{f}_n^z} \mathbf{g}_m^t(\boldsymbol{\rho}) \cdot \left[ \hat{\mathbf{n}}_l(\boldsymbol{\rho}) \times \mathbf{f}_n^z(\boldsymbol{\rho}) \right] d\boldsymbol{\rho}$$
$$- \frac{jk_l}{4} \int_{\mathbf{g}_m^t} \mathbf{g}_m^t(\boldsymbol{\rho}) \cdot \int_{\mathbf{f}_n^z} \mathbf{f}_n^z(\boldsymbol{\rho}') \times \hat{\mathbf{R}}(\boldsymbol{\rho}, \boldsymbol{\rho}') \, H_1^{(2)}(k_l \rho) \, d\boldsymbol{\rho}' \, d\boldsymbol{\rho} \tag{6.120}$$

### 6.2.4 nMFIE: TM Polarization

The nMFIE (3.180) for TM polarization has an $\hat{\mathbf{n}} \times \mathcal{L}$ operator acting on the tangential magnetic current $\mathbf{M}(\boldsymbol{\rho})$, and an $\hat{\mathbf{n}} \times \mathcal{K}$ operator acting on the $\hat{\mathbf{z}}$-directed electric current $\mathbf{J}(\boldsymbol{\rho})$. In region $R_l$, these expressions are

$$\frac{1}{2}\mathbf{J}(\boldsymbol{\rho}) + \frac{jk_l}{4}\hat{\mathbf{n}}(\boldsymbol{\rho}) \times \int_{C'} \mathbf{J}(\boldsymbol{\rho}') \times \hat{\mathbf{R}}(\boldsymbol{\rho}, \boldsymbol{\rho}') \, H_1^{(2)}(k_l \rho) \, d\boldsymbol{\rho}' \tag{6.121}$$

and

$$\frac{\omega \epsilon_l}{4} \, \hat{\mathbf{n}}_l(\boldsymbol{\rho}) \times \int_C \left[ 1 + \frac{1}{k_l^2} \nabla \nabla \cdot \right] \mathbf{M}(\boldsymbol{\rho}') H_0^{(2)}(k_l |\boldsymbol{\rho} - \boldsymbol{\rho}'|) \, d\boldsymbol{\rho}' \tag{6.122}$$

#### 6.2.4.1 MOM Discretization

Testing (6.121) and (6.122) with the $\hat{\mathbf{z}}$-oriented electric basis functions yields the matrix elements (3.189) and (3.190) for the TM case, which are

$$Z_{mn}^{nHJ(l),TM} = nK^{zz}(\mathbf{f}_m^z, \mathbf{f}_n^z) \tag{6.123}$$

and

$$Z_{mn}^{nHM(l),TM} = \frac{\epsilon_l}{\mu_l} nL^{zt}(\mathbf{f}_m^z, \mathbf{g}_n^t) \tag{6.124}$$

The function $nK^{zz}$ was summarized in (6.77). The function $nL^{zt}$ is

$$nL^{zt}(\mathbf{f}_m^z, \mathbf{g}_n^t) = \frac{\omega \mu_l}{4} \int_{\mathbf{f}_m^z} \mathbf{f}_m^z(\boldsymbol{\rho}) \cdot \hat{\mathbf{n}}_l(\boldsymbol{\rho}) \times \int_{\mathbf{g}_n^t} \mathbf{g}_n^t(\boldsymbol{\rho}') \, H_0^{(2)}(k_l|\boldsymbol{\rho} - \boldsymbol{\rho}'|) \, d\boldsymbol{\rho}' \, d\boldsymbol{\rho}$$
$$- \frac{1}{4\omega\epsilon_l} \int_{\mathbf{f}_m^z} \nabla \cdot \left[ \mathbf{f}_m^z(\boldsymbol{\rho}) \times \hat{\mathbf{n}}_l(\boldsymbol{\rho}) \right] \int_{\mathbf{g}_n^t} \nabla' \cdot \mathbf{g}_n^t(\boldsymbol{\rho}') \, H_0^{(2)}(k_l|\boldsymbol{\rho} - \boldsymbol{\rho}'|) \, d\boldsymbol{\rho}' \, d\boldsymbol{\rho} \tag{6.125}$$

where we have used (3.205). Note that the term $\mathbf{f}_m^z(\boldsymbol{\rho}) \times \hat{\mathbf{n}}_l(\boldsymbol{\rho})$ results in a tangentially oriented vector on each segment.

### 6.2.5   EFIE: TE Polarization

The operators in the EFIE for TE polarization are the dual of those from the MFIE in the TM case, where the electric and magnetic currents are exchanged. Thus, we will simply summarize the MOM discretization.

#### 6.2.5.1   MOM Discretization

The matrix elements (3.185) and (3.186) for the TE case are

$$Z_{mn}^{EJ(l),TE} = \mathrm{L}^{tt}(\mathbf{f}_m^t, \mathbf{f}_n^t) \tag{6.126}$$

and

$$Z_{mn}^{EM(l),TE} = \mathrm{K}^{tz}(\mathbf{f}_m^t, \mathbf{g}_n^z) \tag{6.127}$$

### 6.2.6   MFIE: TE Polarization

Likewise, the operators in the MFIE for TE polarization are the dual of those from the EFIE in the TM case, with the electric and magnetic currents exchanged. Thus, we will simply summarize the MOM discretization.

#### 6.2.6.1   MOM Discretization

The matrix elements (3.187) and (3.188) for the TE case are

$$Z_{mn}^{HM(l),TE} = \frac{\epsilon_l}{\mu_l}\mathrm{L}^{zz}(\mathbf{g}_m^z, \mathbf{g}_n^z) \tag{6.128}$$

and

$$Z_{mn}^{HJ(l),TE} = -\mathrm{K}^{zt}(\mathbf{g}_m^z, \mathbf{f}_n^t) \tag{6.129}$$

### 6.2.7   nMFIE: TE Polarization

The expressions for the nMFIE for TE-polarization are functionally identical to those of the nMFIE for TM-polarization, except that $t$ and $z$ are interchanged. Thus, we will simply summarize the MOM discretization.

#### 6.2.7.1   MOM Discretization

The matrix elements (3.189) and (3.190) for the TE case are

$$Z_{mn}^{nHJ(l),TE} = \mathrm{nK}^{tt}(\mathbf{f}_m^t, \mathbf{f}_n^t) \tag{6.130}$$

and

$$Z_{mn}^{nHM(l),TE} = \frac{\epsilon_l}{\mu_l}\mathrm{nL}^{tz}(\mathbf{f}_m^t, \mathbf{g}_n^z) \tag{6.131}$$

The function $nK^{tt}$ was summarized in (6.87). The function $nL^{tz}$ is

$$nL^{tz}(\mathbf{f}_m^t, \mathbf{g}_n^z) = \frac{\omega\mu_l}{4} \int_{\mathbf{f}_m^t} \mathbf{f}_m^t(\boldsymbol{\rho}) \cdot \int_{\mathbf{g}_n^z} \hat{\mathbf{n}}_l(\boldsymbol{\rho}) \times \mathbf{g}_n^z(\boldsymbol{\rho}') \, H_0^{(2)}\left(k_l|\boldsymbol{\rho} - \boldsymbol{\rho}'|\right) \, d\boldsymbol{\rho}' \, d\boldsymbol{\rho}$$

(6.132)

where the second term on the right in (6.125) now goes to zero, since

$$\nabla \cdot \left[\mathbf{f}_m^t(\boldsymbol{\rho}) \times \hat{\mathbf{n}}_l(\boldsymbol{\rho})\right] = 0 \qquad (6.133)$$

## 6.2.8 Numerical Stability

Consider a system matrix that has been constructed from the EFIE and MFIE. In very general terms, it can be represented via the following block structure

$$Z = \left[ \begin{array}{cc} L(\mathbf{f}, \mathbf{f}) & K(\mathbf{f}, \mathbf{g}) \\ -K(\mathbf{g}, \mathbf{f}) & \frac{\epsilon}{\mu}L(\mathbf{g}, \mathbf{g}) \end{array} \right] \qquad (6.134)$$

Note that the matrix block in the lower right is scaled down by $1/\eta^2$ versus the block in the upper left. This disparity in amplitude tends to cause numerical stability issues in the linear system, due to limitations in the floating point representation. As a remedy, we will use the approach from [5], where the magnetic basis function is everywhere scaled by $\eta_0$. That is,

$$\mathbf{g}^{t,z} = \eta_0 \mathbf{f}^{t,z} \qquad (6.135)$$

This modification works very well in practice, and we will incorporate this scale factor into all the magnetic basis functions throughout the remainder of this book.

## 6.2.9 Examples

In this section we apply the expressions developed in this section to the scattering of plane waves by a homogeneous dielectric cylinder and a coated, conducting cylinder. Lossless as well as lossy dielectrics will be considered.

### 6.2.9.1 Dielectric Cylinder

Consider a homogeneous dielectric circular cylinder of radius $a$ with dielectric parameters $(\epsilon_1, \mu_1)$. We divide the problem into an interior region $R_1$ with the aforementioned parameters, and an external free space region $R_0$ with material parameters $(\epsilon_0, \mu_0)$, which are separated by the interface $S_{01}$. Using the PMCHWT approach, we compute the EFIE and MFIE on $S_0$ in $R_0$ and $S_1$ in $R_1$. We then sum the resulting EFIEs and MFIEs on each side of the interface by combining the columns and rows for the basis and testing functions on $S_{01}$.

### 6.2.9.2 Dielectric Cylinder: TM Polarization

For the dielectric cylinder, the series coefficients $A_n$ for TM polarization are [4]

$$A_n = -\frac{(k_1/\mu_1)J_n(k_0 a)J'_n(k_1 a) - (k_0/\mu_0)J'_n(k_0 a)J'_n(k_1 a)}{(k_1/\mu_1)H_n^{(2)}(k_0 a)J'_n(k_1 a) - (k_0/\mu_0)H_n'^{(2)}(k_0 a)J'_n(k_1 a)} \quad (6.136)$$

where the scattered field is given by (6.92).

*MOM Linear System*

For the TM case, the MOM linear system can be written as

$$\begin{bmatrix} Z^{zz,TM} & Z^{zt,TM} \\ Z^{tz,TM} & Z^{tt,TM} \end{bmatrix} \begin{bmatrix} I^J \\ I^M \end{bmatrix} = \begin{bmatrix} V^{z,TM} \\ V^{t,TM} \end{bmatrix} \quad (6.137)$$

where the blocks of the MOM system matrix are

$$Z^{zz,TM} = \mathbf{L}^{zz}\left(\mathbf{f}^z_{S_0}, \mathbf{f}^z_{S_0}, R_0\right) + \mathbf{L}^{zz}\left(\mathbf{f}^z_{S_1}, \mathbf{f}^z_{S_1}, R_1\right) \quad (6.138)$$

$$Z^{tz,TM} = \mathbf{K}^{zt}\left(\mathbf{f}^z_{S_0}, \mathbf{g}^t_{S_0}, R_0\right) + \mathbf{K}^{zt}\left(\mathbf{f}^z_{S_1}, \mathbf{g}^t_{S_1}, R_1\right) \quad (6.139)$$

$$Z^{zt,TM} = -\mathbf{K}^{tz}\left(\mathbf{g}^t_{S_0}, \mathbf{f}^z_{S_0}, R_0\right) - \mathbf{K}^{tz}\left(\mathbf{g}^t_{S_1}, \mathbf{f}^z_{S_1}, R_1\right) \quad (6.140)$$

$$Z^{tt,TM} = \frac{\epsilon_0}{\mu_0}\mathbf{L}^{tt}\left(\mathbf{g}^t_{S_0}, \mathbf{g}^t_{S_0}, R_0\right) + \frac{\epsilon_1}{\mu_1}\mathbf{L}^{tt}\left(\mathbf{g}^t_{S_1}, \mathbf{g}^t_{S_1}, R_1\right) \quad (6.141)$$

The notation $\mathbf{X}\left(\mathbf{a}_{S_l}, \mathbf{b}_{S_m}, R_l\right)$ indicates that X is evaluated in $R_l$ for the testing and basis functions **a** and **b** that reside on the boundaries $S_l$ and $S_m$, respectively. The right-hand side vectors are

$$V^{z,TM} = V^{E,TM}\left(\mathbf{f}^z_{S_0}, R_0\right) \quad (6.142)$$

$$V^{t,TM} = V^{H,TM}\left(\mathbf{g}^t_{S_0}, R_0\right) \quad (6.143)$$

and the solution vectors are

$$I^J = I^{J,z}_{S_{01}} \qquad I^M = I^{M,t}_{S_{01}} \quad (6.144)$$

*Results*

We will consider cylinders of radius $a = 5$, 10, and 20 meters, first with $\epsilon_1 = 2.56$ (lossless) and then with $\epsilon_1 = 2.56 - j.102$ (lossy). Setting $k_0 = 1$, we compute the bistatic RCS for $0 \leq \phi^s \leq 2\pi$ with $\phi^i = 0$. For the MOM we use 10 segments per free space wavelength on the boundary, with a minimum of 64 segments, and for the modal solution $N = 40$. The MOM results are compared to the modal solution in Figure 6.18. The agreement is excellent in all cases. We next compute the RCS in the backscattering and forward-scattering directions for $0 < k_0 a \leq 40$. The MOM uses the same parameters as before, and for the modal solution $N = 60$. For aesthetic purposes, the RCS has been normalized with respect to $\pi a$. The agreement is seen to be very good.

To determine the near fields in $R_0$ and $R_1$, we can use (3.120) and (3.122) and integrate the MOM currents on $S_0$ for points in $R_0$, and the currents on $S_1$ for points in $R_1$. In Figure 6.21a is shown the scattered near electric field in $R_0$ (in V/m) for a lossless dielectric cylinder with radius $1\lambda$ and $\epsilon_1 = 2.56$. In Figure 6.21a are shown the total fields, where in $R_0$ the field is the sum of the incident and scattered electric field, and in $R_1$ it is the transmitted field. We see that the total fields are continuous across the dielectric boundary.

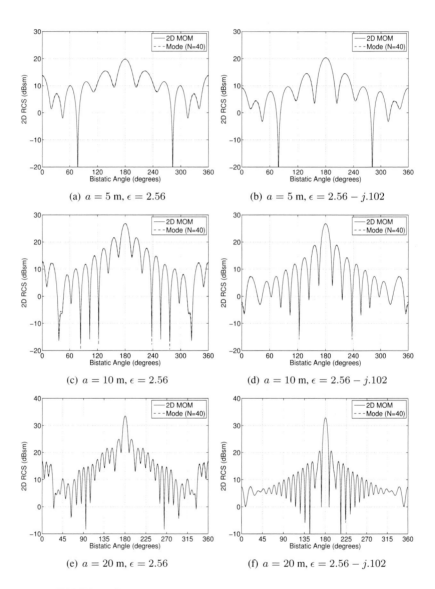

**FIGURE 6.18:** 2D Dielectric Cylinder: Bistatic RCS, TM Polarization

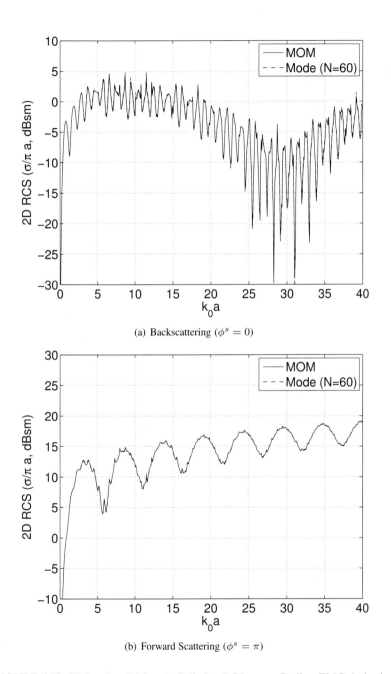

(a) Backscattering ($\phi^s = 0$)

(b) Forward Scattering ($\phi^s = \pi$)

**FIGURE 6.19:** 2D Lossless Dielectric Cylinder: RCS versus Radius, TM Polarization

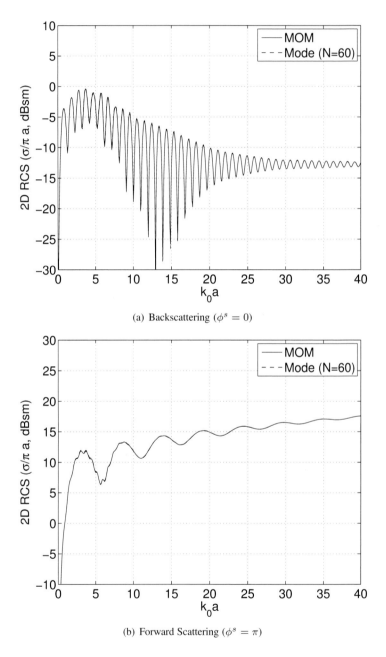

(a) Backscattering ($\phi^s = 0$)

(b) Forward Scattering ($\phi^s = \pi$)

**FIGURE 6.20:** 2D Lossy Dielectric Cylinder: RCS versus Radius, TM Polarization

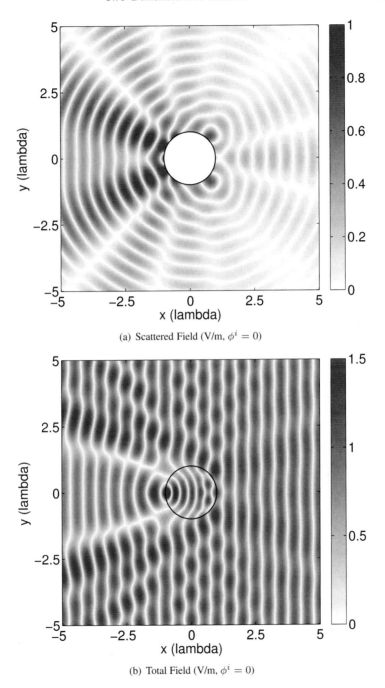

(a) Scattered Field (V/m, $\phi^i = 0$)

(b) Total Field (V/m, $\phi^i = 0$)

**FIGURE 6.21:** 2D Dielectric Cylinder: TM Near Fields

### 6.2.9.3  Dielectric Cylinder: TE Polarization

For the dielectric cylinder, the coefficients $B_n$ for TE polarization are [4]

$$B_n = -\frac{(k_1/\epsilon_1)J_n(k_0a)J_n'(k_1a) - (k_0/\epsilon_0)J_n'(k_0a)J_n(k_1a)}{(k_1/\epsilon_1)H_n^{(2)}(k_0a)J_n'(k_1a) - (k_0/\epsilon_0)H_n'^{(2)}(k_0a)J_n'(k_1a)} \qquad (6.145)$$

where the scattered field is given by (6.97).

*MOM Linear System*

For the TE case, the MOM linear system can be written as

$$\begin{bmatrix} Z^{tt,TE} & Z^{tz,TE} \\ Z^{zt,TE} & Z^{zz,TE} \end{bmatrix} \begin{bmatrix} I^J \\ I^M \end{bmatrix} = \begin{bmatrix} V^{t,TE} \\ V^{z,TE} \end{bmatrix} \qquad (6.146)$$

where the blocks of the system matrix are

$$Z^{tt,TE} = L^{tt}\left(\mathbf{f}_{S_0}^t, \mathbf{f}_{S_0}^t, R_0\right) + L^{tt}\left(\mathbf{f}_{S_1}^t, \mathbf{f}_{S_1}^t, R_1\right) \qquad (6.147)$$

$$Z^{tz,TE} = K^{tz}\left(\mathbf{f}_{S_0}^t, \mathbf{g}_{S_0}^z, R_0\right) + K^{tz}\left(\mathbf{f}_{S_1}^t, \mathbf{g}_{S_1}^z, R_1\right) \qquad (6.148)$$

$$Z^{zt,TE} = -K^{zt}\left(\mathbf{g}_{S_0}^z, \mathbf{f}_{S_0}^t, R_0\right) - K^{zt}\left(\mathbf{g}_{S_1}^z, \mathbf{f}_{S_1}^t, R_1\right) \qquad (6.149)$$

$$Z^{zz,TE} = \frac{\epsilon_0}{\mu_0}L^{zz}\left(\mathbf{g}_{S_0}^z, \mathbf{g}_{S_0}^z, R_0\right) + \frac{\epsilon_1}{\mu_1}L^{zz}\left(\mathbf{g}_{S_1}^z, \mathbf{g}_{S_1}^z, R_1\right) \qquad (6.150)$$

The right-hand side vectors are

$$V^{t,TE} = V^{E,TE}\left(\mathbf{f}_{S_0}^t, R_0\right) \qquad (6.151)$$

$$V^{z,TE} = V^{H,TE}\left(\mathbf{g}_{S_0}^z, R_0\right) \qquad (6.152)$$

and the solution vectors are

$$I^J = I_{S_{01}}^{J,t} \qquad I^M = I_{S_{01}}^{M,z} \qquad (6.153)$$

*Results*

We consider the same examples presented in Section 6.2.9.2, but for TE polarization. The discretization and other parameters remain identical. The results are shown in Figures 6.22–6.25. The comparisons between the MOM and modal solutions are again very good.

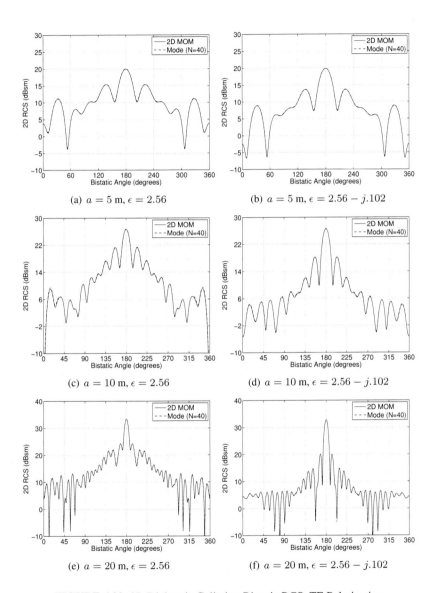

(a) $a = 5$ m, $\epsilon = 2.56$

(b) $a = 5$ m, $\epsilon = 2.56 - j.102$

(c) $a = 10$ m, $\epsilon = 2.56$

(d) $a = 10$ m, $\epsilon = 2.56 - j.102$

(e) $a = 20$ m, $\epsilon = 2.56$

(f) $a = 20$ m, $\epsilon = 2.56 - j.102$

**FIGURE 6.22:** 2D Dielectric Cylinder: Bistatic RCS, TE Polarization

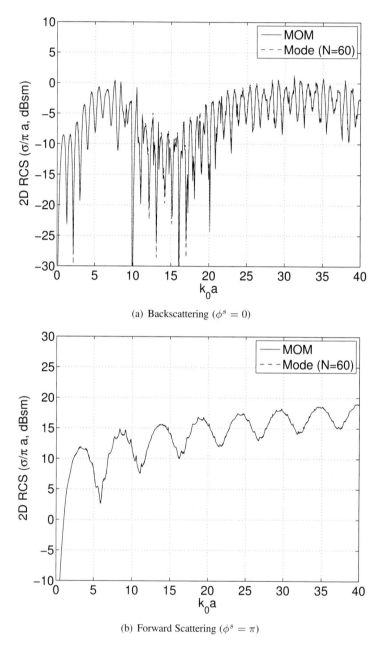

(a) Backscattering ($\phi^s = 0$)

(b) Forward Scattering ($\phi^s = \pi$)

**FIGURE 6.23:** 2D Lossless Dielectric Cylinder: RCS versus Radius, TE Polarization

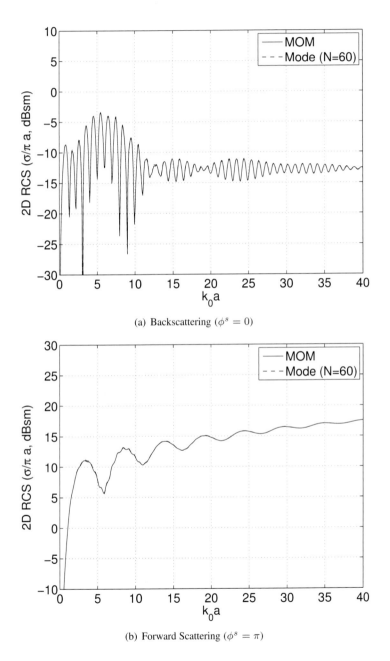

(a) Backscattering ($\phi^s = 0$)

(b) Forward Scattering ($\phi^s = \pi$)

**FIGURE 6.24:** 2D Lossy Dielectric Cylinder: RCS versus Radius, TE Polarization

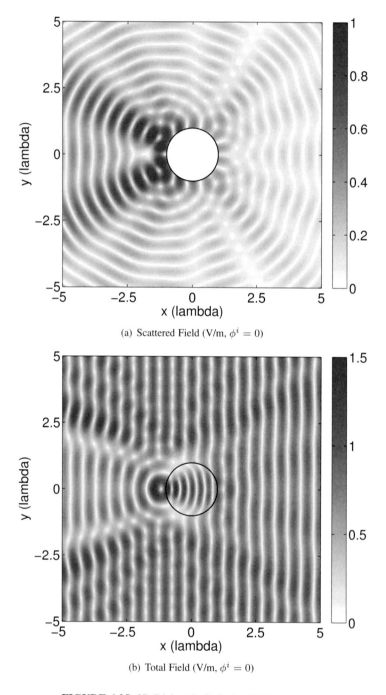

(a) Scattered Field (V/m, $\phi^i = 0$)

(b) Total Field (V/m, $\phi^i = 0$)

**FIGURE 6.25:** 2D Dielectric Cylinder: TE Near Fields

#### 6.2.9.4 Coated Cylinder

We next consider a conducting cylinder of radius $a_1$ with a dielectric coating of thickness $d$. This problem has three regions: the PEC region $R_2$, the dielectric region $R_1$, and the free space region $R_0$. There are two interfaces: the conducting interface $S_{12}$ with radius $a_1$, and the dielectric interface $S_{01}$ with radius $a_0 = a_1 + d$. In this case, we test the EFIE and MFIE on $S_{01}$, and the CFIE on $S_{12}$.

#### 6.2.9.5 Coated Cylinder: TM Polarization

For the coated cylinder, the coefficients $A_n$ for TM polarization are [4]

$$A_n = -\frac{J_n(k_0 a_0) - iZ_n J_n'(k_0 a_0)}{H_n^{(2)}(k_0 a_0) - iZ_n H_n'^{(2)}(k_0 a_0)} \tag{6.154}$$

where

$$iZ_n = \frac{k_0 \mu_1}{k_1 \mu_0}\left[\frac{J_n(k_1 a_0)H_n^{(2)}(k_1 a_1) - H_n^{(2)}(k_1 a_0)J_n(k_1 a_1)}{J_n'(k_1 a_0)H_n^{(2)}(k_1 a_1) - H_n'^{(2)}(k_1 a_0)J_n(k_1 a_1)}\right] \tag{6.155}$$

*MOM Linear System*

For the TM case, the MOM linear system is given by (6.137), where the submatrix blocks are

$$
\begin{aligned}
Z^{zz,TM} &= \begin{bmatrix} \mathrm{L}^{zz}\left(\mathbf{f}_{S_{01}}^z, \mathbf{f}_{S_{01}}^z, R_0\right) + \mathrm{L}^{zz}\left(\mathbf{f}_{S_{10}}^z, \mathbf{f}_{S_{10}}^z, R_1\right) & \mathrm{L}^{zz}\left(\mathbf{f}_{S_{10}}^z, \mathbf{f}_{S_{12}}^z, R_1\right) \\ \alpha \mathrm{L}^{zz}\left(\mathbf{f}_{S_{12}}^z, \mathbf{f}_{S_{10}}^z, R_1\right) & \alpha \mathrm{L}^{zz}\left(\mathbf{f}_{S_{12}}^z, \mathbf{f}_{S_{12}}^z, R_1\right) \end{bmatrix} \\
&+ \begin{bmatrix} 0 & 0 \\ (1-\alpha)\eta_1 \mathrm{n}\mathrm{K}^{zz}\left(\mathbf{f}_{S_{12}}^z, \mathbf{f}_{S_{10}}^z, R_1\right) & (1-\alpha)\eta_1 \mathrm{n}\mathrm{K}^{zz}\left(\mathbf{f}_{S_{12}}^z, \mathbf{f}_{S_{12}}^z, R_1\right) \end{bmatrix}
\end{aligned}
\tag{6.156}
$$

$$
\begin{aligned}
Z^{zt,TM} &= \begin{bmatrix} \mathrm{K}^{zt}\left(\mathbf{f}_{S_{01}}^z, \mathbf{g}_{S_{01}}^t, R_0\right) + \mathrm{K}^{zt}\left(\mathbf{f}_{S_{10}}^z, \mathbf{g}_{S_{10}}^t, R_1\right) \\ \alpha \mathrm{K}^{zt}\left(\mathbf{f}_{S_{12}}^z, \mathbf{g}_{S_{10}}^t, R_1\right) \end{bmatrix} \\
&+ \begin{bmatrix} 0 \\ (1-\alpha)\frac{1}{\eta_1}\mathrm{n}\mathrm{L}^{zt}\left(\mathbf{f}_{S_{12}}^z, \mathbf{g}_{S_{10}}^t, R_1\right) \end{bmatrix}
\end{aligned}
\tag{6.157}
$$

$$Z^{tz,TM} = \begin{bmatrix} -\mathrm{K}^{tz}\left(\mathbf{g}_{S_{01}}^t, \mathbf{f}_{S_{01}}^z, R_0\right) - \mathrm{K}^{tz}\left(\mathbf{g}_{S_{10}}^t, \mathbf{f}_{S_{10}}^z, R_1\right) \mid -\mathrm{K}^{tz}\left(\mathbf{g}_{S_{10}}^t, \mathbf{f}_{S_{12}}^z, R_1\right) \end{bmatrix} \tag{6.158}$$

$$Z^{tt,TM} = \begin{bmatrix} \frac{\epsilon_0}{\mu_0}\mathrm{L}^{tt}\left(\mathbf{g}_{S_{01}}^t, \mathbf{g}_{S_{01}}^t, R_0\right) + \frac{\epsilon_1}{\mu_1}\mathrm{L}^{tt}\left(\mathbf{g}_{S_{10}}^t, \mathbf{g}_{S_{10}}^t, R_1\right) \end{bmatrix} \tag{6.159}$$

The right-hand side vectors are

$$V^{z,TM} = \begin{bmatrix} V^{E,TM}(\mathbf{f}^z_{S_{01}}, R_0) \\ 0 \end{bmatrix} \tag{6.160}$$

and

$$V^{t,TM} = \begin{bmatrix} V^{H,TM}(\mathbf{g}^t_{S_{01}}, R_0) \\ 0 \end{bmatrix} \tag{6.161}$$

and the solution vectors are

$$I^J = \begin{bmatrix} I^{J,z}_{S_{01}} \\ I^{J,z}_{S_{12}} \end{bmatrix} \qquad I^{M,t} = \begin{bmatrix} I^{M,t}_{S_{01}} \\ 0 \end{bmatrix} \tag{6.162}$$

## Results

We will analyze a coated cylinder where $a_0 = 1.1a_1$. Setting $k_0 = 1$, we compute the RCS in the backscattering and forward-scattering directions for $0 < k_0 a \leq 40$. For the MOM solution, both interfaces are discretized using 10 points per $\lambda_0$ with a minimum of 64 segments on each, and for the modal solution, $N = 60$. The MOM results are compared to the modal solution for a lossless ($\epsilon_1 = 2.56$) and lossy ($\epsilon_1 = 2.56 - j.102$) coating in Figures 6.26 and 6.27, respectively. The RCS has again been normalized with respect to $\pi a$. The agreement is seen to be fairly good, except that in the backscattering direction, the MOM results are about $\frac{1}{2}$ dB below the modal results. The reason for this difference is discussed further in Section 6.2.9.7.

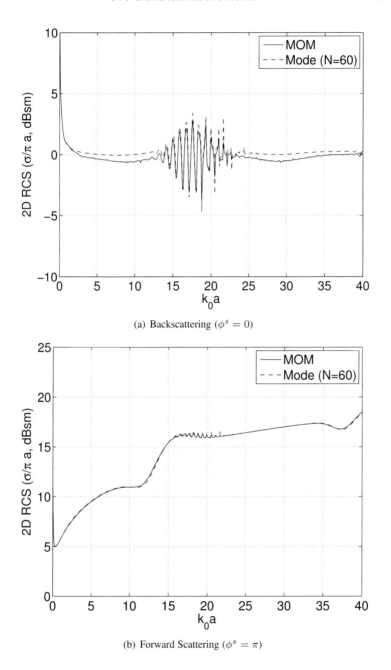

(a) Backscattering ($\phi^s = 0$)

(b) Forward Scattering ($\phi^s = \pi$)

**FIGURE 6.26:** 2D PEC Cylinder with Lossless Coating: RCS versus Radius, TM Polarization

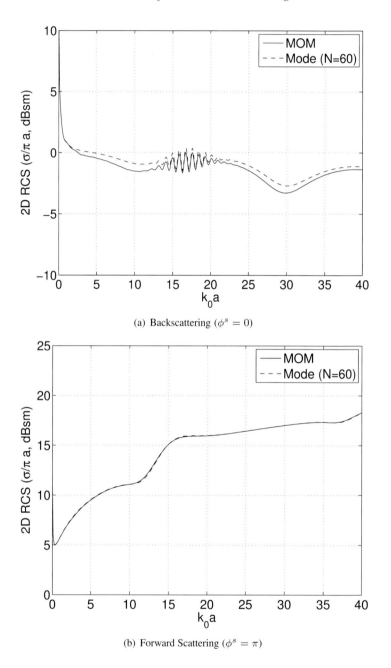

(a) Backscattering ($\phi^s = 0$)

(b) Forward Scattering ($\phi^s = \pi$)

**FIGURE 6.27:** 2D PEC Cylinder with Lossy Coating: RCS versus Radius, TM Polarization

### 6.2.9.6 Coated Cylinder: TE Polarization

For the coated cylinder, the coefficients $B_n$ for TE polarization are [4]

$$B_n = -\frac{J_n(k_0 a_0) - iY_n J_n'(k_0 a_0)}{H_n^{(2)}(k_0 a_0) - iY_n H_n'^{(2)}(k_0 a_0)} \tag{6.163}$$

where

$$iY_n = \frac{k_0 \epsilon_1}{k_1 \epsilon_0}\left[\frac{J_n(k_1 a_0)H_n'^{(2)}(k_1 a_1) - H_n^{(2)}(k_1 a_0)J_n'(k_1 a_1)}{J_n'(k_1 a_0)H_n'^{(2)}(k_1 a_1) - H_n'^{(2)}(k_1 a_0)J_n'(k_1 a_1)}\right] \tag{6.164}$$

*MOM Linear System*

For the TE case, the MOM linear system is given by (6.146), where the sub-matrix blocks are

$$Z^{tt,TE} = \begin{bmatrix} L^{tt}\left(\mathbf{f}_{S_{01}}^t, \mathbf{f}_{S_{01}}^t, R_0\right) + L^{tt}\left(\mathbf{f}_{S_{10}}^t, \mathbf{f}_{S_{10}}^t, R_1\right) & L^{tt}\left(\mathbf{f}_{S_{10}}^t, \mathbf{f}_{S_{12}}^t, R_1\right) \\ \alpha L^{tt}\left(\mathbf{f}_{S_{12}}^t, \mathbf{f}_{S_{10}}^t, R_1\right) & \alpha L^{tt}\left(\mathbf{f}_{S_{12}}^t, \mathbf{f}_{S_{12}}^t, R_1\right) \end{bmatrix}$$
$$+ \begin{bmatrix} 0 & 0 \\ (1-\alpha)\eta_1 \mathrm{n}K^{tt}\left(\mathbf{f}_{S_{12}}^t, \mathbf{f}_{S_{10}}^t, R_1\right) & (1-\alpha)\eta_1 \mathrm{n}K^{tt}\left(\mathbf{f}_{S_{12}}^t, \mathbf{f}_{S_{12}}^t, R_1\right) \end{bmatrix} \tag{6.165}$$

$$Z^{tz,TE} = \begin{bmatrix} K^{tz}\left(\mathbf{f}_{S_{01}}^t, \mathbf{g}_{S_{01}}^z, R_0\right) + K^{tz}\left(\mathbf{f}_{S_{10}}^t, \mathbf{g}_{S_{10}}^z, R_1\right) \\ \alpha K^{tz}\left(\mathbf{f}_{S_{12}}^t, \mathbf{g}_{S_{10}}^z, R_1\right) \end{bmatrix}$$
$$+ \begin{bmatrix} 0 \\ (1-\alpha)\frac{1}{\eta_1}\mathrm{n}L^{tz}\left(\mathbf{f}_{S_{12}}^t, \mathbf{g}_{S_{10}}^z, R_1\right) \end{bmatrix} \tag{6.166}$$

$$Z^{zt,TE} = \begin{bmatrix} -K^{zt}\left(\mathbf{g}_{S_{01}}^z, \mathbf{f}_{S_{01}}^t, R_0\right) - K^{zt}\left(\mathbf{g}_{S_{10}}^z, \mathbf{f}_{S_{10}}^t, R_1\right) & | & -K^{zt}\left(\mathbf{g}_{S_{10}}^z, \mathbf{f}_{S_{12}}^t, R_1\right) \end{bmatrix} \tag{6.167}$$

$$Z^{zz,TE} = \begin{bmatrix} \frac{\epsilon_0}{\mu_0}L^{zz}\left(\mathbf{g}_{S_{01}}^z, \mathbf{g}_{S_{01}}^z, R_0\right) + \frac{\epsilon_1}{\mu_1}L^{zz}\left(\mathbf{g}_{S_{10}}^z, \mathbf{g}_{S_{10}}^z, R_1\right) \end{bmatrix} \tag{6.168}$$

The right-hand side vectors are

$$V^{t,TE} = \begin{bmatrix} V^{E,TE}(\mathbf{f}_{S_{01}}^t, R_0) \\ 0 \end{bmatrix} \tag{6.169}$$

and

$$V^{z,TE} = \begin{bmatrix} V^{H,TE}(\mathbf{g}_{S_{01}}^z, R_0) \\ 0 \end{bmatrix} \tag{6.170}$$

and the solution vectors are

$$
I^J = \begin{bmatrix} I_{S_{01}}^{J,t} \\ I_{S_{12}}^{J,t} \end{bmatrix} \qquad I^{M,t} = \begin{bmatrix} I_{S_{01}}^{M,z} \\ 0 \end{bmatrix} \tag{6.171}
$$

*Results*

We consider again the coated cylinder of Section 6.2.9.5. The MOM results are compared to the modal solution for the lossless ($\epsilon_1 = 2.56$) and lossy ($\epsilon_1 = 2.56 - j.102$) coatings in Figures 6.28 and 6.29, respectively. The amplitude of the MOM results in the backscattering direction is again seen to be about $\frac{1}{2}$ dB below the modal solution.

### 6.2.9.7    Effect of Number of Segments per Wavelength on Accuracy

The MOM results in Sections 6.2.9.5 and 6.2.9.6 used 10 segments per $\lambda_0$ on both interfaces. While this number is sufficient on $S_{10}$, it is not on $S_{12}$, since $\lambda_1 < \lambda_0$. For $\epsilon_1 = 2.56$, $\lambda_1 = \lambda_0/\sqrt{\epsilon_1} = 0.625\lambda_0$. This suggests that $S_{12}$ should have almost twice the number of segments as does $S_{10}$. To demonstrate, consider a coated cylinder with $\epsilon_1 = 2.56$, $k_0 a_1 = 20$ and $k_0 a_0 = 22$. The bistatic RCS obtained using 10 and 20 segments per $\lambda_0$ on $S_{12}$ is shown in Figures 6.30a and 6.30b, respectively, for TE polarization. We see that the difference is eliminated by increasing to 20 segments per $\lambda_0$.

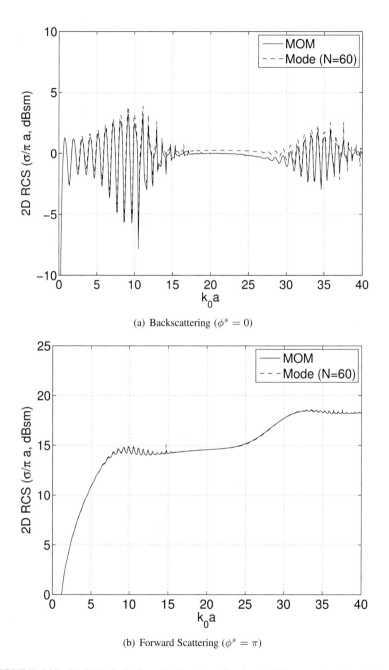

(a) Backscattering ($\phi^s = 0$)

(b) Forward Scattering ($\phi^s = \pi$)

**FIGURE 6.28:** 2D PEC Cylinder with Lossless Coating: RCS versus Radius, TE Polarization

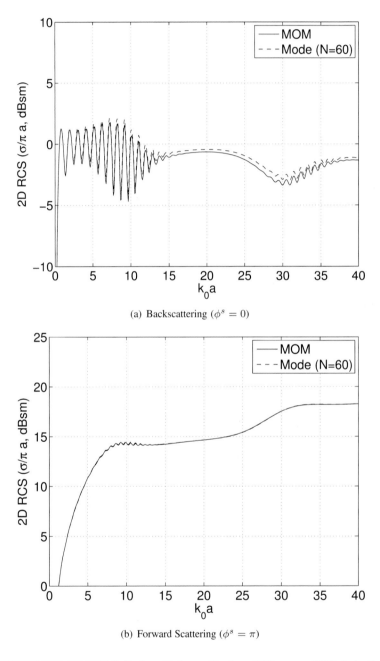

(a) Backscattering ($\phi^s = 0$)

(b) Forward Scattering ($\phi^s = \pi$)

**FIGURE 6.29:** 2D PEC Cylinder with Lossy Coating: RCS versus, TE Polarization

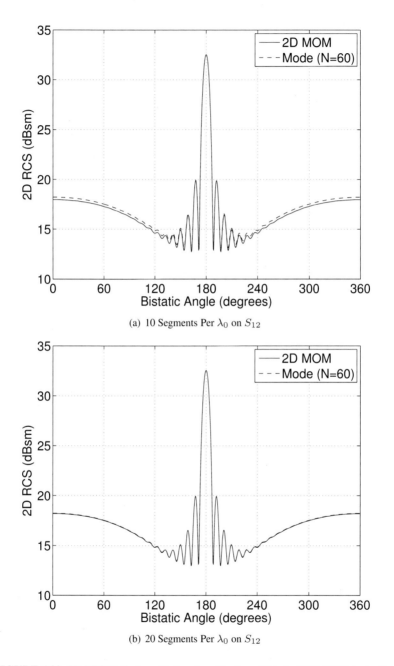

(a) 10 Segments Per $\lambda_0$ on $S_{12}$

(b) 20 Segments Per $\lambda_0$ on $S_{12}$

**FIGURE 6.30:** 2D PEC Cylinder with Lossless Coating: Bistatic RCS, TE Polarization

# References

[1] D. Wilton, S. Rao, A. Glisson, D. Schaubert, M. Al-Bundak, and C. M. Butler, "Potential integrals for uniform and linear source distributions on polygonal and polyhedral domains," *IEEE Trans. Antennas Propagat.*, vol. 32, pp. 276–281, March 1984.

[2] M. Abramowitz and I. Stegun, *Handbook of Mathematical Functions*. National Bureau of Standards, 1966.

[3] H. B. Dwight, *Tables of Integrals and Other Mathematical Data*. The Macmillan Company, 1961.

[4] G. T. Ruck, ed., *Radar Cross Section Handbook*. Plenum Press, 1970.

[5] L. N. Medgyesi-Mitschang and J. M. Putnam, "Electromagnetic scattering from axially inhomogeneous bodies of revolution," *IEEE Trans. Antennas Propagat.*, vol. 32, pp. 796–806, March 1984.

# Chapter 7

## Bodies of Revolution

In this chapter we will use the Method of Moments to solve surface integral equation problems involving bodies of revolution (BORs), which comprise rotationally symmetric objects such as spheres, ellipsoids, and finite cylinders. This is of great practical interest, as such objects are commonly used to benchmark new computational electromagnetics techniques and codes, and to calibrate instruments that measure RCS in a static range. Furthermore, many realistic objects such as the conic reentry vehicle (RV) can be modeled as a body of revolution.

In Chapter 6, we solved the EFIE, MFIE, and nMFIE by formulating the common functional expressions (L, nL, K, and nK), and then assembling the MOM system matrix in a systematic way using those functions. In this chapter, we will use the same methodology to treat BORs of general shape and configuration. We will first discuss modeling of the surface and the surface current expansion, and then derive the functional expressions required in each integral equation. We will also consider the issues with treating the intersection (junction) between three or more dielectric/conducting regions.

## 7.1 BOR Surface Description

Consider a body of revolution such as the cone-sphere in Figure 7.1a. Such an object can be completely described by a bounding curve and an axis of symmetry, where a rotation of the bounding curve by $2\pi$ about the symmetry axis sweeps through all points comprising the surface. Therefore, let us define a BOR in the cylindrical coordinate system as shown in Figure 7.1b. In this system, the axis of symmetry is the $\hat{\mathbf{z}}$ axis, and any point $\mathbf{r}$ on the surface can be identified by its cylindrical components $(\rho, \phi, z)$, or by its corresponding Cartesian components

$$\mathbf{r} = \rho \cos\phi \hat{\mathbf{x}} + \rho \sin\phi \hat{\mathbf{y}} + z\hat{\mathbf{z}} \tag{7.1}$$

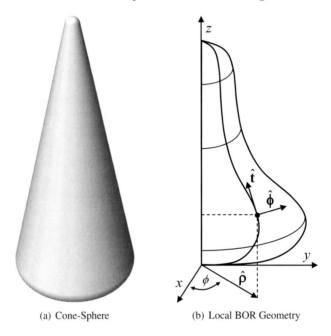

(a) Cone-Sphere            (b) Local BOR Geometry

**FIGURE 7.1:** Body of Revolution (BOR)

The distance $r$ between any two points on the surface can be written as

$$r = |\mathbf{r} - \mathbf{r}'| = \sqrt{(\rho - \rho')^2 + (z - z')^2 + 2\rho\rho'[1 - \cos(\phi - \phi')]} \qquad (7.2)$$

We next define the longitudinal vector $\hat{\mathbf{t}}(\mathbf{r})$, which along with the cylindrical vector $\hat{\phi}(\mathbf{r})$ is everywhere tangent to the surface. Given the surface normal $\hat{\mathbf{n}}(\mathbf{r})$, their relationship is defined as

$$\hat{\mathbf{n}}(\mathbf{r}) = \hat{\phi}(\mathbf{r}) \times \hat{\mathbf{t}}(\mathbf{r}) \qquad (7.3)$$

As $\hat{\mathbf{t}}(\mathbf{r})$ and $\hat{\phi}(\mathbf{r})$ are everywhere normal to each other, they comprise an orthogonal basis set that we will use to expand the surface currents.

## 7.2  Expansion of Surface Currents

Due to the rotational symmetry, we will represent the surface currents $\mathbf{J}$ and $\mathbf{M}$ using $\hat{\mathbf{t}}$- and $\hat{\phi}$-oriented basis functions that are local in the longitudinal dimension and a Fourier series in the azimuthal dimension [1]. For a given region $R_l$ with bounding surface $S_l$, the electric current can be expanded using $N$ basis functions of each type ($2N$ total basis functions), yielding [2]

$$\mathbf{J}_l(\mathbf{r}) = \sum_{\alpha=-\infty}^{\infty} \sum_{n=1}^{N} \left[ I_{\alpha n}^{J,t} \mathbf{f}_{\alpha n}^{t}(\mathbf{r}) - I_{\alpha n}^{J,\phi} \mathbf{f}_{\alpha n}^{\phi}(\mathbf{r}) \right] \tag{7.4}$$

where $a_{\alpha n}^{t}$ and $a_{\alpha n}^{\phi}$ are the longitudinal and azimuthal expansion coefficients of basis function $n$ for Fourier mode $\alpha$. The functions $\mathbf{f}_{\alpha n}^{t,\phi}(\mathbf{r})$ are

$$\mathbf{f}_{\alpha n}^{t}(\mathbf{r}) = f_n(t)\, e^{j\alpha\phi}\, \hat{\mathbf{t}}(\mathbf{r}) \tag{7.5}$$

$$\mathbf{f}_{\alpha n}^{\phi}(\mathbf{r}) = f_n(t)\, e^{j\alpha\phi}\, \hat{\boldsymbol{\phi}}(\mathbf{r}) \tag{7.6}$$

where for the one-dimensional function $f_n(t)$ we will use triangle functions (Section 2.3.2). For convenience, we will treat the integrations in the $t$ dimension in a generic sense in the derivations that follow, and present a more detailed discussion in Section 7.7. Similarly, the magnetic current in $R_l$ is

$$\mathbf{M}_l(\mathbf{r}) = \sum_{\alpha=-\infty}^{\infty} \sum_{n=1}^{N} \left[ I_{\alpha n}^{M,t} \mathbf{g}_{\alpha n}^{t}(\mathbf{r}) - I_{\alpha n}^{M,\phi} \mathbf{g}_{\alpha n}^{\phi}(\mathbf{r}) \right] \tag{7.7}$$

where the functions $\mathbf{g}_{\alpha n}^{t,\phi}(\mathbf{r})$ are

$$\mathbf{g}_{\alpha n}^{t}(\mathbf{r}) = \eta_0\, g_n(t)\, e^{j\alpha\phi}\, \hat{\mathbf{t}}(\mathbf{r}) \tag{7.8}$$

$$\mathbf{g}_{\alpha n}^{\phi}(\mathbf{r}) = \eta_0\, g_n(t)\, e^{j\alpha\phi}\, \hat{\boldsymbol{\phi}}(\mathbf{r}) \tag{7.9}$$

Because it will simplify several expressions later, we define $f_n(t)$ and $g_n(t)$ such that

$$f_n(t) = g_n(t) = \frac{T_n(t)}{\rho(t)} \tag{7.10}$$

where $T_n(t)$ is the triangle function. The minus sign on the second term in (7.4) and (7.7) was chosen as it will yield symmetries in several terms in the sections that follow.

---

## 7.3 EFIE

The EFIE (3.178) contains $\mathcal{L}$ and $\mathcal{K}$ operators, which were discretized and tested via the Moment Method in (3.185) and (3.186). To implement these on a BOR, we require expressions for the testing functions. Following the approach in [1], we will use the complex conjugate of the basis functions as the electric testing functions. Forming the conjugate of the $\hat{\mathbf{t}}$ and $\hat{\boldsymbol{\phi}}$ components from (7.5) and (7.6) yields

$$\mathbf{f}^t_{\beta m}(\mathbf{r}) = f_n(t)\, e^{-j\beta\phi}\, \hat{\mathbf{t}}(\mathbf{r}) \tag{7.11}$$

$$\mathbf{f}^\phi_{\beta m}(\mathbf{r}) = f_n(t)\, e^{-j\beta\phi}\, \hat{\boldsymbol{\phi}}(\mathbf{r}) \tag{7.12}$$

We will also use (7.11) and (7.12) as the testing functions for the nMFIE (Section 7.6).

### 7.3.1  $\mathcal{L}$ Operator

Substitution of the basis (7.5, 7.6) and testing (7.11, 7.12) functions into (3.185) yields a matrix of the form

$$Z^{EJ(l)} = \begin{bmatrix} L(\mathbf{f}^t_\beta, \mathbf{f}^t_\alpha) & L(\mathbf{f}^t_\beta, \mathbf{f}^\phi_\alpha) \\ L(\mathbf{f}^\phi_\beta, \mathbf{f}^t_\alpha) & L(\mathbf{f}^\phi_\beta, \mathbf{f}^\phi_\alpha) \end{bmatrix} = \begin{bmatrix} L^{tt} & L^{t\phi} \\ L^{\phi t} & L^{\phi\phi} \end{bmatrix} \tag{7.13}$$

where in Region $R_l$, the elements of each sub-matrix block are given by

$$\begin{aligned}
L^{**}_{mn} = {}&j\omega\mu_l \int_{\mathbf{f}^{t,\phi}_{\beta m}} \int_{\mathbf{f}^{t,\phi}_{\alpha n}} \mathbf{f}^{t,\phi}_{\beta m}(\mathbf{r}) \cdot \pm\, \mathbf{f}^{t,\phi}_{\alpha n}(\mathbf{r}') \frac{e^{-jk_l r}}{4\pi r}\, d\mathbf{r}'\, d\mathbf{r} \\
&- \frac{j}{\omega\epsilon_l} \int_{\mathbf{f}^{t,\phi}_{\beta m}} \int_{\mathbf{f}^{t,\phi}_{\alpha n}} \Big[\nabla \cdot \mathbf{f}^{t,\phi}_{\beta m}(\mathbf{r})\Big]\Big[\nabla' \cdot \pm\, \mathbf{f}^{t,\phi}_{\alpha n}(\mathbf{r}')\Big] \frac{e^{-jk_l r}}{4\pi r}\, d\mathbf{r}'\, d\mathbf{r}
\end{aligned} \tag{7.14}$$

and we have used (3.204) to re-distribute the differential operators. The $\hat{\mathbf{t}}$- and $\hat{\boldsymbol{\phi}}$-oriented basis functions (7.14) have positive and negative signs, respectively, following the convention in (7.4).

#### 7.3.1.1  L Matrix Elements

To compute the sub-matrix elements in (7.14) we must first determine the divergences of the basis and testing functions. These are

$$\nabla' \cdot \mathbf{f}^t_{\alpha n}(\mathbf{r}') = \frac{1}{\rho(t')}\frac{\partial}{\partial t'}\Big[\rho(t')f_n(t')\Big]e^{j\alpha\phi'} = \frac{1}{\rho(t')}T'_n(t')e^{j\alpha\phi'} \tag{7.15}$$

$$\nabla \cdot \mathbf{f}^t_{\beta m}(\mathbf{r}) = \frac{1}{\rho(t)}\frac{\partial}{\partial t}\Big[\rho(t)f_m(t)\Big]e^{-j\beta\phi} = \frac{1}{\rho(t)}T'_m(t)e^{-j\beta\phi} \tag{7.16}$$

$$\nabla' \cdot \mathbf{f}^\phi_{\alpha n}(\mathbf{r}') = \frac{f_n(t')}{\rho(t')}\frac{\partial}{\partial\phi'}e^{j\alpha\phi'} = \frac{j\alpha}{\rho(t')^2}T_n(t')e^{j\alpha\phi'} \tag{7.17}$$

$$\nabla \cdot \mathbf{f}^\phi_{\beta m}(\mathbf{r}) = \frac{f_m(t)}{\rho(t)}\frac{\partial}{\partial\phi}e^{-j\beta\phi} = -\frac{j\beta}{\rho(t)^2}T_m(t)e^{-j\beta\phi} \tag{7.18}$$

Using (7.15–7.18) and noting that the differential surface element in cylindrical coordinates is $ds = \rho d\phi dt$, the sub-matrix elements can be written as

$$\mathrm{L}_{mn}^{tt} = \int_{T_m} \int_{T_n} \int_0^{2\pi} \int_0^{2\pi} \left[ j\omega\mu_l T_m(t) T_n(t') \, \hat{\mathbf{t}}(\mathbf{r}) \cdot \hat{\mathbf{t}}(\mathbf{r}') \right.$$
$$\left. - \frac{j}{\omega\epsilon_l} T'_m(t) T'_n(t') \right] e^{j(\alpha\phi' - \beta\phi)} \frac{e^{-jk_l r}}{4\pi r} \, d\phi' \, d\phi \, dt' \, dt \qquad (7.19)$$

$$\mathrm{L}_{mn}^{t\phi} = - \int_{T_m} \int_{T_n} \int_0^{2\pi} \int_0^{2\pi} \left[ j\omega\mu_l T_m(t) T_n(t') \, \hat{\mathbf{t}}(\mathbf{r}) \cdot \hat{\boldsymbol{\phi}}(\mathbf{r}') \right.$$
$$\left. + \frac{\alpha}{\omega\epsilon_l \rho(t')} T'_m(t) T_n(t') \right] e^{j(\alpha\phi' - \beta\phi)} \frac{e^{-jk_l r}}{4\pi r} \, d\phi' \, d\phi \, dt' \, dt \qquad (7.20)$$

$$\mathrm{L}_{mn}^{\phi t} = \int_{T_m} \int_{T_n} \int_0^{2\pi} \int_0^{2\pi} \left[ j\omega\mu_l T_m(t) T_n(t') \, \hat{\boldsymbol{\phi}}(\mathbf{r}) \cdot \hat{\mathbf{t}}(\mathbf{r}') \right.$$
$$\left. - \frac{\beta}{\omega\epsilon_l \rho(t)} T_m(t) T'_n(t') \right] e^{j(\alpha\phi' - \beta\phi)} \frac{e^{-jk_l r}}{4\pi r} \, d\phi' \, d\phi \, dt' \, dt \qquad (7.21)$$

$$\mathrm{L}_{mn}^{\phi\phi} = - \int_{T_m} \int_{T_n} \int_0^{2\pi} \int_0^{2\pi} \left[ j\omega\mu_l T_m(t) T_n(t') \, \hat{\boldsymbol{\phi}}(\mathbf{r}) \cdot \hat{\boldsymbol{\phi}}(\mathbf{r}') \right.$$
$$\left. - \frac{j\alpha\beta}{\omega\epsilon_l \rho(t)\rho(t')} T_m(t) T_n(t') \right] e^{j(\alpha\phi' - \beta\phi)} \frac{e^{-jk_l r}}{4\pi r} \, d\phi' \, d\phi \, dt' \, dt \quad (7.22)$$

We note that because integrals are periodic functions of $\phi'$ with period $2\pi$, $\phi'$ can be replaced by $\phi' + \phi$ without altering their values [1]. Making this substitution allows us to write

$$\int_0^{2\pi} e^{j(\alpha\phi' - \beta\phi)} \, d\phi = e^{j\alpha\phi'} \int_0^{2\pi} e^{j(\alpha - \beta)\phi} \, d\phi \qquad (7.23)$$

and noting that

$$\int_0^{2\pi} e^{j(\alpha - \beta)\phi} \, d\phi = \begin{cases} 0 & \alpha \neq \beta \\ 2\pi & \alpha = \beta \end{cases} \qquad (7.24)$$

only those integrals where $\alpha = \beta$ will remain. Additionally, we observe that for a BOR,

$$\frac{e^{jkr(\rho, -\phi, z)}}{r(\rho, -\phi, z)} = \frac{e^{jkr(\rho, \phi, z)}}{r(\rho, \phi, z)} \qquad 0 \leq \phi \leq \pi \qquad (7.25)$$

Let us now define $\hat{\mathbf{t}}(\mathbf{r})$ such that

$$\hat{\mathbf{t}}(\mathbf{r}) = \sin\gamma \hat{\boldsymbol{\rho}} + \cos\gamma \hat{\mathbf{z}} \qquad (7.26)$$

which allows us to write the following dot products

$$\hat{\mathbf{t}}(\mathbf{r}) \cdot \hat{\mathbf{t}}(\mathbf{r}') = \sin\gamma\sin\gamma'\cos(\phi' - \phi) + \cos\gamma\cos\gamma' \tag{7.27}$$

$$\hat{\mathbf{t}}(\mathbf{r}) \cdot \hat{\phi}(\mathbf{r}') = -\sin\gamma\sin(\phi' - \phi) \tag{7.28}$$

$$\hat{\phi}(\mathbf{r}) \cdot \hat{\mathbf{t}}(\mathbf{r}') = \sin\gamma'\sin(\phi' - \phi) \tag{7.29}$$

$$\hat{\phi}(\mathbf{r}) \cdot \hat{\phi}(\mathbf{r}') = \cos(\phi' - \phi) \tag{7.30}$$

Taking into account (7.27–7.30), and using the orthogonality relationships that exist between the trigonometric terms in the various integrands, the sub-matrix elements become

$$L_{mn}^{tt} = \int_{T_m}\int_{T_n}\left[j\omega\mu_l T_m(t)T_n(t')\left(\sin\gamma\sin\gamma'G_2 + \cos\gamma\cos\gamma'G_1\right)\right.$$
$$\left. -\frac{j}{\omega\epsilon_l}T_m'(t)T_n'(t')G_1\right]dt'\,dt \tag{7.31}$$

$$L_{mn}^{t\phi} = -\int_{T_m}\int_{T_n}\left[\omega\mu_l T_m(t)T_n(t')\sin\gamma G_3 + \frac{\alpha}{\omega\epsilon_l\rho(t')}T_m'(t)T_n(t')G_1\right]dt'\,dt \tag{7.32}$$

$$L_{mn}^{\phi t} = -\int_{T_m}\int_{T_n}\left[\omega\mu_l T_m(t)T_n(t')\sin\gamma'G_3 + \frac{\alpha}{\omega\epsilon_l\rho(t)}T_m(t)T_n'(t')G_1\right]dt'\,dt \tag{7.33}$$

$$L_{mn}^{\phi\phi} = -\int_{T_m}\int_{T_n}\left[j\omega\mu_l T_m(t)T_n(t')G_2 - \frac{j\alpha^2}{\omega\epsilon_l\rho(t)\rho(t')}T_m(t)T_n(t')G_1\right]dt'\,dt \tag{7.34}$$

where the integrals $G_1$, $G_2$ and $G_3$ are

$$G_1 = \int_0^\pi \cos(\alpha\phi')\frac{e^{-jk_l r}}{r}d\phi' \tag{7.35}$$

$$G_2 = \int_0^\pi \cos(\phi')\cos(\alpha\phi')\frac{e^{-jk_l r}}{r}d\phi' \tag{7.36}$$

$$G_3 = \int_0^\pi \sin(\phi')\sin(\alpha\phi')\frac{e^{-jk_l r}}{r}d\phi' \tag{7.37}$$

These integrals are evaluated by a one-dimensional Gaussian quadrature. When $\mathbf{r} \neq \mathbf{r}'$, $r$ can be obtained via

$$r = |\mathbf{r} - \mathbf{r}'| = \sqrt{(\rho - \rho')^2 + (z - z')^2 + 2\rho\rho'[1 - \cos(\phi')]} \tag{7.38}$$

When $\mathbf{r} = \mathbf{r}'$, a modification of (7.38) for the above is needed, which is discussed in Section 7.7.

Inspecting the $L$ sub-matrix elements, we find the following relationship between positive and negative modes

$$\begin{bmatrix} \mathbf{L}^{tt} & \mathbf{L}^{t\phi} \\ \mathbf{L}^{\phi t} & \mathbf{L}^{\phi\phi} \end{bmatrix}_{-\alpha} = \begin{bmatrix} \mathbf{L}^{tt} & -\mathbf{L}^{t\phi} \\ -\mathbf{L}^{\phi t} & \mathbf{L}^{\phi\phi} \end{bmatrix}_{\alpha} \tag{7.39}$$

These properties survive inversion.

## 7.3.2 $\mathcal{K}$ Operator

Substitution of the magnetic basis and electric testing functions into (3.186) yields a matrix of the form

$$Z^{EM(l)} = \begin{bmatrix} \mathbf{K}(\mathbf{f}^t_\beta, \mathbf{g}^t_\alpha) & \mathbf{K}(\mathbf{f}^t_\beta, \mathbf{g}^\phi_\alpha) \\ \mathbf{K}(\mathbf{f}^\phi_\beta, \mathbf{g}^t_\alpha) & \mathbf{K}(\mathbf{f}^\phi_\beta, \mathbf{g}^\phi_\alpha) \end{bmatrix} = \begin{bmatrix} \mathbf{K}^{tt} & \mathbf{K}^{t\phi} \\ \mathbf{K}^{\phi t} & \mathbf{K}^{\phi\phi} \end{bmatrix} \tag{7.40}$$

where in Region $R_l$, the elements of each sub-matrix block are given by

$$\begin{aligned} \mathbf{K}^{**}_{mn} &= \frac{1}{2} \int_{\mathbf{f}^{t,\phi}_{\beta m}, \mathbf{g}^{t,\phi}_{\alpha m}} \mathbf{f}^{t,\phi}_{\beta m}(\mathbf{r}) \cdot \left[ \hat{\mathbf{n}}_l(\mathbf{r}) \times \pm \mathbf{g}^{t,\phi}_{\alpha m}(\mathbf{r}) \right] d\mathbf{r} \\ &\quad - \int_{\mathbf{f}^{t,\phi}_{\beta m}} \mathbf{f}^{t,\phi}_{\beta m}(\mathbf{r}) \cdot \int_{\mathbf{g}^{t,\phi}_{\alpha m}} \left[ (\mathbf{r} - \mathbf{r}') \times \pm \mathbf{g}^{t,\phi}_{\alpha m}(\mathbf{r}') \right] \frac{1 + jk_l r}{4\pi r^3} e^{-jk_l r} \, d\mathbf{r}' \, d\mathbf{r} \end{aligned} \tag{7.41}$$

### 7.3.2.1 K Matrix Elements

To evaluate the sub-matrix elements in (7.41), let us first make the following definitions

$$\mathbf{r} = \rho\hat{\boldsymbol{\rho}} + z\hat{\mathbf{z}} \tag{7.42}$$

$$\mathbf{r}' = \rho' \cos(\phi' - \phi)\hat{\boldsymbol{\rho}} + \rho' \sin(\phi' - \phi)\hat{\boldsymbol{\phi}} + z'\hat{\mathbf{z}} \tag{7.43}$$

$$\hat{\mathbf{t}}(\mathbf{r}') = \sin\gamma' \left[ \cos(\phi' - \phi)\hat{\boldsymbol{\rho}} + \sin(\phi' - \phi)\hat{\boldsymbol{\phi}} \right] + \cos\gamma'\hat{\mathbf{z}} \tag{7.44}$$

$$\hat{\boldsymbol{\phi}}(\mathbf{r}') = -\sin(\phi' - \phi)\hat{\boldsymbol{\rho}} + \cos(\phi' - \phi)\hat{\boldsymbol{\phi}} \tag{7.45}$$

and then compute the following cross products

$$\begin{aligned} (\mathbf{r} - \mathbf{r}') \times \hat{\mathbf{t}}(\mathbf{r}') &= \sin(\phi' - \phi)\left[ (z' - z)\sin\gamma' - \rho'\cos\gamma' \right]\hat{\boldsymbol{\rho}} \\ &\quad - \left[ (\rho - \rho'\cos(\phi' - \phi))\cos\gamma' + (z' - z)\sin\gamma'\cos(\phi' - \phi) \right]\hat{\boldsymbol{\phi}} \\ &\quad + \rho\sin\gamma'\sin(\phi' - \phi)\hat{\mathbf{z}} \end{aligned} \tag{7.46}$$

and

$$(\mathbf{r} - \mathbf{r}') \times \hat{\boldsymbol{\phi}}(\mathbf{r}') = (z' - z)\cos(\phi' - \phi)\hat{\boldsymbol{\rho}} + (z' - z)\sin(\phi' - \phi)\hat{\boldsymbol{\phi}}$$
$$+ \left[\rho\cos(\phi' - \phi) - \rho'\right]\hat{\mathbf{z}} \tag{7.47}$$

These definitions then allow us to write

$$\hat{\mathbf{t}}(\mathbf{r}) \cdot \left[(\mathbf{r}-\mathbf{r}') \times \hat{\mathbf{t}}(\mathbf{r}')\right] = \left[(z' - z)\sin\gamma\sin\gamma' - \rho'\cos\gamma'\sin\gamma\right.$$
$$\left. + \rho\cos\gamma\sin\gamma'\right]\sin(\phi' - \phi) \tag{7.48}$$

$$\hat{\mathbf{t}}(\mathbf{r}) \cdot \left[(\mathbf{r}-\mathbf{r}') \times \hat{\boldsymbol{\phi}}(\mathbf{r}')\right] = \left[(z' - z)\sin\gamma\cos(\phi' - \phi)\right.$$
$$\left. + \left[\rho\cos(\phi' - \phi) - \rho'\right]\cos\gamma\right] \tag{7.49}$$

$$\hat{\boldsymbol{\phi}}(\mathbf{r}) \cdot \left[(\mathbf{r}-\mathbf{r}') \times \hat{\mathbf{t}}(\mathbf{r}')\right] = -\left[(z' - z)\sin\gamma'\cos(\phi' - \phi)\right.$$
$$\left. + \left[\rho - \rho'\cos(\phi' - \phi)\right]\cos\gamma'\right] \tag{7.50}$$

$$\hat{\boldsymbol{\phi}}(\mathbf{r}) \cdot \left[(\mathbf{r}-\mathbf{r}') \times \hat{\boldsymbol{\phi}}(\mathbf{r}')\right] = (z' - z)\sin(\phi' - \phi) \tag{7.51}$$

Using (7.48–7.51) as well as the periodic properties of the integrands outlined in Section 7.3.1.1, the system matrix can be re-written as

$$\begin{bmatrix} \mathbf{K}^{tt} & \mathbf{K}^{t\phi} + \tilde{\mathbf{K}}^{t\phi} \\ \mathbf{K}^{\phi t} + \tilde{\mathbf{K}}^{\phi t} & \mathbf{K}^{\phi\phi} \end{bmatrix} \tag{7.52}$$

The sub-matrix elements for mode $\alpha$ are

$$\mathbf{K}_{mn}^{tt} = j\eta_0 \int_{T_m} \int_{T_n} T_m(t) T_n(t') \left([z(t) - z(t')]\sin\gamma(t)\sin\gamma(t')\right.$$
$$+ \sin\gamma(t)\cos\gamma(t')\rho(t') - \sin\gamma(t')\cos\gamma(t)\rho(t)\right)G_6 \, dt' \, dt \tag{7.53}$$

$$\mathbf{K}_{mn}^{t\phi} = -\eta_0 \int_{T_m} \int_{T_n} T_m(t) T_n(t') \cdot \left([z(t) - z(t')]\sin\gamma(t)G_5\right.$$
$$+ \cos\gamma(t)\left[\rho(t')G_4 - \rho(t)G_5\right]\right) dt' \, dt \tag{7.54}$$

$$\mathbf{K}_{mn}^{\phi t} = -\eta_0 \int_{T_m} \int_{T_n} T_m(t) T_n(t') \cdot \left([z(t) - z(t')]\sin\gamma(t')G_5\right.$$
$$+ \cos\gamma(t')\left[\rho(t')G_5 - \rho(t)G_4\right]\right) dt' \, dt \tag{7.55}$$

$$\mathbf{K}_{mn}^{\phi\phi} = -j\eta_0 \int_{T_m} \int_{T_n} T_m(t) T_n(t') [z(t) - z(t')]G_6 \, dt' \, dt \tag{7.56}$$

where the integrations are performed on non-overlapping segments only. For overlapping segments, they are

$$\tilde{K}_{mn}^{t\phi} = \tilde{K}_{mn}^{\phi t} = \tilde{K}_{mn} = -\pi\eta_0 \int_{T_m, T_n} \frac{T_m(t)T_n(t)}{\rho(t)} dt \qquad (7.57)$$

The integrals $G_4$, $G_5$, and $G_6$ are

$$G_4 = \int_0^\pi \cos(\alpha\phi') \frac{1 + jk_l r}{r^3} e^{-jk_l r} d\phi' \qquad (7.58)$$

$$G_5 = \int_0^\pi \cos(\phi') \cos(\alpha\phi') \frac{1 + jk_l r}{r^3} e^{-jk_l r} d\phi' \qquad (7.59)$$

$$G_6 = \int_0^\pi \sin(\phi') \sin(\alpha\phi') \frac{1 + jk_l r}{r^3} e^{-jk_l r} d\phi' \qquad (7.60)$$

where $r$ was defined in (7.38).

Inspecting the K sub-matrix elements, we find the following relationship between positive and negative modes

$$\begin{bmatrix} K^{tt} & K^{t\phi} \\ K^{\phi t} & K^{\phi\phi} \end{bmatrix}_{-\alpha} = \begin{bmatrix} -K^{tt} & K^{t\phi} \\ K^{\phi t} & -K^{\phi\phi} \end{bmatrix}_{\alpha} \qquad (7.61)$$

These properties survive inversion.

### 7.3.3 Excitation

Right-hand side vector elements for the EFIE can be obtained by substituting the testing functions $\mathbf{f}_{\alpha m}^{t,\phi}(\mathbf{r})$ into (3.191). This yields a column vector of the form

$$V^{E(l)} = \begin{bmatrix} V^{E(l),t} \\ V^{E(l),\phi} \end{bmatrix} \qquad (7.62)$$

The individual elements are given by

$$V_m^{E(l),(t,\phi)} = \int_{T_m} \int_0^{2\pi} \mathbf{f}_{\alpha m}^{t,\phi}(\mathbf{r}) \cdot \mathbf{E}^i(\mathbf{r}) \, \rho \, d\phi \, dt \qquad (7.63)$$

where $\mathbf{E}^i(\mathbf{r})$ is the impressed electric field. In general, the above integral must be evaluated numerically, however for plane waves it can be evaluated in the $\phi$ dimension analytically.

#### 7.3.3.1 Plane Wave Excitation

An incident plane wave in the free space region $R_0$ has a phase component at a point $\mathbf{r}$ given by

$$e^{jk_0 \hat{\mathbf{r}}^i \cdot \mathbf{r}} = e^{jk_0 z \cos\theta^i} e^{jk_0 \sin\theta^i (x \cos\phi^i + y \sin\phi^i)} \qquad (7.64)$$

where $(\theta^i, \phi^i)$ are the incident spherical angles, and $\hat{\mathbf{r}}^i$ the direction of incidence. In cylindrical coordinates, (7.64) can be written as

$$e^{jk_0 z \cos \theta^i} e^{jk_0 \sin \theta^i (x \cos \phi^i + y \sin \phi^i)} = e^{jk_0 z \cos \theta^i} e^{jk_0 \sin \theta^i \rho \cos(\phi - \phi^i)} \quad (7.65)$$

To obtain a complete scattering matrix, we must consider $\hat{\boldsymbol{\theta}}^i$- and $\hat{\boldsymbol{\phi}}^i$-polarized incident waves, so we will treat each polarization separately. Using the fact that

$$\hat{\boldsymbol{\theta}}^i = \cos \theta^i \cos \phi^i \hat{\mathbf{x}} + \cos \theta^i \sin \phi^i \hat{\mathbf{y}} - \sin \theta^i \hat{\mathbf{z}} \quad (7.66)$$

and

$$\hat{\boldsymbol{\phi}}^i = -\sin \phi^i \hat{\mathbf{x}} + \cos \phi^i \hat{\mathbf{y}} \quad (7.67)$$

we can take the dot products with the $\hat{\mathbf{t}}$ and $\hat{\boldsymbol{\phi}}$ component vectors to yield

$$\hat{\mathbf{t}} \cdot \hat{\boldsymbol{\theta}}^i = \cos \theta^i \sin \gamma \cos(\phi - \phi^i) - \sin \theta^i \cos \gamma \quad (7.68)$$

$$\hat{\boldsymbol{\phi}} \cdot \hat{\boldsymbol{\theta}}^i = -\cos \theta^i \sin(\phi - \phi^i) \quad (7.69)$$

$$\hat{\mathbf{t}} \cdot \hat{\boldsymbol{\phi}}^i = \sin \gamma \sin(\phi - \phi^i) \quad (7.70)$$

$$\hat{\boldsymbol{\phi}} \cdot \hat{\boldsymbol{\phi}}^i = \cos(\phi - \phi^i) \quad (7.71)$$

Testing the $\hat{\boldsymbol{\theta}}^i$-polarized electric field with the $\hat{\mathbf{t}}$ and $\hat{\boldsymbol{\phi}}$-oriented electric testing functions yields a right-hand side vector of the form

$$V^{E,\theta^i} = \begin{bmatrix} V^{E,t\theta^i} \\ V^{E,\phi\theta^i} \end{bmatrix} \quad (7.72)$$

with elements given by

$$V_m^{E,t\theta^i} = \int_{T_m} T_m(t) e^{jk_0 z(t) \cos \theta^i} \int_0^{2\pi} \left[ \cos \theta^i \sin \gamma(t) \cos \phi - \sin \theta^i \cos \gamma(t) \right] \cdot$$

$$e^{jk_0 \sin \theta^i \rho(t) \cos \phi - j\alpha\phi} \, d\phi \, dt \quad (7.73)$$

and

$$V_m^{E,\phi\theta^i} = -\int_{T_m} T_m(t) e^{jk_0 z(t) \cos \theta^i} \int_0^{2\pi} \cos \theta^i \sin \phi \cdot$$

$$e^{jk_0 \sin \theta^i \rho(t) \cos \phi - j\alpha\phi} \, d\phi \, dt \quad (7.74)$$

Similarly, testing the $\hat{\boldsymbol{\phi}}^i$-polarized electric field yields a right-hand side vector of the form

$$V^{E,\phi^i} = \begin{bmatrix} V^{E,t\phi^i} \\ V^{E,\phi\phi^i} \end{bmatrix} \quad (7.75)$$

with elements given by

$$V_m^{E,t\phi^i} = \int_{T_m} T_m(t)e^{jk_0 z(t)\cos\theta^i} \int_0^{2\pi} \sin\gamma(t)\sin\phi\, e^{jk_0\sin\theta^i\rho(t)\cos\phi - j\alpha\phi}\, d\phi\, dt \tag{7.76}$$

$$V_m^{E,\phi\phi^i} = \int_{T_m} T_m(t)e^{jk_0 z(t)\cos\theta^i} \int_0^{2\pi} \cos\phi\, e^{jk_0\sin\theta^i\rho(t)\cos\phi - j\alpha\phi}\, d\phi\, dt \tag{7.77}$$

Due to the rotational symmetry, we have set $\phi^i$ to zero for simplicity. If we need to solve bistatic scattering problems, we can simply replace $\phi^s$ by $\phi^s - \phi^i$ when computing the scattered field in Section 7.5.2.1. Let us now consider the following integral from $V_m^{E,t\theta^i}$

$$I = \cos\theta^i \sin\gamma(t) \int_0^{2\pi} \cos\phi e^{jk_0\sin\theta^i\rho(t)\cos\phi - j\alpha\phi}\, d\phi$$
$$- \sin\theta^i \cos\gamma(t) \int_0^{2\pi} e^{jk_0\sin\theta^i\rho(t)\cos\phi - j\alpha\phi}\, d\phi \tag{7.78}$$

Using the identity [3]

$$2\pi j^\alpha J_\alpha(z) = \int_0^{2\pi} e^{jz\cos\phi}\cos(\alpha\phi) \tag{7.79}$$

the second integral in (7.78) is

$$\int_0^{2\pi} e^{jk_0\sin\theta^i\rho(t)\cos\phi - j\alpha\phi}\, d\phi = 2\pi j^\alpha J_\alpha(x) \tag{7.80}$$

where $x = k_0\sin\theta^i\rho(t)$. We can then write (7.78) as

$$I = \cos\theta^i \sin\gamma(t) \int_0^{2\pi} \cos\phi e^{jk_0\sin\theta^i\rho_p\cos\phi - j\alpha\phi}\, d\phi - 2\pi j^\alpha J_\alpha(x)\sin\theta^i\cos\gamma(t) \tag{7.81}$$

To evaluate the remaining integral, we make the substitution

$$\int_0^{2\pi} \cos\phi e^{jk\sin\theta^i\rho(t)\cos\phi - j\alpha\phi}\, d\phi = \int_0^{2\pi} (e^{j\phi} - j\sin\phi)e^{jk\sin\theta^i\rho(t)\cos\phi - j\alpha\phi}\, d\phi \tag{7.82}$$

The first term is evaluated as before, yielding

$$2\pi j^{\alpha-1} J_{\alpha-1}(x) - j\int_0^{2\pi} \sin\phi e^{jk\sin\theta^i\rho(t)\cos\phi - j\alpha\phi}\, d\phi \tag{7.83}$$

and the second term can be integrated by parts, yielding

$$2\pi j^{\alpha-1} J_{\alpha-1}(x) + j\left[\frac{2\pi\alpha}{x} j^{\alpha} J_{\alpha}(x)\right]$$ (7.84)

Using the recurrence relationship

$$\frac{2\alpha}{x} J_{\alpha}(x) = J_{\alpha+1}(x) + J_{\alpha-1}(x)$$ (7.85)

we can write (7.84) as

$$\pi j^{\alpha} j\left[J_{\alpha+1}(x) - J_{\alpha-1}(x)\right]$$ (7.86)

and $V_m^{E,t\theta^i}$ becomes

$$\begin{aligned} V_m^{E,t\theta^i} =\pi j^{\alpha} \int_{T_m} T_m(t)^{jkz(t)\cos\theta^i} &\left[j\cos\theta^i \sin\gamma(t)\left[J_{\alpha+1}(x) - J_{\alpha-1}(x)\right]\right. \\ &\left. -2\sin\theta^i \cos\gamma(t) J_{\alpha}(x)\right] dt \end{aligned}$$ (7.87)

The other elements are obtained similarly, yielding

$$V_m^{E,\phi\theta^i} = \pi j^{\alpha} \int_{T_m} T_m(t) e^{jkz(t)\cos\theta^i} \cos\theta^i \left[J_{\alpha+1}(x) + J_{\alpha-1}(x)\right] dt$$ (7.88)

$$V_m^{E,t\phi^i} = -\pi j^{\alpha} \int_{T_m} T_m(t) e^{jkz(t)\cos\theta^i} \sin\gamma(t) \left[J_{\alpha+1}(x) + J_{\alpha-1}(x)\right] dt$$ (7.89)

$$V_m^{E,\phi\phi^i} = \pi j^{\alpha+1} \int_{T_m} T_m(t) e^{jkz(t)\cos\theta^i} \left[J_{\alpha+1}(x) - J_{\alpha-1}(x)\right] dt$$ (7.90)

Inspecting (7.87–7.90), we observe the following relationship between positive and negative modes

$$\begin{bmatrix} V^{E,t\theta^i} \\ V^{E,\phi\theta^i} \end{bmatrix}_{-\alpha} = \begin{bmatrix} V^{E,t\theta^i} \\ -V^{E,\phi\theta^i} \end{bmatrix}_{\alpha}$$ (7.91)

and

$$\begin{bmatrix} V^{E,t\phi^i} \\ V^{E,\phi\phi^i} \end{bmatrix}_{-\alpha} = \begin{bmatrix} -V^{E,t\phi^i} \\ V^{E,\phi\phi^i} \end{bmatrix}_{\alpha}$$ (7.92)

## 7.4 MFIE

The MFIE (3.179) also has $\mathcal{K}$ and $\mathcal{L}$ operators, which were discretized and tested via the Moment Method in (3.187) and (3.188). For the testing functions, we will use the complex conjugates of the magnetic basis functions (7.8) and (7.9). Substitution of the electric basis and magnetic testing functions into (3.187) yields in Region $R_l$ a matrix of the form

$$Z_{mn}^{HJ(l)} = -\begin{bmatrix} K(g_\beta^t, f_\alpha^t) & K(g_\beta^t, f_\alpha^\phi) \\ K(g_\beta^\phi, f_\alpha^t) & K(g_\beta^\phi, f_\alpha^\phi) \end{bmatrix} = -\begin{bmatrix} K^{tt} & K^{t\phi} \\ K^{\phi t} & K^{\phi\phi} \end{bmatrix} \tag{7.93}$$

and substitution of the magnetic basis and testing functions into (3.188) yields

$$Z_{mn}^{HM(l)} = \frac{\epsilon_l}{\mu_l}\begin{bmatrix} L(g_\beta^t, g_\alpha^t) & L(g_\beta^t, g_\alpha^\phi) \\ L(g_\beta^\phi, g_\alpha^t) & L(g_\beta^\phi, g_\alpha^\phi) \end{bmatrix} = \eta_0^2 \frac{\epsilon_l}{\mu_l}\begin{bmatrix} L^{tt} & L^{t\phi} \\ L^{\phi t} & L^{\phi\phi} \end{bmatrix} \tag{7.94}$$

where the expressions for the L and K sub-matrix elements were derived in Sections 7.3.1.1 and 7.3.2.1, respectively.

### 7.4.1 Excitation

Right-hand side vector elements for the MFIE can be obtained by substituting the testing functions $g_{\alpha m}^{t,\phi}(\mathbf{r})$ into (3.192). This yields a column vector of length $2N$ of the form

$$V^{H(l)} = \begin{bmatrix} V^{H(l),t} \\ V^{H(l),\phi} \end{bmatrix} \tag{7.95}$$

The individual elements are given by

$$V_m^{H(l),(t,\phi)} = \int_{T_m} \int_0^{2\pi} g_{\alpha m}^{t,\phi}(\mathbf{r}) \cdot \mathbf{H}^i(\mathbf{r})\, \rho\, d\phi\, dt \tag{7.96}$$

where $\mathbf{H}^i(\mathbf{r})$ is the impressed magnetic field. As with the EFIE, this can be evaluated analytically in the $\phi$ dimension for plane waves.

#### 7.4.1.1 Plane Wave Excitation

Given $\hat{\theta}^i$- and $\hat{\phi}^i$-polarized incident electric fields, the corresponding magnetic fields are $-\hat{\phi}^i$- and $\hat{\theta}^i$-polarized, respectively. Thus, we can re-use the components of $V^{E(l)}$, yielding MFIE right-hand side vectors given by

$$V^{H,\theta^i} = \begin{bmatrix} V^{H,t\theta^i} \\ V^{H,\phi\theta^i} \end{bmatrix} = \begin{bmatrix} -V^{E,t\phi^i} \\ -V^{E,\phi\phi^i} \end{bmatrix} \tag{7.97}$$

and

$$V^{H,\phi^i} = \begin{bmatrix} V^{H,t\phi^i} \\ V^{H,\phi\phi^i} \end{bmatrix} = \begin{bmatrix} V^{E,t\theta^i} \\ V^{E,\phi\theta^i} \end{bmatrix} \tag{7.98}$$

## 7.5  Solution

Assembling the sub-matrix blocks for the EFIE and MFIE, the system matrix (3.184) for a general BOR has the form

$$\begin{bmatrix} Z^{EJ,tt} & Z^{EJ,t\phi} & Z^{EM,tt} & Z^{EM,t\phi} \\ Z^{EJ,\phi t} & Z^{EJ,\phi\phi} & Z^{EM,\phi t} & Z^{EM,\phi\phi} \\ Z^{HJ,tt} & Z^{HJ,t\phi} & Z^{HM,tt} & Z^{HM,t\phi} \\ Z^{HJ,\phi t} & Z^{HJ,\phi\phi} & Z^{HM,\phi t} & Z^{HM,\phi\phi} \end{bmatrix} \begin{bmatrix} I^{J,t} \\ I^{J,\phi} \\ I^{M,t} \\ I^{M,\phi} \end{bmatrix} = \begin{bmatrix} V^{E,t} \\ V^{E,\phi} \\ V^{H,t} \\ V^{H,\phi} \end{bmatrix} \tag{7.99}$$

where the nMFIE is omitted for now.

### 7.5.1  Plane Wave Solution

Substitution of the matrix and vector sub-blocks for $\hat{\theta}^i$-polarization into (7.99) yields as the solution for mode $\alpha$

$$\begin{bmatrix} I^{J,t\theta^i} \\ I^{J,\phi\theta^i} \\ I^{M,t\theta^i} \\ I^{M,\phi\theta^i} \end{bmatrix}_\alpha = \begin{bmatrix} L_E^{tt} & L_E^{t\phi} & K_E^{tt} & K_E^{t\phi} \\ L_E^{\phi t} & L_E^{\phi\phi} & K_E^{\phi t} & K_E^{\phi\phi} \\ K_H^{tt} & K_H^{t\phi} & L_H^{tt} & L_H^{t\phi} \\ K_H^{\phi t} & K_H^{\phi\phi} & L_H^{\phi t} & L_H^{\phi\phi} \end{bmatrix}_\alpha^{-1} \begin{bmatrix} V^{E,t\theta^i} \\ V^{E,\phi\theta^i} \\ -V^{E,t\phi^i} \\ -V^{E,\phi\phi^i} \end{bmatrix}_\alpha \tag{7.100}$$

Using (7.39) and (7.61), together with (7.91) and (7.92), the solution (7.100) for negative modes can be written as

$$\begin{bmatrix} I^{J,t\theta^i} \\ I^{J,\phi\theta^i} \\ I^{M,t\theta^i} \\ I^{M,\phi\theta^i} \end{bmatrix}_{-\alpha} = \begin{bmatrix} L_E^{tt} & -L_E^{t\phi} & -K_E^{tt} & K_E^{t\phi} \\ -L_E^{\phi t} & L_E^{\phi\phi} & K_E^{\phi t} & -K_E^{\phi\phi} \\ -K_H^{tt} & K_H^{t\phi} & L_H^{tt} & -L_H^{t\phi} \\ K_H^{\phi t} & -K_H^{\phi\phi} & -L_H^{\phi t} & L_H^{\phi\phi} \end{bmatrix}_\alpha^{-1} \begin{bmatrix} V^{E,t\theta^i} \\ -V^{E,\phi\theta^i} \\ V^{E,t\phi^i} \\ -V^{E,\phi\phi^i} \end{bmatrix}_\alpha \tag{7.101}$$

thus

$$\begin{bmatrix} I^{J,t\theta^i} \\ I^{J,\phi\theta^i} \\ I^{M,t\theta^i} \\ I^{M,\phi\theta^i} \end{bmatrix}_{-\alpha} = \begin{bmatrix} I^{J,t\theta^i} \\ -I^{J,\phi\theta^i} \\ -I^{M,t\theta^i} \\ I^{M,\phi\theta^i} \end{bmatrix}_\alpha \tag{7.102}$$

where only $I^{J,t\theta^i}$ and $I^{M,\phi\theta^i}$ are nonzero for $\alpha = 0$. Similarly, for $\hat{\phi}^i$-polarization, the solution to (7.99) for mode $\alpha$ is

$$
\begin{bmatrix} I^{J,t\phi^i} \\ I^{J,\phi\phi^i} \\ I^{M,t\phi^i} \\ I^{M,\phi\phi^i} \end{bmatrix}_\alpha = \begin{bmatrix} L_E^{tt} & L_E^{t\phi} & K_E^{tt} & K_E^{t\phi} \\ L_E^{\phi t} & L_E^{\phi\phi} & K_E^{\phi t} & K_E^{\phi\phi} \\ K_H^{tt} & K_H^{t\phi} & L_H^{tt} & L_H^{t\phi} \\ K_H^{\phi t} & K_H^{\phi\phi} & L_H^{\phi t} & L_H^{\phi\phi} \end{bmatrix}_\alpha^{-1} \begin{bmatrix} V^{E,t\phi^i} \\ V^{E,\phi\phi^i} \\ V^{E,t\theta^i} \\ V^{E,\phi\theta^i} \end{bmatrix}_\alpha
\tag{7.103}
$$

and for negative modes it is

$$
\begin{bmatrix} I^{J,t\phi^i} \\ I^{J,\phi\phi^i} \\ I^{M,t\phi^i} \\ I^{M,\phi\phi^i} \end{bmatrix}_{-\alpha} = \begin{bmatrix} L_E^{tt} & -L_E^{t\phi} & -K_E^{tt} & K_E^{t\phi} \\ -L_E^{\phi t} & L_E^{\phi\phi} & K_E^{\phi t} & -K_E^{\phi\phi} \\ -K_H^{tt} & K_H^{t\phi} & L_H^{tt} & -L_H^{t\phi} \\ K_H^{\phi t} & -K_H^{\phi\phi} & -L_H^{\phi t} & L_H^{\phi\phi} \end{bmatrix}_\alpha^{-1} \begin{bmatrix} -V^{E,t\phi^i} \\ V^{E,\phi\phi^i} \\ V^{E,t\theta^i} \\ -V^{E,\phi\theta^i} \end{bmatrix}_\alpha
\tag{7.104}
$$

thus

$$
\begin{bmatrix} I^{J,t\phi^i} \\ I^{J,\phi\phi^i} \\ I^{M,t\phi^i} \\ I^{M,\phi\phi^i} \end{bmatrix}_{-\alpha} = \begin{bmatrix} -I^{J,t\phi^i} \\ I^{J,\phi\phi^i} \\ I^{M,t\phi^i} \\ -I^{M,\phi\phi^i} \end{bmatrix}_\alpha
\tag{7.105}
$$

where only $I^{J,\phi\phi^i}$ and $I^{M,t\phi^i}$ are nonzero for $\alpha = 0$.

### 7.5.1.1 Currents

For $\hat{\theta}^i$-polarization, the electric and magnetic currents for basis function $m$ can be written as

$$
\mathbf{J}_m^{\theta^i}(\mathbf{r}) = \sum_{\alpha=-\infty}^{\infty} \left[ I_{\alpha m}^{J,t\theta^i} f_m(t)\hat{\mathbf{t}}(\mathbf{r}) - I_{\alpha m}^{J,\phi\theta^i} f_m(t)\hat{\phi}(\mathbf{r}) \right] e^{j\alpha\phi}
\tag{7.106}
$$

and

$$
\mathbf{M}_m^{\theta^i}(\mathbf{r}) = \sum_{\alpha=-\infty}^{\infty} \left[ I_{\alpha m}^{M,t\theta^i} g_m(t)\hat{\mathbf{t}}(\mathbf{r}) - I_{\alpha m}^{M,\phi\theta^i} g_m(t)\hat{\phi}(\mathbf{r}) \right] e^{j\alpha\phi}
\tag{7.107}
$$

Using (7.102), these currents can be re-written using only positive modes as

$$
\mathbf{J}_m^{\theta^i}(\mathbf{r}) = I_{0m}^{J,t\theta^i} f_m(t)\hat{\mathbf{t}}(\mathbf{r})
$$
$$
+ 2\sum_{\alpha=1}^{\infty} \left[ I_{\alpha m}^{J,t\theta^i} f_m(t)\cos(\alpha\phi)\hat{\mathbf{t}}(\mathbf{r}) - j I_{\alpha m}^{J,\phi\theta^i} f_m(t)\sin(\alpha\phi)\hat{\phi}(\mathbf{r}) \right]
$$

$$
\tag{7.108}
$$

and

$$\mathbf{M}_m^{\theta^i}(\mathbf{r}) = -I_{0m}^{M,\phi\theta^i}\,g_m(t)\hat{\phi}(\mathbf{r})$$
$$+2\sum_{\alpha=1}^{\infty}\left[jI_{\alpha m}^{M,t\theta^i}\,g_m(t)\sin(\alpha\phi)\hat{\mathbf{t}}(\mathbf{r}) - I_{\alpha m}^{M,\phi\theta^i}\,g_m(t)\cos(\alpha\phi)\hat{\phi}(\mathbf{r})\right]$$

$$(7.109)$$

Similarly, for $\hat{\phi}^i$-polarization, the currents are

$$\mathbf{J}_m^{\phi^i}(\mathbf{r}) = -I_{0m}^{J,\phi\phi^i}\,f_m(t)\hat{\phi}(\mathbf{r})$$
$$+2\sum_{\alpha=1}^{\infty}\left[jI_{\alpha m}^{J,t\phi^i}\,f_m(t)\sin(\alpha\phi)\hat{\mathbf{t}}(\mathbf{r}) - I_{\alpha m}^{J,\phi\phi^i}\,f_m(t)\cos(\alpha\phi)\hat{\phi}(\mathbf{r})\right]$$

$$(7.110)$$

and

$$\mathbf{M}_m^{\phi^i}(\mathbf{r}) = I_{0m}^{M,t\phi^i}\,g_m(t)\hat{\mathbf{t}}(\mathbf{r})$$
$$+2\sum_{\alpha=1}^{\infty}\left[I_{\alpha m}^{M,t\phi^i}\,g_m(t)\cos(\alpha\phi)\hat{\mathbf{t}}(\mathbf{r}) - jI_{\alpha m}^{M,\phi\phi^i}\,g_m(t)\sin(\alpha\phi)\hat{\phi}(\mathbf{r})\right]$$

$$(7.111)$$

### 7.5.2   Scattered Field

The scattered near electric field in Region $R_l$ can be obtained by substituting (7.4) and (7.7) into (3.108) and (3.112), and performing the integration over all basis functions in $R_l$. For general problems, the near field integrals must be performed numerically in the $t$ and $\phi$ dimensions for each mode.

#### 7.5.2.1   Scattered Far Fields

Using (7.108–7.111), the scattered far electric fields in the free space region $R_0$ are obtained using (3.133), with the components of $\mathbf{A}(\mathbf{r})$ and $\mathbf{F}(\mathbf{r})$ given by

$$A_{**}(\mathbf{r}) = \mu_0\frac{e^{-jk_0 r}}{4\pi r}\sum_{m=1}^{N}\int_{t_m}(\hat{\boldsymbol{\theta}}^s,\hat{\boldsymbol{\phi}}^s)\cdot\int_0^{2\pi}\mathbf{J}_m^{\theta^i,\phi^i}(\mathbf{r})e^{jk_0\hat{\mathbf{r}}^s\cdot\mathbf{r}'}\rho\,d\phi\,dt \quad (7.112)$$

and

$$F_{**}(\mathbf{r}) = \epsilon_0\frac{e^{-jk_0 r}}{4\pi r}\sum_{m=1}^{N}\int_{t_m}(\hat{\boldsymbol{\theta}}^s,\hat{\boldsymbol{\phi}}^s)\cdot\int_0^{2\pi}\mathbf{M}_m^{\theta^i,\phi^i}(\mathbf{r})e^{jk_0\hat{\mathbf{r}}^s\cdot\mathbf{r}'}\rho\,d\phi\,dt$$

$$(7.113)$$

where $(\theta^s, \phi^s)$ are the spherical scattering angles, and $\hat{\mathbf{r}}^s$ the direction of scattering. The integration is performed over all basis functions in $R_0$. As before, the integrations in the $\phi$ dimension in (7.112) and (7.113) can be performed analytically.

Let us focus on the innermost integral in (7.112) for basis function $m$, which has the form

$$(\hat{\boldsymbol{\theta}}^s, \hat{\boldsymbol{\phi}}^s) \cdot \int_0^{2\pi} \mathbf{J}_m^{\theta^i, \phi^i}(\mathbf{r}) e^{jk_0 \hat{\mathbf{r}}^s \cdot \mathbf{r}'} \rho \, d\phi \qquad (7.114)$$

The exponential can be written as

$$e^{jk_0 \hat{\mathbf{r}}^s \cdot \mathbf{r}'} = e^{jk_0 z \cos \theta^s} e^{jk_0 \sin \theta^s \rho \cos(\phi - \phi^s)} \qquad (7.115)$$

Applying the dot products from before, the $(\theta^s, \theta^i)$ term for modes $\alpha > 0$ is

$$I_{\alpha m}^{J, t\theta^i} \int_0^{2\pi} e^{jk_0 \sin \theta^s \rho(t) \cos \phi} \cos(\alpha\phi + \alpha\phi^s) \big[ \cos \theta^s \sin \gamma(t) \cos \phi - \sin \theta^s \cos \gamma(t) \big] \, d\phi$$

$$+ I_{\alpha m}^{J, \phi\theta^i} \int_0^{2\pi} j e^{jk_0 \sin \theta^s \rho(t) \cos \phi} \sin(\alpha\phi + \alpha\phi^s) \sin \phi \cos \theta^s \, d\phi \qquad (7.116)$$

where we have made the substitution $\phi = \phi + \phi^s$ due to the $2\pi$ periodicity. Now consider the first integral in (7.116). Using the identity

$$\cos(\alpha\phi + \alpha\phi^s) = \cos(\alpha\phi) \cos(\alpha\phi^s) - \sin(\alpha\phi) \sin(\alpha\phi^s) \qquad (7.117)$$

this integral becomes

$$I_1 = \cos(\alpha\phi^s) \int_0^{2\pi} e^{jk_0 \sin \theta^s \rho(t) \cos \phi} \cos(\alpha\phi) \big[ \cos \theta^s \sin \gamma(t) \cos \phi - \sin \theta^s \cos \gamma(t) \big] \, d\phi$$

$$- \sin(\alpha\phi^s) \int_0^{2\pi} e^{jk_0 \sin \theta^s \rho(t) \cos \phi} \sin(\alpha\phi) \big[ \cos \theta^s \sin \gamma(t) \cos \phi - \sin \theta^s \cos \gamma(t) \big] \, d\phi$$

$$(7.118)$$

The second term goes to zero by inspection. Using (7.79), the integral can then be written as

$$I_1 = \pi j^\alpha \Big[ \cos \theta^s \sin \gamma(t) j \big[ J_{\alpha+1}(x) - J_{\alpha-1}(x) \big] - 2 \sin \theta^s \cos \gamma(t) J_\alpha(x) \Big] \cos(\alpha\phi^s)$$

$$(7.119)$$

where $J_\alpha(x) = J_\alpha(k_0 \rho_p \sin \theta^s)$. Using the identity

$$\sin(\alpha\phi + \alpha\phi^s) = \sin(\alpha\phi) \cos(\alpha\phi^s) + \cos(\alpha\phi) \sin(\alpha\phi^s) \qquad (7.120)$$

the second integral in (7.116) becomes

$$I_2 = \cos(\alpha\phi^s) \cos \theta^s \int_0^{2\pi} j e^{jk_0 \sin \theta^s \rho(t) \cos \phi} \sin(\alpha\phi) \sin \phi \, d\phi$$

$$+ \sin(\alpha\phi^s) \cos \theta^s \int_0^{2\pi} j e^{jk_0 \sin \theta^s \rho(t) \cos \phi} \cos(\alpha\phi) \sin \phi \, d\phi \qquad (7.121)$$

The second term is again seen to be zero. The first term can be integrated by parts yielding

$$I_2 = \cos(\alpha\phi^s)\cos\theta^s\pi j^\alpha \frac{2\alpha}{x} J_\alpha(x) \tag{7.122}$$

and again using (7.85) this becomes

$$I_2 = \pi j^\alpha \left[J_{\alpha+1}(x) + J_{\alpha-1}(x)\right]\cos\theta^s\cos(\alpha\phi^s) \tag{7.123}$$

Combining the results, we see that the integrals have the same form as (7.87) and (7.88), where $\theta^i$ is replaced by $\theta^s$. Thus, $A_{\theta\theta}$ can be written as

$$A_{\theta\theta}(\mathbf{r}) = \frac{C_A}{2}I_{0m}^{J,t\theta^i}V_0^{E,t\theta^s} + C_A\sum_{\alpha=1}^{\infty}\left[I_{\alpha m}^{J,t\theta^i}V_\alpha^{E,t\theta^s} + I_{\alpha m}^{J,\phi\theta^i}V_\alpha^{E,\phi\theta^s}\right]\cos(\alpha\phi^s) \tag{7.124}$$

The other terms follow via similarly, and are

$$A_{\phi\theta}(\mathbf{r}) = -jC_A\sum_{\alpha=1}^{\infty}\left[I_{\alpha n}^{J,t\theta^i}V_\alpha^{E,t\phi^s} + I_{\alpha n}^{J,\phi\theta^i}V_\alpha^{E,\phi\phi^s}\right]\sin(\alpha\phi^s) \tag{7.125}$$

$$A_{\theta\phi}(\mathbf{r}) = jC_A\sum_{\alpha=1}^{\infty}\left[I_{\alpha n}^{J,t\phi^i}V_\alpha^{E,t\theta^s} + I_{\alpha n}^{J,\phi\phi^i}V_\alpha^{E,\phi\theta^s}\right]\sin(\alpha\phi^s) \tag{7.126}$$

$$A_{\phi\phi}(\mathbf{r}) = -\frac{C_A}{2}I_{0m}^{J,\phi\phi^i}V_0^{E,\phi\phi^s} - C_A\sum_{\alpha=1}^{\infty}\left[I_{\alpha n}^{J,t\phi^i}V_\alpha^{E,t\phi^s} + I_{\alpha n}^{J,\phi\phi^i}V_\alpha^{E,\phi\phi^s}\right]\cos(\alpha\phi^s) \tag{7.127}$$

where $C_A$ is

$$C_A = \mu_0\frac{e^{-jk_0 r}}{2\pi r} \tag{7.128}$$

Similarly, the components of $\mathbf{F}(\mathbf{r})$ are

$$F_{\theta\theta}(\mathbf{r}) = jC_F\sum_{\alpha=1}^{\infty}\left[I_{\alpha n}^{M,t\theta^i}V_\alpha^{E,t\theta^s} + I_{\alpha n}^{M,\phi\theta^i}V_\alpha^{E,\phi\theta^s}\right]\sin(\alpha\phi^s) \tag{7.129}$$

$$F_{\phi\theta}(\mathbf{r}) = -\frac{C_F}{2}I_{0m}^{M,\phi\theta^i}V_0^{E,\phi\phi^s} - C_F\sum_{\alpha=1}^{\infty}\left[I_{\alpha n}^{M,t\theta^i}V_\alpha^{E,t\phi^s} + I_{\alpha n}^{M,\phi\theta^i}V_\alpha^{E,\phi\phi^s}\right]\cos(\alpha\phi^s) \tag{7.130}$$

$$F_{\theta\phi}(\mathbf{r}) = \frac{C_F}{2}I_{0m}^{M,t\phi^i}V_0^{E,t\theta^s} + C_F\sum_{\alpha=1}^{\infty}\left[I_{\alpha m}^{M,t\phi^i}V_\alpha^{E,t\theta^s} + I_{\alpha m}^{M,\phi\phi^i}V_\alpha^{E,\phi\theta^s}\right]\cos(\alpha\phi^s) \tag{7.131}$$

$$F_{\phi\phi}(\mathbf{r}) = -jC_F \sum_{\alpha=1}^{\infty} \left[ I_{\alpha n}^{M,t\phi^i} V_{\alpha}^{E,t\phi^s} + I_{\alpha n}^{M,\phi\phi^i} V_{\alpha}^{E,\phi\phi^s} \right] \sin(\alpha\phi^s) \quad (7.132)$$

where $C_F$ is

$$C_F = \epsilon_0 \eta_0 \frac{e^{-jk_0 r}}{2\pi r} \quad (7.133)$$

Note that the right-hand side vectors $V^{E,\theta^i}$ and $V^{E,\phi^i}$ contain no references to the incident or observation angles $\phi^i$ or $\phi^s$. For monostatic calculations, these vectors can be computed once per right-hand side and used to obtain the currents as well as the scattered fields.

---

## 7.6   nMFIE

The nMFIE (3.180) contains the $\hat{\mathbf{n}} \times \mathcal{L}$ and $\hat{\mathbf{n}} \times \mathcal{K}$ operators, which were discretized and tested via the Moment Method in (3.189) and (3.190). For the testing functions, we will use the electric testing functions in (7.11) and (7.12).

### 7.6.1   $\hat{\mathbf{n}} \times \mathcal{L}$ Operator

Substitution of the basis (7.8, 7.9) and testing (7.11, 7.12) functions into (3.190) yields a $2N \times 2N$ matrix of the form

$$Z^{nHM(l)} = \begin{bmatrix} \mathrm{nL}(\mathbf{f}_\beta^t, \mathbf{g}_\alpha^t) & \mathrm{nL}(\mathbf{f}_\beta^t, \mathbf{g}_\alpha^\phi) \\ \mathrm{nL}(\mathbf{f}_\beta^\phi, \mathbf{g}_\alpha^t) & \mathrm{nL}(\mathbf{f}_\beta^\phi, \mathbf{g}_\alpha^\phi) \end{bmatrix} = \begin{bmatrix} \mathrm{nL}^{tt} & \mathrm{nL}^{t\phi} \\ \mathrm{nL}^{\phi t} & \mathrm{nL}^{\phi\phi} \end{bmatrix} \quad (7.134)$$

The elements of each sub-matrix block are given by

$$\mathrm{nL}_{mn}^{**} = j\omega\epsilon_l \int_{\mathbf{f}_{\beta m}^{t,\phi}} \int_{\mathbf{g}_{\alpha n}^{t,\phi}} \left[ \mathbf{f}_{\beta m}^{t,\phi}(\mathbf{r}) \times \hat{\mathbf{n}}_l(\mathbf{r}) \right] \cdot \int_{\mathbf{g}_{\alpha n}^{t,\phi}} \pm \mathbf{g}_{\alpha n}^{t,\phi}(\mathbf{r}') \frac{e^{-jk_l r}}{4\pi r} \, d\mathbf{r}' \right] d\mathbf{r}$$

$$- \frac{j}{\omega\mu_l} \int_{\mathbf{f}_{\beta m}^{t,\phi}} \nabla \cdot \left[ \mathbf{f}_{\beta m}^{t,\phi}(\mathbf{r}) \times \hat{\mathbf{n}}_l(\mathbf{r}) \right] \int_{\mathbf{g}_{\alpha n}^{t,\phi}} \left[ \nabla' \cdot \pm \mathbf{g}_{\alpha n}^{t,\phi}(\mathbf{r}') \right] \frac{e^{-jk_l r}}{4\pi r} \, d\mathbf{r}' \, d\mathbf{r}$$

$$(7.135)$$

where we have used (3.205) to re-distribute the cross products and differential operators.

#### 7.6.1.1　nL Matrix Elements

Noting that

$$\hat{t}(\mathbf{r}) \times \hat{\mathbf{n}}_l(\mathbf{r}) = \hat{\phi}(\mathbf{r}) \tag{7.136}$$

and

$$\hat{\phi}(\mathbf{r}) \times \hat{\mathbf{n}}_l(\mathbf{r}) = -\hat{t}(\mathbf{r}) \tag{7.137}$$

we see that (7.135) has the same form as (7.14). This allows us to write the nL sub-matrix elements in terms of the L sub-matrix elements as

$$\begin{bmatrix} \mathrm{nL}^{tt} & \mathrm{nL}^{t\phi} \\ \mathrm{nL}^{\phi t} & \mathrm{nL}^{\phi\phi} \end{bmatrix} = \eta_0 \frac{\epsilon_l}{\mu_l} \begin{bmatrix} \mathrm{L}^{\phi t} & \mathrm{L}^{\phi\phi} \\ -\mathrm{L}^{tt} & -\mathrm{L}^{t\phi} \end{bmatrix} \tag{7.138}$$

Inspection of (7.138) reveals the following relationship between positive and negative modes

$$\begin{bmatrix} \mathrm{nL}^{tt} & \mathrm{nL}^{t\phi} \\ \mathrm{nL}^{\phi t} & \mathrm{nL}^{\phi\phi} \end{bmatrix}_{-\alpha} = \begin{bmatrix} -\mathrm{nL}^{tt} & \mathrm{nL}^{t\phi} \\ \mathrm{nL}^{\phi t} & -\mathrm{nL}^{\phi\phi} \end{bmatrix}_{\alpha} \tag{7.139}$$

which are the same properties as in the K matrix.

### 7.6.2　$\hat{\mathbf{n}} \times \mathcal{K}$ Operator

Substitution of the basis (7.5, 7.6) and testing (7.11, 7.12) functions into (3.189) yields a matrix of the form

$$Z^{nHJ(l)} = \begin{bmatrix} \mathrm{nK}(\mathbf{f}_\beta^t, \mathbf{f}_\alpha^t) & \mathrm{nK}(\mathbf{f}_\beta^t, \mathbf{f}_\alpha^\phi) \\ \mathrm{nK}(\mathbf{f}_\beta^\phi, \mathbf{f}_\alpha^t) & \mathrm{nK}(\mathbf{f}_\beta^\phi, \mathbf{f}_\alpha^\phi) \end{bmatrix} = \begin{bmatrix} \mathrm{nK}^{tt} & \mathrm{nK}^{t\phi} \\ \mathrm{nK}^{\phi t} & \mathrm{nK}^{\phi\phi} \end{bmatrix} \tag{7.140}$$

where in Region $R_l$, the elements of each sub-matrix block are given by

$$\begin{aligned} \mathrm{nK}_{mn}^{**} =& \frac{1}{2} \int_{\mathbf{f}_{\beta m}^{t,\phi}, \mathbf{f}_{\alpha n}^{t,\phi}} \mathbf{f}_{\beta m}^{t,\phi}(\mathbf{r}) \cdot \pm \mathbf{f}_{\alpha n}^{t,\phi}(\mathbf{r}) \, d\mathbf{r} \\ &+ \int_{\mathbf{f}_m^{t,\phi}} \mathbf{f}_{\beta m}^{t,\phi}(\mathbf{r}) \cdot \hat{\mathbf{n}}(\mathbf{r}) \times \int_{\mathbf{f}_{\alpha n}^{t,\phi}} \left[(\mathbf{r} - \mathbf{r}') \times \pm \mathbf{f}_{\alpha n}^{t,\phi}(\mathbf{r}')\right] \frac{1 + jk_l r}{4\pi r^3} e^{-jk_l r} \, d\mathbf{r}' d\mathbf{r} \end{aligned} \tag{7.141}$$

#### 7.6.2.1　nK Matrix Elements

In evaluating the sub-matrix elements in (7.141), we again use (7.42–7.47) and note that

$$\hat{t}(\mathbf{r}) \cdot \hat{\mathbf{n}}(\mathbf{r}) \times \left[(\mathbf{r} - \mathbf{r}') \times \mathbf{f}(\mathbf{r}')\right] = \hat{\phi}(\mathbf{r}) \cdot \left[(\mathbf{r} - \mathbf{r}') \times \mathbf{f}(\mathbf{r}')\right] \tag{7.142}$$

and

$$\hat{\phi}(\mathbf{r}) \cdot \hat{\mathbf{n}}(\mathbf{r}) \times \left[(\mathbf{r} - \mathbf{r}') \times \mathbf{f}(\mathbf{r}')\right] = -\hat{t}(\mathbf{r}) \cdot \left[(\mathbf{r} - \mathbf{r}') \times \mathbf{f}(\mathbf{r}')\right] \tag{7.143}$$

which have the same form as (7.48–7.51). Thus, we can write the nK sub-matrix elements in terms of the K sub-matrix elements as

$$\begin{bmatrix} nK^{tt} & nK^{t\phi} \\ nK^{\phi t} & nK^{\phi\phi} \end{bmatrix} = \frac{1}{\eta_0} \begin{bmatrix} -K^{\phi t} - \tilde{K} & -K^{\phi\phi} \\ K^{tt} & K^{t\phi} + \tilde{K} \end{bmatrix} \tag{7.144}$$

Inspection of (7.144) reveals the following relationship between positive and negative modes

$$\begin{bmatrix} nK^{tt} & nK^{t\phi} \\ nK^{\phi t} & nK^{\phi\phi} \end{bmatrix}_{-\alpha} = \begin{bmatrix} nK^{tt} & -nK^{t\phi} \\ -nK^{\phi t} & nK^{\phi\phi} \end{bmatrix}_{\alpha} \tag{7.145}$$

which are the same properties found in the L matrix.

### 7.6.3 Excitation

Right-hand side vector elements for the nMFIE can be obtained by substituting the testing functions $\mathbf{f}_{\alpha m}^{t,\phi}(\mathbf{r})$ into (3.193). This yields a column vector of the form

$$V^{nH(l)} = \begin{bmatrix} V^{nH(l),t} \\ V^{nH(l),\phi} \end{bmatrix} \tag{7.146}$$

The individual elements are given by

$$V_m^{nH(l),(t,\phi)} = \int_{T_m} \int_0^{2\pi} \mathbf{f}_{\alpha m}^{t,\phi}(\mathbf{r}) \cdot \left[ \hat{\mathbf{n}}_l(\mathbf{r}) \times \mathbf{H}^i(\mathbf{r}) \right] \rho \, d\phi \, dt \tag{7.147}$$

where $\mathbf{H}^i(\mathbf{r})$ is the impressed magnetic field. As with the EFIE and MFIE, this can be evaluated analytically in the $\phi$ dimension for incident plane waves.

#### 7.6.3.1 Plane Wave Excitation

While the components of the incident magnetic field are the same as in Section 7.4.1.1, we must now take into account the presence of $\hat{\mathbf{n}}(\mathbf{r})$ in (7.147). Making the following substitutions

$$\hat{\mathbf{t}}(\mathbf{r}) = \sin\gamma\cos\phi\hat{\mathbf{x}} + \sin\gamma\sin\phi\hat{\mathbf{y}} + \cos\gamma\hat{\mathbf{z}} \tag{7.148}$$

$$\hat{\boldsymbol{\phi}}(\mathbf{r}) = -\sin\phi\hat{\mathbf{x}} + \cos\phi\hat{\mathbf{y}} \tag{7.149}$$

$$\hat{\mathbf{n}}(\mathbf{r}) = \cos\gamma\cos\phi\hat{\mathbf{x}} + \cos\gamma\sin\phi\hat{\mathbf{y}} - \sin\gamma\hat{\mathbf{z}} \tag{7.150}$$

and using (7.66) and (7.67), we find that

$$\hat{\mathbf{t}}(\mathbf{r}) \cdot \left[ \hat{\mathbf{n}}(\mathbf{r}) \times \hat{\boldsymbol{\theta}}^i \right] = -\cos\theta^i \sin\phi \tag{7.151}$$

$$\hat{\boldsymbol{\phi}}(\mathbf{r}) \cdot \left[ \hat{\mathbf{n}}(\mathbf{r}) \times \hat{\boldsymbol{\theta}}^i \right] = \cos\gamma\sin\theta^i - \sin\gamma\cos\theta^i \cos\phi \tag{7.152}$$

$$\hat{t}(\mathbf{r}) \cdot \left[ \hat{n}(\mathbf{r}) \times -\hat{\phi}^i \right] = -\cos\phi \tag{7.153}$$

$$\hat{\phi}(\mathbf{r}) \cdot \left[ \hat{n}(\mathbf{r}) \times -\hat{\phi}^i \right] = \sin\gamma\sin\phi \tag{7.154}$$

where we have again set $\phi^i = 0$ for convenience. We immediately note that these results are identical to those in (7.68–7.71), thus the elements of $V^{nH(l)}$ can be written in terms of $V^{E(l)}$. In particular,

$$V^{nH,\theta^i} = \begin{bmatrix} V^{nH,t\theta^i} \\ V^{nH,\phi\theta^i} \end{bmatrix} = \frac{1}{\eta_0} \begin{bmatrix} -V^{E,\phi\phi^i} \\ V^{E,t\phi^i} \end{bmatrix} \tag{7.155}$$

and

$$V^{nH,\phi^i} = \begin{bmatrix} V^{nH,t\phi^i} \\ V^{nH,\phi\phi^i} \end{bmatrix} = \frac{1}{\eta_0} \begin{bmatrix} V^{E,\phi\theta^i} \\ -V^{E,t\theta^i} \end{bmatrix} \tag{7.156}$$

### 7.6.3.2  Plane Wave Solution

Referring again to (7.87–7.90), we observe the following relationship between positive and negative modes

$$\begin{bmatrix} V^{nH,t\theta^i} \\ V^{nH,\phi\theta^i} \end{bmatrix}_{-\alpha} = \begin{bmatrix} V^{nH,t\theta^i} \\ -V^{nH,\phi\theta^i} \end{bmatrix}_{\alpha} \tag{7.157}$$

and

$$\begin{bmatrix} V^{nH,t\phi^i} \\ V^{nH,\phi\phi^i} \end{bmatrix}_{-\alpha} = \begin{bmatrix} -V^{nH,t\phi^i} \\ V^{nH,\phi\phi^i} \end{bmatrix}_{\alpha} \tag{7.158}$$

which are the same properties observed in (7.91) and (7.92). Noting that the nL and nK matrix have the same behavior as the K and L matrix for negative modes, as well as the EFIE and nMFIE right-hand side vectors, the EFIE and nMFIE can be combined in (7.99) to form the CFIE (3.181) with no additional modifications.

## 7.7  Numerical Discretization

A common way to describe a BOR is a piecewise linear approximation of the bounding curve in the $xz$ plane ($\phi = 0$). The curve is divided into $N$ segments with $N + 1$ unique endpoints, where the segments need not be of equal length. Each segment, when rotated 360 degrees about $\hat{z}$, sweeps out an annulus on the surface. To define the longitudinal and normal vectors on each

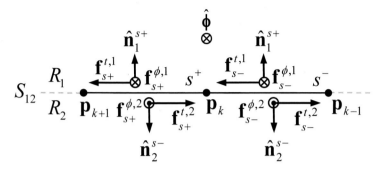

**FIGURE 7.2:** Basis Function Orientation for BOR Problems

segment, we again use a right-handed rule (see Section 6.17). Consider the triangle function with support on segments $s+$ and $s-$, whose endpoints comprise $(\mathbf{p}_{k-1}, \mathbf{p}_k, \mathbf{p}_{k+1})$. These segments reside on the interface $S_{12}$ between dielectric regions $R_1$ and $R_2$, as shown in Figure 7.2. Let us define the unit vectors

$$\hat{\mathbf{t}}^{s+} = \frac{\mathbf{p}_{k+1} - \mathbf{p}_k}{|\mathbf{p}_{k+1} - \mathbf{p}_k|} \tag{7.159}$$

$$\hat{\mathbf{t}}^{s-} = \frac{\mathbf{p}_k - \mathbf{p}_{k-1}}{|\mathbf{p}_k - \mathbf{p}_{k-1}|} \tag{7.160}$$

Following Section 6.2.1, we define $R_1$ as the "positive" side of the interface. On this side, the vectors normal to each segment are then

$$\hat{\mathbf{n}}_1^{s+} = \hat{\phi} \times \hat{\mathbf{t}}^{s+} \tag{7.161}$$

$$\hat{\mathbf{n}}_1^{s-} = \hat{\phi} \times \hat{\mathbf{t}}^{s-} \tag{7.162}$$

We then defined the longitudinal basis vectors on each segment in $R_1$ as

$$\hat{\mathbf{f}}_{s+}^{t,1} = \hat{\mathbf{t}}^{s+} \tag{7.163}$$

$$\hat{\mathbf{f}}_{s-}^{t,1} = \hat{\mathbf{t}}^{s-} \tag{7.164}$$

and the azimuthal basis vectors as

$$\hat{\mathbf{f}}_{s+}^{\phi,1} = \hat{\phi} \tag{7.165}$$

$$\hat{\mathbf{f}}_{s-}^{\phi,1} = \hat{\phi} \tag{7.166}$$

On the opposite or "negative" side of the interface in $R_2$, the direction of the normal vectors and the basis functions in $R_1$ are then reversed, as shown in

Figure 7.2. All currents have opposite orientations on each side of the interface, and for the longitudinally oriented currents, Kirchoff's law is satisfied at $p_k$. To facilitate this orientation, the $(x, z)$ locations of the segment endpoints are typically stored in a *boundary* or *points* structure, such that the longitudinal vectors are piecewise continuous between segments.

Other treatments [1, 2, 4] employ triangle functions that span $2d$ boundary segments with $d$ segments per half-triangle. Triangles are assigned a constant value on each segment, and integrations in the $t$ dimension are evaluated using a centroidal approximation. Therein, the use of two segments per half-triangle is suggested, though this choice was likely due to the computer limitations of the time. The treatment herein uses triangle functions as defined in Section 2.3.2, where triangles are assigned to no more than two segments. This assignment is illustrated on a simple, closed ellipse in Figure 7.3a. We will continue to use the centroidal approximation, so we sub-divide each segment into $N$ sub-segments of length $\Delta_s = \Delta_l/N$, and perform the integrations at the midpoint of each sub-segment.

As mentioned previously, when computing the integrals (7.35–7.37) and

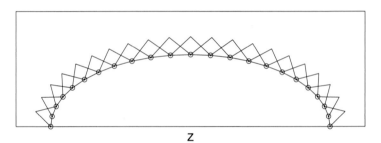

(a) Assignment of Triangle Functions

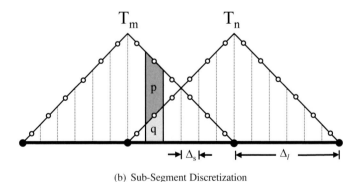

(b) Sub-Segment Discretization

**FIGURE 7.3:** Triangle Functions on a BOR

(7.58–7.60), special treatment is needed when $\mathbf{r} = \mathbf{r}'$. This occurs when the source and testing sub-segments $p$ and $q$ coincide, as outlined in Figure 7.3b. In this case, the following value of $r$ should be used [1]

$$r = \sqrt{(\Delta_s/4)^2 + 2\rho^2[1 - \cos\phi']} \tag{7.167}$$

## 7.8 Notes on Software Implementation

### 7.8.1 Parallelization

Because of the relatively modest system matrix size obtained with the MOM/BOR formulation, it is convenient to allocate matrix storage space for each thread. Unique modes can then be assigned to each thread in the pool, with each thread performing the matrix fill, factorization, and calculation of the fields for all excitations and scattering angles for the assigned mode. Because the calculations from each mode are not dependent on the others, there is minimal inter-process communication between threads except for the assignment of mode and frequency and reporting of results. This scheme is equally well suited for shared memory and distributed memory (MPI) systems.

### 7.8.2 Convergence

While it is simple to assign a maximum mode number and allow the code to calculate the fields for each mode, it is likely that the calculation may converge to a fairly accurate solution at a lower mode number. One way to determine the point where this happens is to set a threshold on the amplitude of the radiated/scattered field for each angle. The software then tracks the magnitude for each angle and when this threshold is reached, calculations are terminated at that angle for all subsequent modes. If all angles converge before the maximum mode number is reached, the loop over modes is stopped and the code can move on to the next frequency. This method is especially useful in monostatic RCS computations due to the many right-hand side operations that can be skipped, increasing the efficiency of the code for larger problems.

## 7.9 Examples

In this section we will use the formulations developed in this chapter to compute the scattering by several test articles. We will first consider a series of examples involving conducting, dielectric, and coated spheres, then a set of benchmark targets that were constructed and measured by the Electromagnetic Code Consortium (EMCC) [5]. We will then look at more complicated configurations, such as conducting and partially conducting (coated) reentry vehicles (RVs). The numerical results in this section were computed using a software called *Galaxy*, which is part of the *lucernhammer* suite of radar cross section codes. *Galaxy* is written in the C++ programming language, and is parallelized versus modes via threads. Unless otherwise specified, the BOR treatment employs 5 subsegments per half-triangle for integration in the $t$ dimension, and the CFIE is used on closed conductors with $\alpha = 0.5$.

### 7.9.1 Spheres

Spheres are excellent benchmark geometries for computational electromagnetic codes, as analytic solutions for their scattering of plane waves are available. Let us assume that there is a plane wave traveling in the $-\hat{\mathbf{z}}$ direction with an $\hat{\mathbf{x}}$-oriented electric field of unit amplitude. The exact value for the scattered far electric field is [6]

$$\mathbf{E}^s(\theta^s, \phi^s) = \frac{e^{-jk_0 r}}{k_0 r}\left[\cos \phi^s S_1(\theta^s)\hat{\boldsymbol{\theta}} - \sin \phi^s S_2(\theta^s)\hat{\boldsymbol{\phi}}\right] \tag{7.168}$$

where

$$S_1(\theta) = \sum_{n=1}^{\infty} j^{n+1}\left[A_n \frac{P_n^1(\cos \theta)}{\sin \theta} - jB_n \frac{d}{d\theta}P_n^1(\cos \theta)\right] \tag{7.169}$$

$$S_2(\theta) = \sum_{n=1}^{\infty} j^{n+1}\left[A_n \frac{d}{d\theta}P_n^1(\cos \theta) - jB_n \frac{P_n^1(\cos \theta)}{\sin \theta}\right] \tag{7.170}$$

and $P_n^1(\cos \theta)$ is the associated Legendre polynomial of degree 1 and order $n$. In practice, the sums in (7.169) and (7.170) are truncated after a finite number of modes, and the recurrence relationships in [3] used to compute the associated Legendre polynomials of higher order. The coefficients $A_n$ and $B_n$ differ depending on whether the sphere is conducting, or fully or partially dielectric. We will consider examples of each in this section.

### 7.9.1.1 Conducting Sphere

For a conducting sphere of radius $a$, the coefficients $A_n$ and $B_n$ are

$$A_n = -j^n \frac{2n+1}{n(n+1)} \frac{j_n(k_0 a)}{h_n^{(2)}(k_0 a)} \tag{7.171}$$

and

$$B_n = j^{n+1} \frac{2n+1}{n(n+1)} \frac{\left[k_0 a\, j_n(k_0 a)\right]'}{\left[k_0 a\, h_n^{(2)}(k_0 a)\right]'} \tag{7.172}$$

where $j_n(k_0 a)$ and $h_n^{(2)}(k_0 a)$ are the spherical Bessel function of the first kind, and spherical Hankel function of the second kind, of order $n$, and the primes indicate differentiation with respect to $k_0 a$ [6]. Spherical Bessel functions can be written as

$$f_n(x) = \sqrt{\frac{\pi}{2x}} F_{n+1/2}(x) \tag{7.173}$$

where $f_n(x)$ is a spherical Bessel function of the first or second kind ($j_n(x)$ and $y_n(x)$, respectively), and $F_n(x)$ is a regular Bessel function of the first or second kind ($J_n(x)$ and $Y_n(x)$, respectively). The spherical Hankel function is

$$h_n^{(2)}(x) = j_n(x) - jy_n(x) \tag{7.174}$$

The derivatives of the spherical Bessel functions are

$$\frac{d}{dx}\left[x f_n(x)\right] = \sqrt{\frac{\pi}{2x}}\left[x F_{n-1/2}(x) - n F_{n+1/2}(x)\right] \tag{7.175}$$

*MOM System Matrix*

The conducting sphere comprises a single bounding curve $S_{01}$ between the free space region $R_0$ and the interior conducting region $R_1$. Applying the CFIE yields a linear system of the form

$$\left[\alpha_c Z^{EJ} + (1 - \alpha_c) Z^{nHJ}\right]\left[I^J\right] = \left[\alpha_c V^E + (1 - \alpha_c) V^{nH}\right] \tag{7.176}$$

where

$$Z^{EJ} = \begin{bmatrix} L\left(\mathbf{f}_{S_{01}}^t, \mathbf{f}_{S_{01}}^t, R_0\right) & L\left(\mathbf{f}_{S_{01}}^t, \mathbf{f}_{S_{01}}^\phi, R_0\right) \\ L\left(\mathbf{f}_{S_{01}}^\phi, \mathbf{f}_{S_{01}}^t, R_0\right) & L\left(\mathbf{f}_{S_{01}}^\phi, \mathbf{f}_{S_{01}}^\phi, R_0\right) \end{bmatrix} \tag{7.177}$$

and

$$Z^{nHJ} = \begin{bmatrix} nK\left(\mathbf{f}_{S_{01}}^t, \mathbf{f}_{S_{01}}^t, R_0\right) & nK\left(\mathbf{f}_{S_{01}}^t, \mathbf{f}_{S_{01}}^\phi, R_0\right) \\ nK\left(\mathbf{f}_{S_{01}}^\phi, \mathbf{f}_{S_{01}}^t, R_0\right) & nK\left(\mathbf{f}_{S_{01}}^\phi, \mathbf{f}_{S_{01}}^\phi, R_0\right) \end{bmatrix} \tag{7.178}$$

The notation $X\left(\mathbf{a}_{S_l}, \mathbf{b}_{S_m}, R_l\right)$ indicates that $X$ is evaluated in $R_l$ for the testing and basis functions **a** and **b** that reside on the boundaries $S_l$ and $S_m$, respectively. The right-hand side vectors are

$$V^E = \begin{bmatrix} V^E(\mathbf{f}_{S_0}^t, R_0) \\ V^E(\mathbf{f}_{S_0}^\phi, R_0) \end{bmatrix} \tag{7.179}$$

and

$$V^{nH} = \begin{bmatrix} V^{nH}(\mathbf{f}_{S_0}^t, R_0) \\ V^{nH}(\mathbf{f}_{S_0}^\phi, R_0) \end{bmatrix} \tag{7.180}$$

and the solution vector is

$$I^J = \begin{bmatrix} I_{S_{01}}^{J,t} \\ I_{S_{01}}^{J,\phi} \end{bmatrix} \tag{7.181}$$

*Monostatic RCS versus Frequency*

We first compute the monostatic RCS of a conducting sphere of radius 0.5 meters for frequencies between 10 MHz and 2 GHz. For the Mie terms (7.169) and (7.170) we truncate the summation after 40 modes, and for the BOR bounding curve, we use 180 segments of equal length. In Figures 7.4a and 7.4b we compare the Mie series RCS to that of the EFIE and nMFIE, respectively. The comparisons are good at lower frequencies, though at higher ones discrepancies begin to appear, particularly in the nMFIE results. These correspond to internal resonances of the discretized BOR. In Figure 7.5a are plotted the results obtained using the CFIE, where the comparison is very good and the resonance problems have been eliminated, as expected.

*Range Profile*

We next compare the *range profiles* of the sphere using the backscattered field from the CFIE and Mie series. The range profile is a time domain representation of the electromagnetic signature and can be used to localize the scattering centers of a target in the down-range dimension. Given an array of complex-valued backscattered electric field samples $S$ over a range of evenly spaced frequencies $f_{min}$ to $f_{max}$, the range profile $R$ is

$$R = \mathrm{IFFT}(S) \tag{7.182}$$

which is known in radar signal processing as *pulse compression* or *matched filtering*. The range resolution of the time-domain waveform is

$$\Delta_R = \frac{c}{2B} \tag{7.183}$$

where the bandwidth $B$ is

$$B = f_{max} - f_{min} \qquad (7.184)$$

A window function is typically applied to the array before performing the FFT to reduce the sidelobes in the range domain. A commonly used window function is the *Hamming window*, which comprises the $N$ element array $W$ with elements given by

$$W_{k+1} = 0.54 - 0.46 \cos\left[\frac{2\pi k}{N-1}\right] \qquad k = 0, \ldots, N-1 \qquad (7.185)$$

The windowing operation decreases the sidelobes at the expense of decreased range resolution. Figure 7.5b depicts the monostatic range profiles of the CFIE and Mie series computed from the complex-valued scattered field computed from 1 to 2 GHz. A total of 32 data points were used, and the 1 GHz of bandwidth results in a range resolution of approximately 15 cm. A Hamming window is applied to the data, which decreases the range resolution by approximately 50 percent. Positive range in this figure represents the down-range direction. Immediately seen at -0.5 m is the bright specular reflection from the sphere. Slightly further down-range is the well-known creeping wave response, caused by electromagnetic energy that originates from the shadow boundary and travels around the shadowed side of the sphere, eventually coming back toward the transmitter. The results are virtually indistinguishable.

*Bistatic RCS*

The bistatic RCS of the sphere is another commonly used benchmark. We retain the same geometry and simulation parameters as before, and consider the bistatic RCS at 0.3, 1.0 and 2.0 GHz. The results from the CFIE are compared to the Mie series in Figures 7.6a–7.6f. The comparison is fairly good at 0.3 GHz, and at 1.0 and 2.0 GHz, the results are indistinguishable.

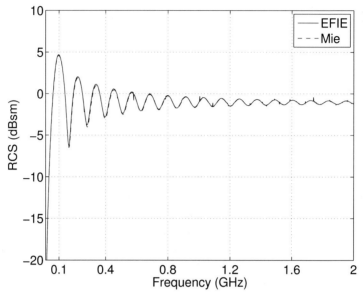

(a) RCS vs. Frequency: EFIE vs. Mie

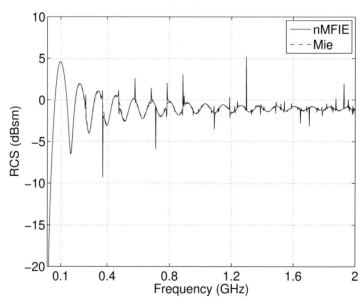

(b) RCS vs. Frequency: nMFIE vs. Mie

**FIGURE 7.4:** Conducting Sphere: EFIE and nMFIE vs. Mie

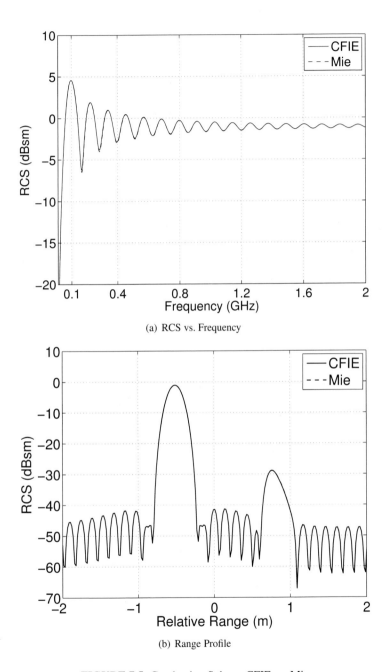

(a) RCS vs. Frequency

(b) Range Profile

**FIGURE 7.5:** Conducting Sphere: CFIE vs. Mie

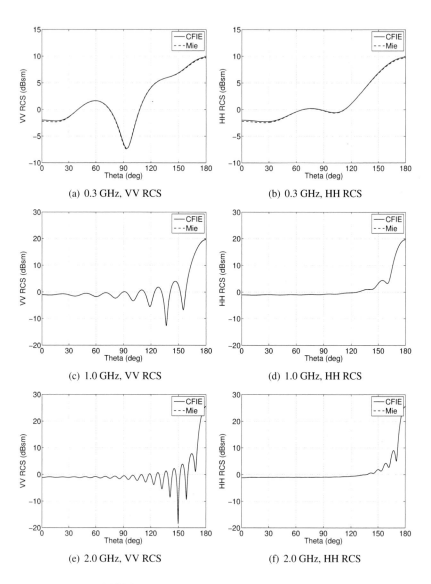

(a) 0.3 GHz, VV RCS

(b) 0.3 GHz, HH RCS

(c) 1.0 GHz, VV RCS

(d) 1.0 GHz, HH RCS

(e) 2.0 GHz, VV RCS

(f) 2.0 GHz, HH RCS

**FIGURE 7.6:** Conducting Sphere: Bistatic RCS

### 7.9.1.2 Stratified Sphere

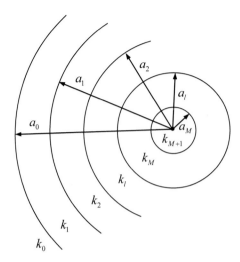

**FIGURE 7.7:** Geometry of an $M$-Layer Stratified Sphere

We next consider the radially stratified geometry of Figure 7.7, where the core may or may not be conducting. The Mie series coefficients [1] for this configuration are [6]

$$A_n = -j^n \frac{2n+1}{n(n+1)} \frac{k_0 a_0 j_n(k_0 a_0) + j Z_n(k_0 a_0) \left[k_0 a_0 j_n(k_0 a_0)\right]'}{k_0 a_0 h_n^{(2)}(k_0 a_0) + j Z_n(k_0 a_0) \left[k_0 a_0 h_n^{(2)}(k_0 a_0)\right]'} \tag{7.186}$$

and

$$B_n = j^{n+1} \frac{2n+1}{n(n+1)} \frac{k_0 a_0 j_n(k_0 a_0) + j Y_n(k_0 a_0) \left[k_0 a_0 j_n(k_0 a_0)\right]'}{k_0 a_0 h_n^{(2)}(k_0 a_0) + j Y_n(k_0 a_0) \left[k_0 a_0 h_n^{(2)}(k_0 a_0)\right]'} \tag{7.187}$$

where $a_0$ is the outermost radius of the sphere. $Z_n(k_0 a_0)$ and $Y_n(k_0 a_0)$ are called the modal surface impedances and admittance, respectively, which differ depending on the composition of the sphere. We consider the general case in this section, which can be applied to dielectric as well as coated spheres with many layers. Calculation of these coefficients is as follows: First, we define the surface impedance at the interface of radius $a_l$ as $Z_n^l$. The impedance $Z_n^M$ between the core and the first layer is then determined, and from $Z_n^M$, the impedance $Z_n^{M-1}$ at the second interface from the center is found. This process

---

[1]There is an error in the leading coefficient on the right-hand side of (3.4-2) in [6], which has been corrected here.

is repeated until $Z_n^0$ at the outer surface is obtained. Then,

$$Z_n(k_0 a_0) = \frac{j}{\eta_0} Z_n^0 \qquad (7.188)$$

and

$$Y_n(k_0 a_0) = j\eta_0 Y_n^0 \qquad (7.189)$$

where the admittance $Y_n^0$ is found from $Y_n^M$ using a similar procedure. For the iterative procedure, we first make the following definitions

$$P_n(x) = \frac{x j_n(x)}{[x j_n(x)]'} \qquad (7.190)$$

$$Q_n(x) = \frac{x h_n^{(2)}(x)}{[x h_n^{(2)}(x)]'} \qquad (7.191)$$

$$U_n^l = \frac{[k_{l+1} a_{l+1} j_n(k_{l+1} a_{l+1})]'}{[k_{l+1} a_l j_n(k_{l+1} a_l)]'} \frac{[k_{l+1} a_l h_n^{(2)}(k_{l+1} a_l)]'}{[k_{l+1} a_{l+1} h_n^{(2)}(k_{l+1} a_{l+1})]'} \qquad (7.192)$$

$$V_n^l = \frac{[k_{l+1} a_{l+1} j_n(k_{l+1} a_{l+1})]}{[k_{l+1} a_l j_n(k_{l+1} a_l)]} \frac{[k_{l+1} a_l h_n^{(2)}(k_{l+1} a_l)]}{[k_{l+1} a_{l+1} h_n^{(2)}(k_{l+1} a_{l+1})]} \qquad (7.193)$$

$$Z_n^M = \eta_{M+1} P_n(k_{M+1} a_M) \qquad Y_n^M = \frac{1}{\eta_{M+1}} P_n(k_{M+1} a_M) \qquad (7.194)$$

At the interface at radius $a_l$, the impedance and admittance are

$$Z_n^{(l)} = \eta_{l+1} P_n(k_{l+1} a_l) \left[ 1 - V_n^{(l)} \frac{1 - Z_n^{(l+1)}/[\eta_{l+1} P_n(k_{l+1} a_{l+1})]}{1 - Z_n^{(l+1)}/[\eta_{l+1} Q_n(k_{l+1} a_{l+1})]} \right] \cdot$$
$$\left[ 1 - U_n^{(l)} \frac{1 - \eta_{l+1} P_n(k_{l+1} a_{l+1})/Z_n^{(l+1)}}{1 - \eta_{l+1} Q_n(k_{l+1} a_{l+1})/Z_n^{(l+1)}} \right]^{-1} \qquad (7.195)$$

and

$$Y_n^{(l)} = \frac{P_n(k_{l+1} a_l)}{\eta_{l+1}} \left[ 1 - V_n^{(l)} \frac{1 - \eta_{l+1} Y_n^{(l+1)}/P_n(k_{l+1} a_{l+1})}{1 - \eta_{l+1} Y_n^{(l+1)}/Q_n(k_{l+1} a_{l+1})} \right] \cdot$$
$$\left[ 1 - U_n^{(l)} \frac{1 - P_n(k_{l+1} a_{l+1})/[\eta_{l+1} Y_n^{(l+1)}]}{1 - Q_n(k_{l+1} a_{l+1})/[\eta_{l+1} Y_n^{(l+1)}]} \right]^{-1} \qquad (7.196)$$

where $l$ begins at $M-1$ and ends at $l=0$. Note that if the core of the sphere is conducting, $Z_n^M \to 0$ and $Y_n^M \to \infty$, thus

$$Z_n^{(M-1)} = \eta_M P_n(k_M a_{M-1}) \frac{1 - V_n^{(M-1)}}{1 - U_n^{(M-1)} P_n(k_M a_M)/Q_n(k_M a_M)}$$
(7.197)

and

$$Y_n^{(M-1)} = \frac{P_n(k_M a_{M-1})}{\eta_M} \frac{1 - V_n^{(M-1)} Q_n(k_M a_M)/P_n(k_M a_M)}{1 - U_n^{(M-1)}}$$
(7.198)

We use these expressions to compute the field scattered by dielectric and coated conducting spheres in the following sections.

### 7.9.1.3 Dielectric Sphere

A dielectric sphere comprises a single bounding curve on the interface $S_{01}$ between the free space region $R_0$ and the interior dielectric region $R_1$. Applying the PMCHWT formulation on $S_{01}$ yields a linear system of the form

$$\begin{bmatrix} Z^{EJ} & Z^{EM} \\ Z^{HJ} & Z^{HM} \end{bmatrix} \begin{bmatrix} I^J \\ I^M \end{bmatrix} \begin{bmatrix} V^E \\ V^H \end{bmatrix}$$
(7.199)

where the sub-matrix blocks are

$$Z^{EJ} = \mathrm{L}\left(\mathbf{f}_{S_{01}}, \mathbf{f}_{S_{01}}, R_0\right) + \mathrm{L}\left(\mathbf{f}_{S_{10}}, \mathbf{f}_{S_{10}}, R_1\right)$$
(7.200)

$$Z^{EM} = \mathrm{K}\left(\mathbf{f}_{S_{01}}, \mathbf{g}_{S_{01}}, R_0\right) + \mathrm{K}\left(\mathbf{f}_{S_{10}}, \mathbf{g}_{S_{10}}, R_1\right)$$
(7.201)

$$Z^{HJ} = - \mathrm{K}\left(\mathbf{g}_{S_{01}}, \mathbf{f}_{S_{01}}, R_0\right) - \mathrm{K}\left(\mathbf{g}_{S_{10}}, \mathbf{f}_{S_{10}}, R_1\right)$$
(7.202)

$$Z^{HM} = \frac{\epsilon_0}{\mu_0} \mathrm{L}\left(\mathbf{g}_{S_{01}}, \mathbf{g}_{S_{01}}, R_0\right) + \frac{\epsilon_1}{\mu_1} \mathrm{L}\left(\mathbf{g}_{S_{10}}, \mathbf{g}_{S_{10}}, R_1\right)$$
(7.203)

Each block is understood to contain the $\hat{\mathbf{t}}$ and $\hat{\phi}$ components of all basis and testing functions. The right-hand side vectors are

$$V^E = V^E\left(\mathbf{f}_{S_{01}}, R_0\right)$$
(7.204)

and

$$V^H = V^H\left(\mathbf{g}_{S_{01}}, R_0\right)$$
(7.205)

and the solution vectors are

$$I^J = I_{S_{01}}^J \qquad I^M = I_{S_{01}}^M$$
(7.206)

*RCS versus Frequency*

We first consider the RCS of a dielectric sphere of radius 0.5 meters for frequencies between 10 MHz and 2 GHz. For the Mie terms (7.169) and (7.170), we truncate the summation after 40 modes, and the BOR bounding curve comprises 180 segments of equal length. In Figure 7.8a we compare the Mie series to the MOM for a lossless sphere ($\epsilon_r = 2.56$) in backscattering ($\theta^i = \theta^s = 0$). The comparison is very good. A similar comparison is made in Figure 7.8b for a lossy sphere ($\epsilon_r = 2.56 - j.102$), where the comparison is again very good. The forward scattering ($\theta^i = 0, \theta^s = \pi$) for each case is summarized in Figures 7.9a and 7.9b, respectively, where the comparison is again quite good.

*Bistatic RCS*

Using the same parameters as before, we next consider the bistatic RCS at 0.3, 1.0 and 2.0 GHz. The MOM results are compared to those of the Mie series for the lossless sphere in Figures 7.10a–7.10f. The comparison is very good. The results for the lossy sphere are compared in Figures 7.11a–7.11f, where the comparison is again very good.

### 7.9.1.4  Coated Sphere

A coated conducting sphere comprises two bounding curves: the interface $S_{01}$ between the free space region $R_0$ and dielectric region $R_1$, and the interface $S_{12}$ between the dielectric region $R_1$ and the core conducting region $R_2$. Applying the CFIE/PMCHWT formulation to this problem yields a linear system of the form

$$\begin{bmatrix} Z^{EJ} + Z^{nHJ} & Z^{EM} + Z^{nHM} \\ Z^{HJ} & Z^{HM} \end{bmatrix} \begin{bmatrix} I^J \\ I^M \end{bmatrix} \begin{bmatrix} V^E \\ V^H \end{bmatrix} \qquad (7.207)$$

where the sub-matrix blocks are

$$Z^{EJ} = \begin{bmatrix} L(\mathbf{f}_{S_{01}}, \mathbf{f}_{S_{01}}, R_0) + L(\mathbf{f}_{S_{10}}, \mathbf{f}_{S_{10}}, R_1) & L(\mathbf{f}_{S_{10}}, \mathbf{f}_{S_{12}}, R_1) \\ \alpha_c L(\mathbf{f}_{S_{12}}, \mathbf{f}_{S_{10}}, R_1) & \alpha_c L(\mathbf{f}_{S_{12}}, \mathbf{f}_{S_{12}}, R_1) \end{bmatrix}$$
$$(7.208)$$

$$Z^{nHJ} = \begin{bmatrix} 0 & 0 \\ (1 - \alpha_c)\eta_1 \mathrm{nK}(\mathbf{f}_{S_{12}}, \mathbf{f}_{S_{10}}, R_1) & (1 - \alpha_c)\eta_1 \mathrm{nK}(\mathbf{f}_{S_{12}}, \mathbf{f}_{S_{12}}, R_1) \end{bmatrix}$$
$$(7.209)$$

$$Z^{EM} = \begin{bmatrix} K(\mathbf{f}_{S_{01}}, \mathbf{g}_{S_{01}}, R_0) + K(\mathbf{f}_{S_{10}}, \mathbf{g}_{S_{10}}, R_1) \\ \alpha_c K(\mathbf{f}_{S_{12}}, \mathbf{g}_{S_{10}}, R_1) \end{bmatrix}$$
$$(7.210)$$

$$Z^{nHM} = \begin{bmatrix} 0 \\ (1 - \alpha_c)\eta_1 nL(\mathbf{f}_{S_{12}}, \mathbf{g}_{S_{10}}, R_1) \end{bmatrix} \qquad (7.211)$$

$$Z^{HJ} = \begin{bmatrix} -K(\mathbf{g}_{S_{01}}, \mathbf{f}_{S_{01}}, R_0) - K(\mathbf{g}_{S_{10}}, \mathbf{f}_{S_{10}}, R_1) & | & -K(\mathbf{g}_{S_{10}}, \mathbf{f}_{S_{12}}, R_1) \end{bmatrix} \qquad (7.212)$$

$$Z^{HM} = \begin{bmatrix} \frac{\epsilon_0}{\mu_0} L(\mathbf{g}_{S_{01}}, \mathbf{g}_{S_{01}}, R_0) + \frac{\epsilon_1}{\mu_1} L(\mathbf{g}_{S_{10}}, \mathbf{g}_{S_{10}}, R_1) \end{bmatrix} \qquad (7.213)$$

where each operator is understood to contain the $\hat{\mathbf{t}}$ and $\hat{\phi}$ components of all basis and testing functions. The right-hand side vectors are

$$V^E = \begin{bmatrix} V^E(\mathbf{f}_{S_{01}}, R_0) \\ 0 \end{bmatrix} \qquad (7.214)$$

and

$$V^H = \begin{bmatrix} V^H(\mathbf{g}_{S_{01}}, R_0) \\ 0 \end{bmatrix} \qquad (7.215)$$

and the solution vectors are

$$I^J = \begin{bmatrix} I^J_{S_{01}} \\ I^J_{S_{12}} \end{bmatrix} \qquad I^M = \begin{bmatrix} I^M_{S_{01}} \\ 0 \end{bmatrix} \qquad (7.216)$$

### RCS versus Frequency

We now consider a conducting sphere with a dielectric coating for frequencies between 10 MHz and 2 GHz. The conducting core of the sphere has a radius of 0.5 meters, and the dielectric coating has a thickness of 5 centimeters, yielding an outermost radius of 0.55 meters. The bounding curve of the dielectric interface is divided into 180 segments, and the bounding curve of the PEC core is divided into 360 segments. The extra density on the innermost interface is chosen due to the compression of the wavelength inside the dielectric coating. The remainder of the run parameters remain the same. In Figure 7.12a we compare the Mie series to the MOM in backscattering for a lossless coating ($\epsilon_r = 2.56$). The comparison is very good. In Figure 7.12b is a similar comparison for a lossy coating ($\epsilon_r = 2.56 - j.102$). The forward-scattering results for each case are compared in 7.13a and 7.13b, respectively, where the comparison is again quite good.

*Bistatic RCS*

Using the same parameters as before, we next consider the bistatic RCS at 0.3, 1.0, and 2.0 GHz. The MOM results are compared to those of the Mie series for the lossless coating in Figures 7.14a–7.14f. The comparison is very good. The results for the lossy coating are compared in Figures 7.15a–7.15f, where the comparison is again very good.

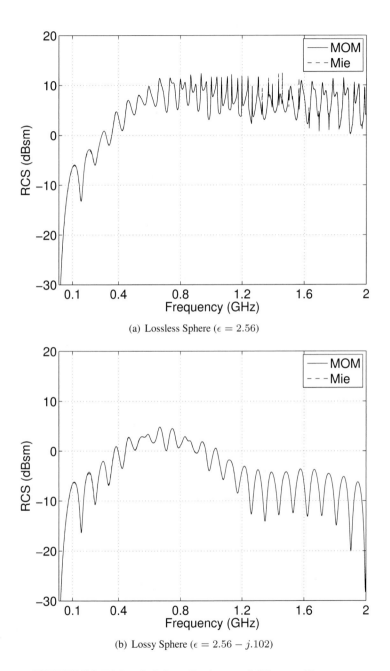

(a) Lossless Sphere ($\epsilon = 2.56$)

(b) Lossy Sphere ($\epsilon = 2.56 - j.102$)

**FIGURE 7.8:** Dielectric Sphere: Backscatter RCS versus Frequency

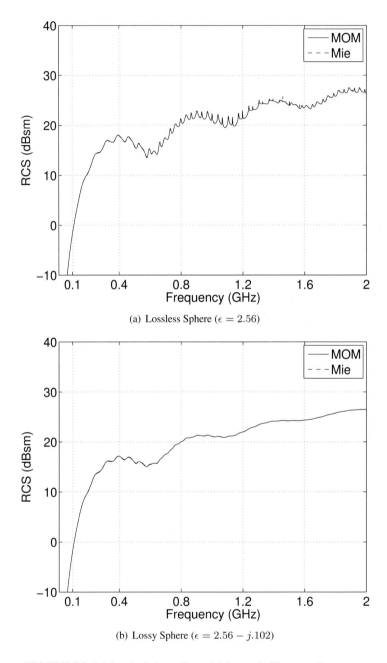

(a) Lossless Sphere ($\epsilon = 2.56$)

(b) Lossy Sphere ($\epsilon = 2.56 - j.102$)

**FIGURE 7.9:** Dielectric Sphere: Forward Scatter RCS versus Frequency

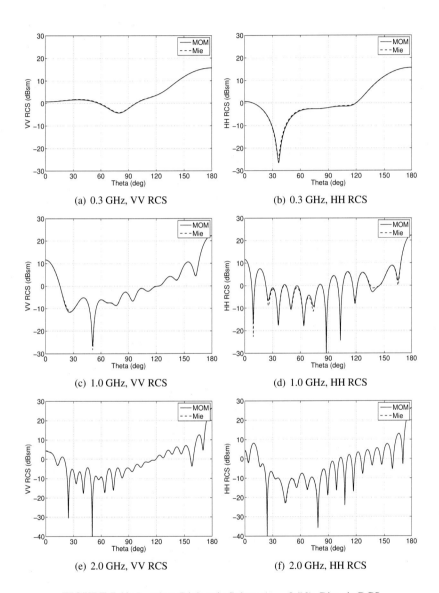

(a) 0.3 GHz, VV RCS

(b) 0.3 GHz, HH RCS

(c) 1.0 GHz, VV RCS

(d) 1.0 GHz, HH RCS

(e) 2.0 GHz, VV RCS

(f) 2.0 GHz, HH RCS

**FIGURE 7.10:** Lossless Dielectric Sphere ($\epsilon = 2.56$): Bistatic RCS

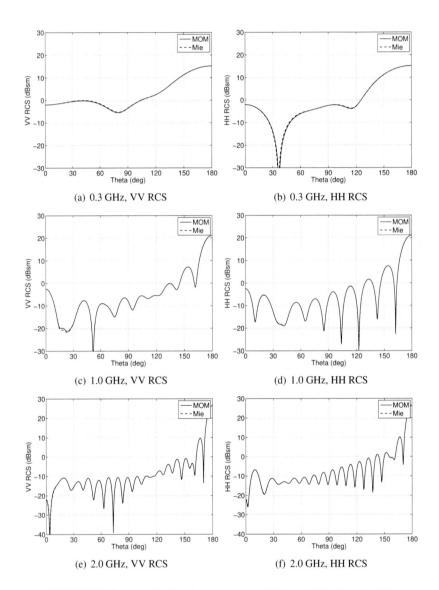

**FIGURE 7.11:** Lossy Dielectric Sphere ($\epsilon = 2.56 - j.102$): Bistatic RCS

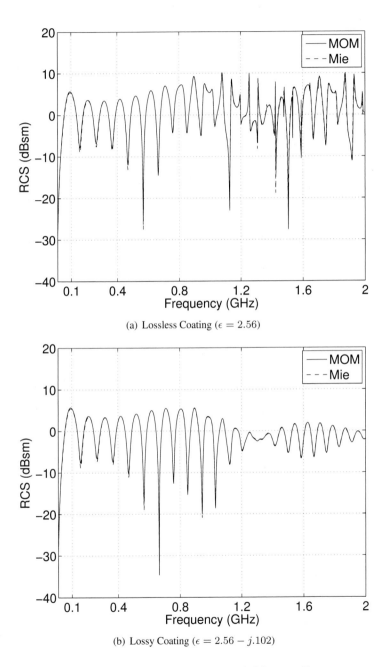

(a) Lossless Coating ($\epsilon = 2.56$)

(b) Lossy Coating ($\epsilon = 2.56 - j.102$)

**FIGURE 7.12:** Coated Sphere: Backscatter RCS versus Frequency

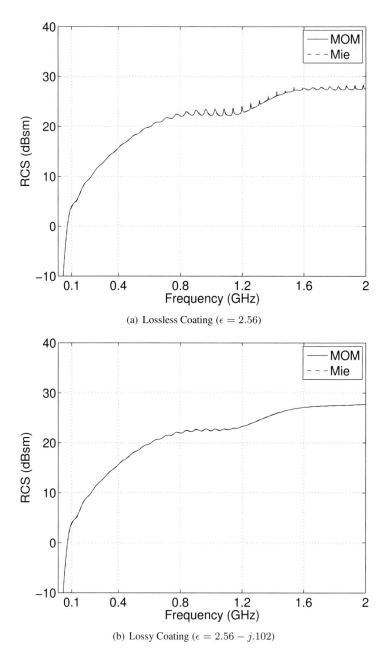

(a) Lossless Coating ($\epsilon = 2.56$)

(b) Lossy Coating ($\epsilon = 2.56 - j.102$)

**FIGURE 7.13:** Coated Sphere: Forward Scatter RCS versus Frequency

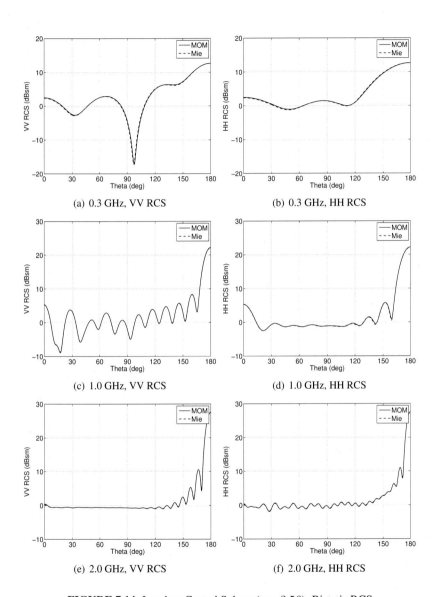

(a) 0.3 GHz, VV RCS

(b) 0.3 GHz, HH RCS

(c) 1.0 GHz, VV RCS

(d) 1.0 GHz, HH RCS

(e) 2.0 GHz, VV RCS

(f) 2.0 GHz, HH RCS

**FIGURE 7.14:** Lossless Coated Sphere ($\epsilon = 2.56$): Bistatic RCS

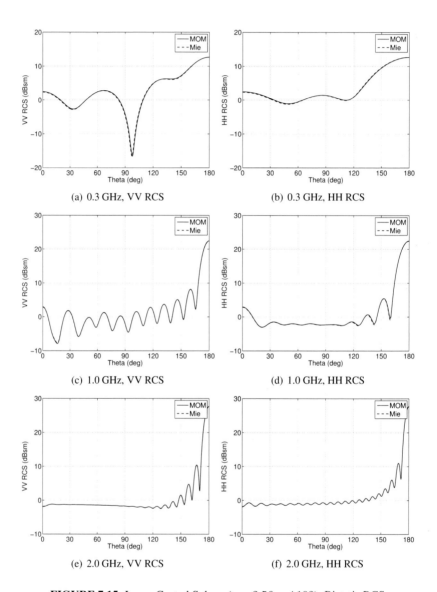

**FIGURE 7.15:** Lossy Coated Sphere ($\epsilon = 2.56 - j.102$): Bistatic RCS

## 7.9.2 EMCC Benchmark Targets

We will next use the BOR/MOM to compute the RCS of several benchmark radar targets described and measured by the EMCC in [7]. The objects considered are the ogive, double ogive, cone-sphere, and cone-sphere with gap. The first three are illustrated in Figures 7.16a–7.16c. The test articles used for the measurements were fabricated from aluminum using a numerically controlled mill. In the plots that follow, the measurement data have been shifted slightly in angle to better align them with the computed results. The comparisons are made using the measurements made at 9.0 GHz. Note that in the range, the objects were positioned on a rotating platform in the $xy$ plane. Therefore, computed $\hat{\theta}$-polarization will be mapped to measured horizontal polarization, and computed $\hat{\phi}$-polarization mapped to measured vertical polarization.

### 7.9.2.1 EMCC Ogive

The 10-inch EMCC ogive of Figure 7.16a can be expressed for $0 \leq \phi \leq 2\pi$ as

$$f(x) = \sqrt{1 - \left(\frac{x}{5}\right)^2 \sin^2(22.62°)} - \cos(22.62°) \qquad (7.217)$$

$$y(x) = \frac{f(x)\cos\phi}{1 - \cos(22.62°)} \qquad (7.218)$$

$$z(x) = \frac{f(x)\sin\phi}{1 - \cos(22.62°)} \qquad (7.219)$$

for $-5 \leq x \leq 5$ inches. The BOR bounding curve has 64 segments of equal length, and the MOM calculation is halted after 30 modes. The numerical results are compared to the EMCC measurements in Figures 7.17a and 7.17b for vertical and horizontal polarizations, respectively. The agreement is excellent, even at near-tip angles where the cross section is very low.

### 7.9.2.2 EMCC Double Ogive

The 7.5-inch EMCC double ogive of Figure 7.16b can be expressed for $0 \leq \phi \leq 2\pi$ as

$$g(x) = \sqrt{1 - \left(\frac{x}{2.5}\right)^2 \sin^2(46.6°)} - \cos(46.6°) \qquad (7.220)$$

$$y(x) = \frac{g(x)\cos\phi}{1 - \cos(46.4°)} \qquad (7.221)$$

$$z(x) = \frac{g(x)\sin\phi}{1 - \cos(46.4°)} \qquad (7.222)$$

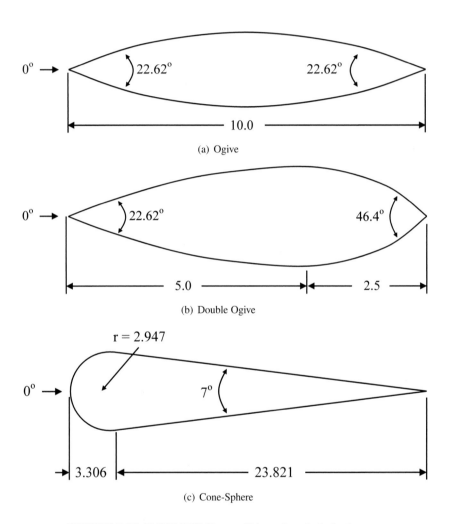

(a) Ogive

(b) Double Ogive

(c) Cone-Sphere

**FIGURE 7.16:** EMCC BOR Targets (Dimensions in Inches)

for $-2.5 \leq x \leq 0$ inches, and

$$f(x) = \sqrt{1 - \left(\frac{x}{5}\right)^2 \sin^2(22.62°)} - \cos(22.62°) \tag{7.223}$$

$$y(x) = \frac{f(x)\cos\phi}{1 - \cos(22.62°)} \tag{7.224}$$

$$z(x) = \frac{f(x)\sin\phi}{1 - \cos(22.62°)} \tag{7.225}$$

for $0 \leq x \leq 5$ inches. The BOR bounding curve has 64 segments of equal length, and the MOM calculation is halted after 30 modes. The numerical results are compared to the EMCC measurements in Figures 7.18a and 7.18b for vertical and horizontal polarizations, respectively. The agreement is again excellent.

### 7.9.2.3 EMCC Cone-Sphere

The 27-inch EMCC cone-sphere of Figure 7.16c can be expressed for $0 \leq \phi \leq 2\pi$ as

$$y(x) = 0.87145(x + 23.821)\cos\phi \tag{7.226}$$

$$z(x) = 0.87145(x + 23.821)\sin\phi \tag{7.227}$$

for $-23.821 \leq x \leq 0$ inches, and

$$y(x) = 2.947\sqrt{1 - \left(\frac{x - 0.359}{2.947}\right)^2}\cos\phi \tag{7.228}$$

$$z(x) = 2.947\sqrt{1 - \left(\frac{x - 0.359}{2.947}\right)^2}\sin\phi \tag{7.229}$$

for $0 \leq x \leq 3.306$ inches. The BOR bounding curve has 225 segments of equal length, and the MOM calculation is halted after 30 modes. The numerical results are compared to the EMCC measurements in Figures 7.19a and 7.19b for vertical and horizontal polarizations, respectively. The agreement is fairly good at all angles. The side specular in the measured data near 100 degrees is noticeably broader and of lesser amplitude than in the computed data. The reason for this differences is not known, and it was not mentioned in [7].

### 7.9.2.4  EMCC Cone-Sphere with Gap

The EMCC cone-sphere with gap is identical to the cone-sphere except that a square groove $1/4$ inch deep is cut into the target near the cone-sphere junction point. The gap can be expressed for $0 \leq \phi \leq 2\pi$ as

$$y(x) = 2.697 \cos \phi \qquad (7.230)$$

$$z(x) = 2.697 \sin \phi \qquad (7.231)$$

for $0 \leq x \leq 0.25$ inches. At 9.0 GHz, the dimensions of the gap are approximately $\lambda/5$. To generate the bounding curve, an additional 25 segments were added to the cone-sphere model in the vicinity of the gap, yielding a total of 250 boundary segments. The numerical results are compared to the EMCC measurements in Figures 7.20a and 7.20b for vertical and horizontal polarizations, respectively. The agreement is again excellent. The presence of the gap has a significant effect on the backscattered field, especially for horizontal polarization.

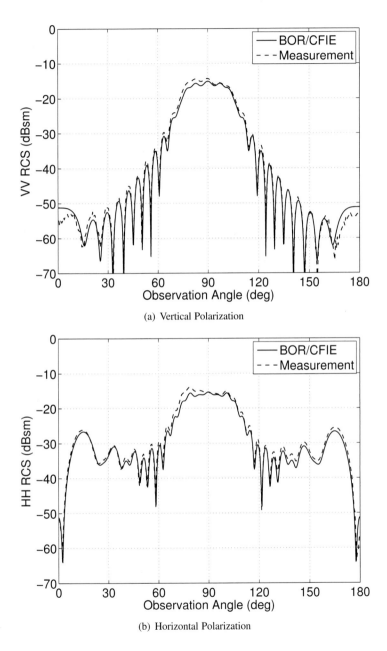

(a) Vertical Polarization

(b) Horizontal Polarization

**FIGURE 7.17:** EMCC Ogive: BOR versus Measurement at 9.0 GHz

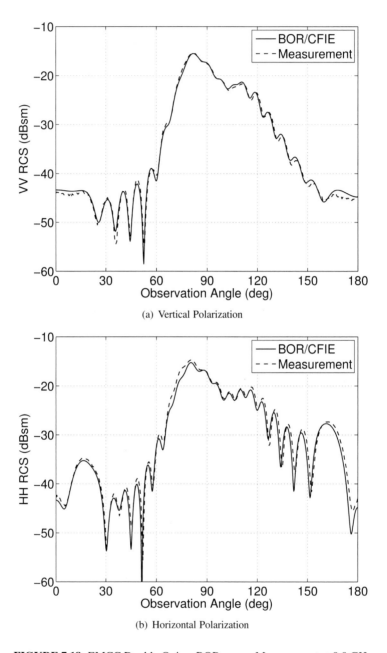

(a) Vertical Polarization

(b) Horizontal Polarization

**FIGURE 7.18:** EMCC Double Ogive: BOR versus Measurement at 9.0 GHz

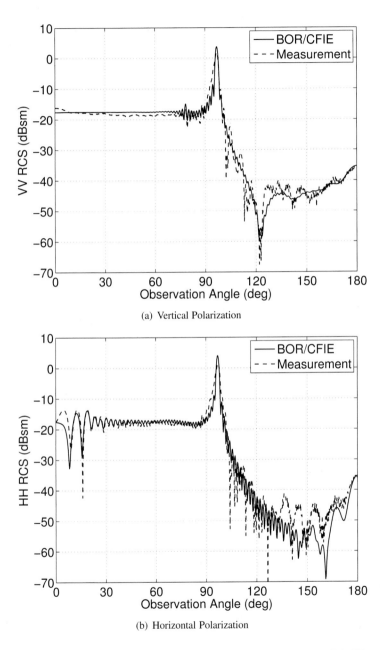

(a) Vertical Polarization

(b) Horizontal Polarization

**FIGURE 7.19:** EMCC Cone-Sphere: BOR versus Measurement at 9.0 GHz

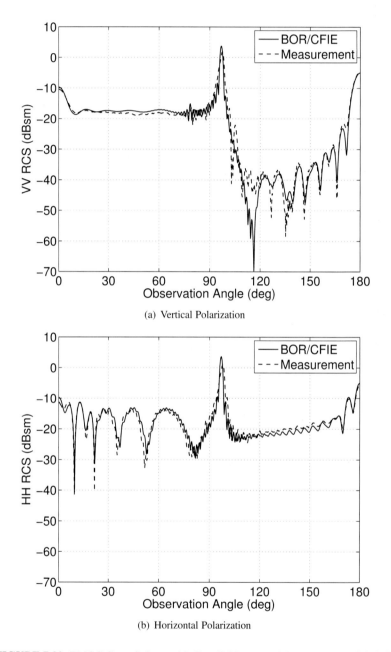

(a) Vertical Polarization

(b) Horizontal Polarization

**FIGURE 7.20:** EMCC Cone-Sphere with Gap: BOR versus Measurement at 9.0 GHz

### 7.9.3 Biconic Reentry Vehicle

We next consider a biconic reentry vehicle (RV) whose dimensions are outlined in Figure 7.21a. The RV has a large nose and recessed rear cavity that should be significant sources of backscatter. The three-dimensional shape of the RV is illustrated at forward and rear aspects in Figures 7.21b and 7.21c, respectively. The length of the BOR bounding curve for this object is approximately 3.2 meters.

In the study of a target's scattering behavior, it is often useful to consider its range profile over a wide range of monostatic angles. This allows for the isolation of target scattering features. Therefore, we will compute the complex-

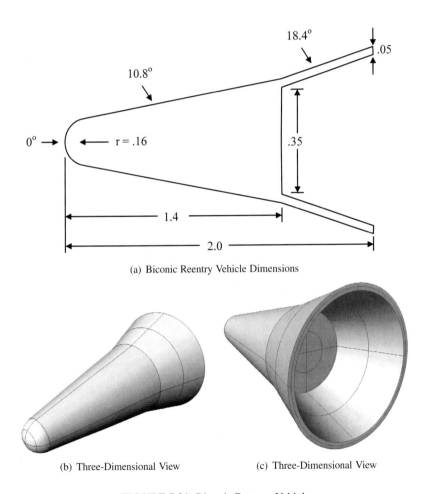

(a) Biconic Reentry Vehicle Dimensions

(b) Three-Dimensional View    (c) Three-Dimensional View

**FIGURE 7.21:** Biconic Reentry Vehicle

valued far electric field of the RV for frequencies ranging from 5 to 6 GHz (C-Band), for observation angles between 0 to 180 degrees. At 6 GHz, the length of the bounding curve of the RV is approximately 64 wavelengths. Our numerical model uses 577 boundary segments, which is approximately 10 segments per wavelength.

For visualization purposes, we compute the range profiles across all incident angles and stack them together to create a two-dimensional figure called a *range-aspect-intensity* (RAI) plot. RAI plots for vertical and horizontal polarizations are shown in Figures 7.22a and 7.22b, respectively. The choice of bandwidth results in a down-range resolution of approximately 15 centimeters, and a Hamming window is used to reduce the sidelobes. Clearly visible at forward-scattering angles are the large scattering responses from the nose as well as the base edge. At rear aspects, the multiple bounce returns from inside the cavity clearly dominate the response, giving rise to significant returns from down-range locations.

We can further isolate the scattering centers of the target by generating a two-dimensional *range-Doppler* image. Such an image is generated by computing the FFT of each range line over a small range of angles. If we assume that the Doppler frequency of each bin remains constant across the integration interval, then we can achieve a cross-range resolution given by [8]

$$\Delta_{cr} = \frac{\lambda_o}{2\Delta_\theta} \tag{7.232}$$

where $\lambda_o$ is the wavelength at the center of the waveband, and $\Delta_\theta$ is the angular extent of the integration interval. If we integrate over approximately 6 degrees, the cross-range resolution is approximately 0.26 meters at 5.5 GHz. Range-Doppler images centered at observation angles of 3 and 138 degrees are shown in Figures 7.23a and 7.23b, respectively, for vertical polarization. A Hamming window is used in the cross-range dimension to reduce sidelobes. Superimposed onto each image is the two-dimensional outline of the RV, allowing us to register the scattering features. At the forward aspect, the scattering centers at the nose and stationary phase points of the base edge (sometimes referred to as slipping scatterers) are clearly visible. The joint between the two frustums is also visible, as well as the faint double-diffraction response located slightly past the target. At the rear aspect, the delayed returns dominate the image. These scattering centers appear to originate from off-body locations, obscuring the true shape of the target. Such features are often detrimental to automated target recognition (ATR) algorithms that use the object's shape as a discrimination feature. Real-world reentry vehicles also have hoses, propellant tanks, wiring, and assorted electronics inside the rear cavity that give rise to additional multiple-bounce behavior. As a result, radar observations of such a target could result in an apparent length several times that of the true object.

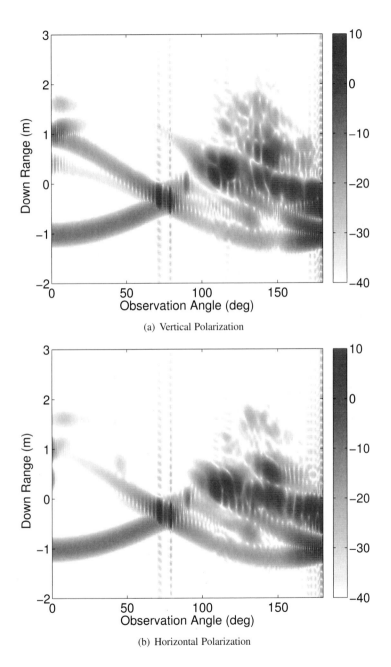

(a) Vertical Polarization

(b) Horizontal Polarization

**FIGURE 7.22:** Biconic Reentry Vehicle: C-Band Range Profiles (dBsm)

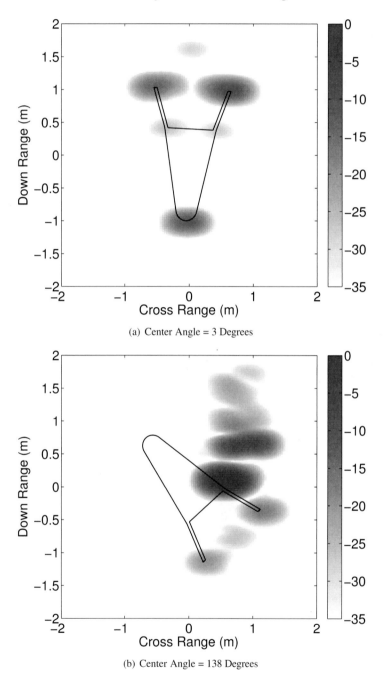

(a) Center Angle = 3 Degrees

(b) Center Angle = 138 Degrees

**FIGURE 7.23:** Biconic Reentry Vehicle: C-Band Range-Doppler Images (dBsm)

## 7.10   Treatment of Junctions

So far, our treatment has considered interfaces between no more than two regions. Points where three or more regions meet are called *junctions*, and the boundary conditions and orientation of vector basis functions must be considered more carefully. In this section, we will consider the treatment, placement, and orientation of the longitudinal and azimuthal vectors at several types of junctions in BOR problems. We refer the reader to Section 8.6 for an in-depth discussion of the treatment of the boundary conditions and enforcement of integral equations at generalized junctions.

### 7.10.1   Orientation of Basis Functions

In Section 7.7, we examined the assignment and orientation of basis functions at an interface between two regions. We will now generalize this further by considering junctions between $N$ regions, where one or more of these regions may conducting. In a BOR geometry, junctions occur where an endpoint is shared by three or more boundary segments.

#### 7.10.1.1   Longitudinal Basis Vectors

According to Section 7.7, the longitudinally oriented basis functions assigned to a point on the interface between two dielectric regions must have opposing directions on each side, as shown in Figure 7.24a. Consider now the interface between three or more dielectric regions, as shown in Figure 7.24b. As Kirchoff's laws must be satisfied in each region, the basis vectors must flow past the common point in a clockwise (or counterclockwise) manner, as shown. Furthermore, the boundary conditions stipulate that these basis functions will

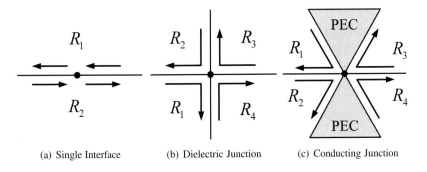

(a) Single Interface          (b) Dielectric Junction          (c) Conducting Junction

**FIGURE 7.24:** Orientation of Longitudinal Basis Vectors

have the same coefficient. If we now make one or more of the regions conducting, as in Figure 7.24c, this assignment will still be valid for dielectric interfaces that lie between conductors. Thus, the oriented basis functions in $R_1$ and $R_2$ will be combined into one unknown, as will the basis functions in $R_3$ and $R_4$. This concept will be further generalized in Section 8.6.

### 7.10.1.2  Azimuthal Basis Vectors

At the intersection between two regions, as shown in Figure 7.25a, the azimuthally oriented basis functions assigned to the common point have a $+\hat{\phi}$ orientation in $R_1$, and a $-\hat{\phi}$ orientation in $R_2$. Under this arrangement, the boundary conditions at the interface are satisfied, and the basis functions on each side of the interface can be combined into a single unknown. If we now increase the number of dielectric regions meeting at this point to any odd number greater than or equal to three, this arrangement is no longer possible. This is illustrated in Figure 7.25b for a junction between three regions, where the boundary condition is not satisfied at the interface between $R_1$ and $R_3$. This implies that at the intersection of an odd number of dielectric regions, we must use half-triangles on each segment and assign separate coefficients to each. Increasing the number of regions to four, as in Figure 7.25c, we see that the boundary conditions are again satisfied, and we can again use full triangle functions and combine the unknowns into a single unknown. This approach is similar to that in [4], except that the approach taken there suggests using half-triangles for all azimuthal *and* longitudinal functions at a junction, regardless of the number of regions meeting there.

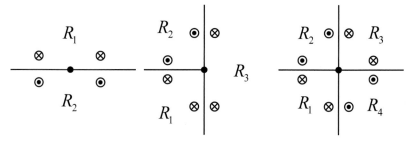

(a) Two Regions (Good)    (b) Three Regions (Not Good)    (c) Four Regions (Good)

**FIGURE 7.25:** Orientation of Azimuthal Basis Vectors

## 7.10.2 Examples with Junctions

We now consider several examples involving dielectric and dielectric/conducting objects with junctions.

### 7.10.2.1 Dielectric Sphere with Septum

We will first consider the dielectric sphere depicted in Figure 7.26a, where the sphere has been split into two hemispheres of equal size. There are now three dielectric regions: the free space region $R_0$ and the two dielectric halves $R_1$ and $R_2$. This object has a single junction at the intersection of the three regions.

*Results*

A good way to test the effectiveness of the junction treatment is to treat $R_1$ and $R_2$ as independent regions having identical dielectric properties. This effectively models a homogeneous dielectric sphere, which we can compare against the Mie series solution. The outermost radius of the sphere is $a_0 = 1\lambda_0$. The dielectric parameters of both regions is $\epsilon = 2.56$, and for the Mie solution, $N = 40$. The bounding curves are subdivided into 10 segments per wavelength in each region. The MOM results are compared to the Mie series in Figures 7.27a and 7.27b for vertical and horizontal polarization, respectively, where the comparison is very good.

### 7.10.2.2 Coated Sphere with Septum

We next consider a conducting sphere with a dielectric coating, depicted in Figure 7.26b. The coating is again separated into two partial hemispheres of equal size. There are three dielectric regions: the free space region $R_0$, and the two hemispherical dielectric regions $R_1$ and $R_2$. The conducting core comprises an additional region, $R_3$, which is not labeled. This object has two junc-

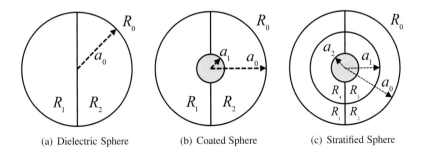

(a) Dielectric Sphere      (b) Coated Sphere      (c) Stratified Sphere

**FIGURE 7.26:** Spheres with a Hemispherical Division

tions: A junction at the intersection of $R_0$, $R_1$, and $R_2$, and another at the intersection of $R_1$, $R_2$, and $R_3$.

*Results*

$R_1$ and $R_2$ are again treated as independent regions with identical dielectric properties. The radius of the conducting core is $a_1 = 0.25\lambda_0$, and the radius of the coating is $a_0 = 1\lambda_0$. The dielectric parameter of both regions is $\epsilon = 5$, and for the Mie solution, $N = 40$. The bounding curves are subdivided into 10 segments per wavelength in each region. The MOM results are compared to the Mie series in Figures 7.27c and 7.27d for vertical and horizontal polarization, respectively, where the comparison is again very good.

### 7.10.2.3 Stratified Sphere with Septum

We next consider a conducting sphere with a stratified dielectric coating, depicted in Figure 7.26c. The stratified coatings are split into two partial hemispheres of equal size. There are five dielectric regions: the free space region $R_0$, the hemispherical regions of the outermost coating ($R_1$ and $R_2$) and the hemispherical regions of the innermost coating ($R_3$ and $R_4$). The conducting core comprises an additional region, $R_5$, which is not labeled. This object has three junctions: A junction at the intersection of $R_0$, $R_1$, and $R_2$, one at the intersection of $R_1$, $R_2$, $R_3$, and $R_4$, and another at the intersection of $R_3$, $R_4$, and $R_5$.

*Results*

For this example, regions $R_1$ and $R_2$ are assigned a dielectric parameter of $\epsilon = 5.0$, and regions $R_3$ and $R_4$ are assigned $\epsilon = 7.0$. The radius of the conducting core is $a_2 = 0.125\lambda_0$, the radius of the innermost coating is $a_1 = 0.6\lambda_0$, and the radius of the outermost coating is $a_0 = 1\lambda_0$. The bounding curves are subdivided into 10 segments per wavelength in each region, and for the Mie solution, $N = 40$. The MOM results are compared to the Mie series in Figures 7.27e and 7.27f for vertical and horizontal polarization, respectively, where the comparison is again very good.

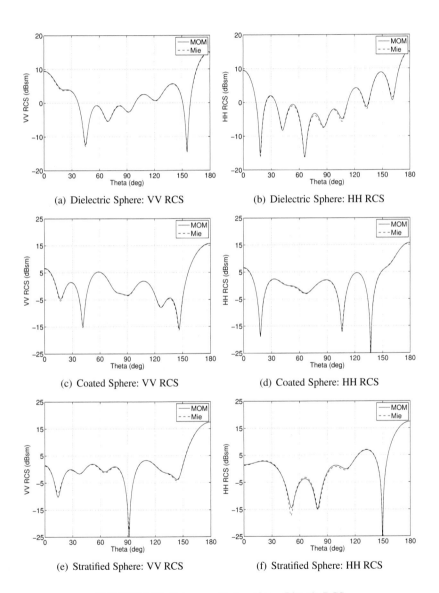

(a) Dielectric Sphere: VV RCS

(b) Dielectric Sphere: HH RCS

(c) Coated Sphere: VV RCS

(d) Coated Sphere: HH RCS

(e) Stratified Sphere: VV RCS

(f) Stratified Sphere: HH RCS

**FIGURE 7.27:** Spheres with Junctions: Bistatic RCS

### 7.10.2.4   Monoconic Reentry Vehicle with Dielectric Nose

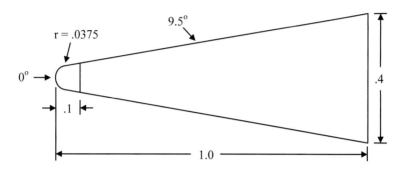

**FIGURE 7.28:** Monoconic Reentry Vehicle Dimensions (meters)

Finally, we examine a monoconic reentry vehicle (RV) whose dimensions are outlined in Figure 7.28. The RV has a length of 1 meter and a base diameter of 0.4 meters, with a cone angle of 9.5 degrees. The dielectric section is 10 centimeters in length, and the nose tip has a radius of curvature of 3.75 centimeters. This problem has three regions: the free space region $R_0$, the dielectric (nose tip) region $R_1$, and the conducting (frustum) region $R_2$. As reentry vehicles typically have nose tips made of an ablative material, this is a somewhat realistic example. The BOR boundaries are constructed as follows: 223 segments for conducting frustum ($S_{02}$), 25 segments for the nose ($S_{01}$), and 19 segments for the interface (septum) $S_{12}$ between the frustum and nose.

### Results

We compute the monostatic RCS of the RV at 5.0 GHz, and consider the nose tip with lossless ($\epsilon = 5$) and lossy ($\epsilon = 5.0 - j0.25$) dielectric parameters. The RV with a lossless nose tip is compared to a perfectly conducting RV of the same dimensions in Figures 7.29a and 7.29b for vertical and horizontal polarizations, respectively. The RV with the dielectric nose has a significantly higher RCS at many aspect angles, particularly below 30 degrees. This is due to the many multiple-bounces inside the nose. A similar comparison is done for the RV with lossy nose in Figures 7.29a and 7.29b. The RCS has been significantly reduced from the lossless case, although it is still significantly higher at nose-forward aspect angles, particularly in horizontal polarization.

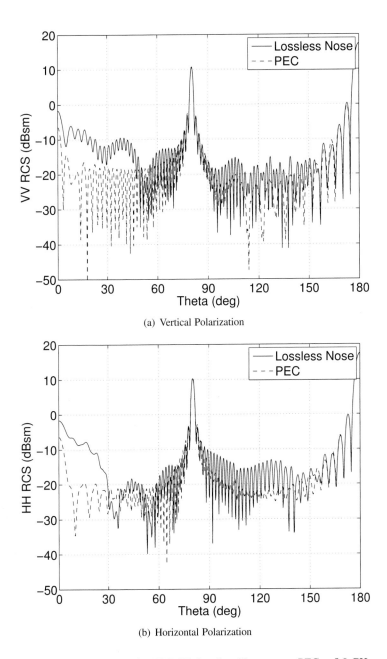

(a) Vertical Polarization

(b) Horizontal Polarization

**FIGURE 7.29:** Monoconic RV RCS: Lossless Nose versus PEC at 5.0 GHz

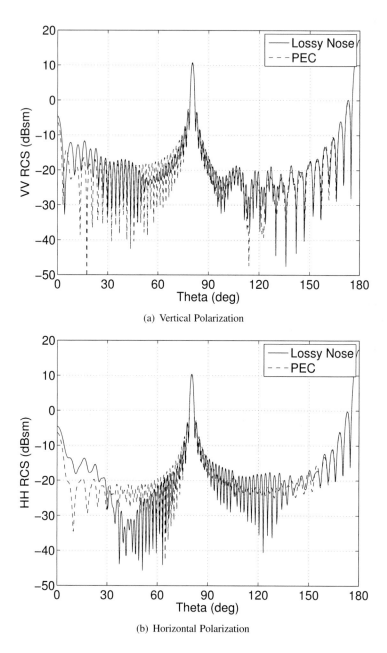

(a) Vertical Polarization

(b) Horizontal Polarization

**FIGURE 7.30:** Monoconic RV RCS: Lossy Nose versus PEC at 5.0 GHz

# References

[1] R. Mautz and R. Harrington, "H-field, E-field and combined solutions for bodies of revolution," Tech. Rep. Report RADC-TR-77-109, Rome Air Development Center, Griffiss AFB, N.Y., March 1977.

[2] L. N. Medgyesi-Mitschang and J. M. Putnam, "Electromagnetic scattering from axially inhomogeneous bodies of revolution," *IEEE Trans. Antennas Propagat.*, vol. 32, pp. 796–806, March 1984.

[3] M. Abramowitz and I. Stegun, *Handbook of Mathematical Functions*. National Bureau of Standards, 1966.

[4] L. N. Medgyesi-Mitschang and J. M. Putnam, "Combined field integral equation formulation for inhomogeneous two- and three-dimensional bodies: The junction problem," *IEEE Trans. Antennas Propagat.*, vol. 39, pp. 667–672, May 1991.

[5] K. Faison, "The Electromagnetics Code Consortium," *IEEE Antennas Propagat. Magazine*, vol. 30, pp. 19–23, February 1990.

[6] G. T. Ruck, ed., *Radar Cross Section Handbook*. Plenum Press, 1970.

[7] A. C. Woo, H. T. G. Wang, M. J. Schuh, and M. L. Sanders, "Benchmark radar targets for the validation of computational electromagnetics programs," *IEEE Antennas Propagat. Magazine*, vol. 35, pp. 84–89, February 1993.

[8] A. Ausherman, A. Kozma, J. Waker, H. M. Jones, and E. C. Poggio, "Developments in radar imaging," *IEEE Trans. Aerospace Electron. Syst.*, vol. 20, pp. 363–400, July 1984.

# Chapter 8

## *Three-Dimensional Problems*

In this chapter we will use the Method of Moments to solve surface integral equation problems involving three-dimensional objects of arbitrary shape. This will allow us to treat many problems of practical interest such as those found in electromagnetic interference (EMI), electronics packaging, scattering, radar cross section, and antenna design. This area has received significant attention in the literature in the last thirty years, and many different methods of treating 3D surfaces of general shape have been considered. We will consider one of the most commonly used approaches, where the surfaces are described using flat, planar triangles, and the currents expanded using the well-known Rao-Wilton-Glisson (RWG) basis functions. In Chapter 7, we treated several dielectric and dielectric/conducting junctions of limited type. In this chapter, we will use the method outlined in [1] to treat junctions with a much more general configuration.

Until recently, most 3D problems were limited to those of small electrical size, due to the limitations in processor time and system memory. Solving larger problems typically required access to supercomputer-class equipment, which was not available to most people. With the advances in processor speed and system memory over the last fifteen years, larger problems can now be solved on most desktop computers. This has opened up areas of analysis that were not attempted or considered before. The development of the Fast Multipole Method (FMM) has even further increased the tractability of larger problems, which we will discuss in detail in Chapter 9.

The approach in this chapter comprises three general steps: We will first consider the discretized EFIE, MFIE, and nMFIE of Section 3.6.2.2 in a region-wise context, and derive expressions for computing the required matrix elements. Next, we enforce the boundary conditions along each interface and at junctions, such that all the linearly dependent unknowns are combined. The overdetermined matrix system will then be reduced by applying the integral equation formulations described in Section (3.6.3.1). We will then consider several examples.

## 8.1 Modeling of Three-Dimensional Surfaces

Let us consider the digital construction and representation of three-dimensional objects. First, a description of the object to be modeled is required. If the object is conceptual, line drawings or engineering blueprints are drafted, or if the actual object is available, drawings are generated through physical measurement. Next, a model of the object is constructed using a computer-aided design (CAD) software. It is common for this step to often overlap with the first, as the conceptual design and surface modeling of many objects is now done completely inside the CAD program, with no pen or paper required. The result of this step is a curved surface model, which is highly accurate. Most modern modeling programs such as Rhinoceros 3D [2] represent free-form objects using a mathematically exact description such as non-uniform rational B-splines (NURBS) [3]. This approach preserves necessary information such as the surface normal and radius of curvature at every point. NURBS is a part of many industry-wide standard CAD file formats and modeling engines such as IGES, STEP, ACIS, and Parasolid. Once the curved surface model is developed, it must be reduced to a set of geometric primitives that can be processed. One of the most commonly used primitives are planar triangles (or *facets*), which are flexible and reasonably good at following the curvature of most realistic shapes [4]. Furthermore, numerical quadratures have been developed for the triangle (see Section 10.2), and analytic solutions to potential integrals exist for triangular subdomains [5, 6, 7, 8]. The model is then discretized or *tesselated* into a set of triangle meshes conforming to the surfaces of interest, and most CAD programs can generate meshes of exceptionally high quality. Due to their flexibility, complex shapes such as ground and air vehicles can be represented with great precision by facet models. Figures 8.1a, 8.1b, and 8.1c depict a sphere, a missile, and a main battle tank represented entirely by triangular facets.

### 8.1.1 Facet File

A common way to describe a triangular surface mesh is through the finite element connectivity file (or *facet file*) [9]. This file comprises a *node list* which contains the $(x, y, z)$ components of all nodes in the mesh, and a *facet list* which contains three integers for each triangle. These integers define which nodes comprise the triangle corners. Additional information regarding each triangle, such as its interface assignment, can be added to the facet list. An example is provided in Figure 8.2, which comprises a square plate in the $xy$ plane constructed from 9 nodes and 8 triangles.

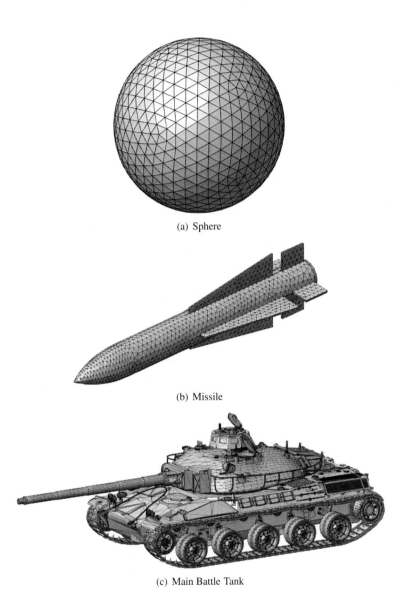

(a) Sphere

(b) Missile

(c) Main Battle Tank

**FIGURE 8.1:** Example 3D Facet Models

```
9
-1.0  1.0  0.0
 0.0  1.0  0.0
 1.0  1.0  0.0
-1.0  0.0  0.0
 0.0  0.0  0.0
 1.0  0.0  0.0
-1.0 -1.0  0.0
 0.0 -1.0  0.0
 1.0 -1.0  0.0
8
4 2 1
4 5 2
5 3 2
5 6 3
7 5 4
7 8 5
8 6 5
8 9 6
```

**FIGURE 8.2:** Example Facet File

## 8.1.2   Edge-Finding Algorithm

For a given surface mesh, we need an algorithm that will identify and register all unique triangle edges. To illustrate an edge-finding algorithm, we will use the facet file in Figure 8.2, which comprises a flat plate in the $xy$ plane. The numbering of the nodes and facets in this model is illustrated in Figures 8.3a, and 8.3b, respectively. Our first task is to identify the connections between each node and each facet that references it. This will allow us to determine which facets share an edge. To do this, we create two lists for each node, a *node connectivity list* and a *facet connectivity list*. The first list contains all unique connections between a node and other nodes, and the second list contains the facets having that node as a vertex. We next loop over individual triangles and identify the nodes that make up the triangle, and sort them into ascending order. We then add the index of the current facet to the facet connectivity list of each node. For the node of lowest index, we add to its list the two higher node indexes if they are not already in the list. For the second node, the index of the third node is added to its list if not already present, and nothing is done for the third node. After this operation is completed, the node and facet connectivity lists are as shown in Table 8.1. Note that the node connectivity list contains no redundant links; this results in an empty list for node 9 since it is already referenced in the list for nodes 6 and 8.

At this point, the connectivity lists contain all the information needed to

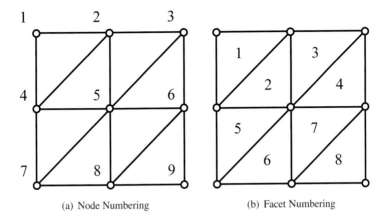

FIGURE 8.3: Simple Flat Plate Geometry

TABLE 8.1: Node and Facet Connectivity Lists

| Node # | Connected Nodes | Connected Facets |
|--------|-----------------|------------------|
| 1 | 2 4 | 1 |
| 2 | 4 5 3 | 1 2 3 |
| 3 | 5 6 | 3 4 |
| 4 | 5 7 | 1 2 5 |
| 5 | 6 7 8 | 2 3 4 5 6 7 |
| 6 | 8 9 | 4 7 8 |
| 7 | 8 | 5 6 |
| 8 | 9 | 6 7 8 |
| 9 | - | 8 |

find all unique edges in the model. The remaining task is to go through the connectivity list for each node and create an edge for each entry. The facets common to the endpoints of each edge are recorded. If an edge belongs to only one triangle, it is called a *boundary edge*. Edges shared by two triangles are called *regular* edges (or just edges), and those shared by three or more facets are *junction* edges. Additional information regarding each edge can be stored by the programmer at this step, such as its length, the facets that share it, its endpoints, and the node that is opposite the edge on each triangle sharing it.

### 8.1.2.1 Shared Nodes

The edge-finding algorithm requires that facets that are connected to each other use common node indexes, otherwise there is no logical connectivity and the algorithm will not find any edges. Some CAD programs generate redundant nodes for each triangle, and some triangle file formats such as Stereolithography (.stl) store the nodes of each triangle separately. The CAD modeler should ensure that these redundant nodes are collapsed into single nodes. Some CAD programs have a function that will *join* or *weld* meshes to remove coincident nodes before finalizing the geometry, and some advanced moment method codes will do this automatically when reading the facet file.

## 8.2   Expansion of Surface Currents

On each dielectric interface, the electric and magnetic currents are expanded using a sum of weighted basis functions as in (3.182) and (3.183). In the expansion, we use the Rao-Wilton-Glisson (RWG) triangular basis function [10], which has become the most commonly used basis function in 3D problems since its introduction. An RWG function is defined as

$$\mathbf{f}_n(\mathbf{r}) = \frac{L_n}{2A_n^+} \boldsymbol{\rho}_n^+(\mathbf{r}) \qquad \mathbf{r} \text{ in } T_n^+ \tag{8.1}$$

$$\mathbf{f}_n(\mathbf{r}) = \frac{L_n}{2A_n^-} \boldsymbol{\rho}_n^-(\mathbf{r}) \qquad \mathbf{r} \text{ in } T_n^- \tag{8.2}$$

$$\mathbf{f}_n(\mathbf{r}) = 0 \qquad \text{otherwise} \tag{8.3}$$

where $T_n^+$ and $T_n^-$ are a pair of triangles that share edge $n$, and $L_n$ is the length of edge $n$. On $T_n^+$, the vector $\boldsymbol{\rho}_n^+(\mathbf{r})$ points *toward* the vertex $\mathbf{v}^+$ opposite the edge, and is

$$\boldsymbol{\rho}_n^+(\mathbf{r}) = \mathbf{v}^+ - \mathbf{r} \quad \mathbf{r} \text{ in } T_n^+ \tag{8.4}$$

and on the $T_n^-$, $\boldsymbol{\rho}_n^-(\mathbf{r})$ points *away* from the opposite vertex $\mathbf{v}^-$, and is

$$\boldsymbol{\rho}_n^-(\mathbf{r}) = \mathbf{r} - \mathbf{v}^- \quad \mathbf{r} \text{ in } T_n^- \tag{8.5}$$

These basis functions are assigned only to *interior* edges shared by two or more adjacent triangles (we will discuss how RWG functions are assigned at junctions shortly). The RWG function has no component normal to any edge other than the one it is assigned to, and the component of the function normal to that edge is unity. Kirchoff's law is satisfied along each edge, and as boundary edges have no current flowing across them, they are not used. For the Galerkin testing procedure, we will use as testing functions the same RWG functions used to expand the current.

RWG functions are often referred to as *first-order* or *linear* basis functions. Functions of higher order are also possible; interested readers can consult references such as [11] for more information.

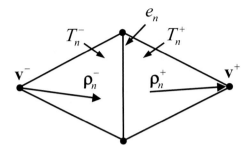

**FIGURE 8.4:** RWG Basis Function

## 8.2.1 Divergence of the RWG Function

To compute the divergence of the RWG function, let us consider $\mathbf{f}_n(\mathbf{r})$ on $T_n^-$. If $T_n^-$ is in a cylindrical coordinate system with $\mathbf{v}^-$ at the origin, $\mathbf{f}_n(\mathbf{r})$ can be written as a vector with a radial component only, yielding

$$\mathbf{f}_n(\mathbf{r}) = \frac{L_n}{2A_n} r \, \hat{\mathbf{r}} \tag{8.6}$$

Computing the divergence in cylindrical coordinates yields

$$\nabla \cdot \mathbf{f}_n(\mathbf{r}) = \frac{1}{r} \frac{\partial}{\partial r} \left[ \frac{L_n}{2A_n} r^2 \right] = \frac{L_n}{A} \quad \text{in } T_n^- \tag{8.7}$$

By a similar calculation,

$$\nabla \cdot \mathbf{f}_n(\mathbf{r}) = -\frac{L_n}{A} \quad \text{in } T_n^+ \tag{8.8}$$

As the divergence of the current is proportional to the surface charge density through the equation of continuity, we see from (8.7) and (8.8) that the total charge density associated with adjacent triangle pairs is zero. As there is no accumulation of charges on an edge, the RWG function is said to be *divergence conforming* [11].

## 8.2.2 Assignment and Orientation of Basis Functions

Consider a problem of general configuration, comprising multiple dielectric and conducting regions, where there may be one or more junctions between regions. Let us now assign basis functions to the various interfaces on a region-wise basis. If $e_n$ is a regular edge on the interface $S_{lm}$ between the two dielectric regions $R_l$ and $R_m$, we assign one electric and magnetic basis function to the edge in each region, and the functions are oriented so that they flow in opposite directions on each side of the interface, as illustrated in Figure 7.24a. If $e_n$ is a junction edge, a similar assignment is made, except that the basis functions are oriented such that the current flows past the edge in a counterclockwise (or clockwise) manner, as shown in Figure 7.24b. These assignments ensure that Kirchoff's law is satisfied at $e_n$ in each region. If the edge lies on a conducting interface, or at the junction of multiple interfaces, where one or more of those interfaces are conducting, we omit the magnetic basis function, as magnetic currents vanish on conductors. Similarly, no electric or magnetic basis functions are assigned inside a closed conducting region. Doing this for all non-boundary edges leads to the discretized matrix system in (3.194), where the basis functions have oriented correctly but the linear dependencies at the interfaces are not yet taken into account. We will consider the boundary conditions in more detail in Section 8.6.

---

## 8.3    EFIE

The EFIE (3.178) contains $\mathcal{L}$ and $\mathcal{K}$ operators, which were discretized and tested via the Moment Method in (3.185) and (3.186). We will consider these in the context of three-dimensional problems and RWG functions in this section.

### 8.3.1    $\mathcal{L}$ Operator

Substitution of the electric basis and testing functions into (3.185) yields in Region $R_l$ matrix elements given by

$$Z_{mn}^{EJ(l)} = \mathrm{L}(\mathbf{f}_m, \mathbf{f}_n) \tag{8.9}$$

where

$$\mathrm{L}(\mathbf{f}_m, \mathbf{f}_n) = \int_{\mathbf{f}_m} \int_{\mathbf{f}_n} \left[ j\omega\mu_l\, \mathbf{f}_m(\mathbf{r})\cdot\mathbf{f}_n(\mathbf{r}') - \frac{j}{\omega\epsilon_l}\nabla\cdot\mathbf{f}_m(\mathbf{r})\,\nabla'\cdot\mathbf{f}_n(\mathbf{r}') \right] \frac{e^{-jk_l r}}{4\pi r}\, d\mathbf{r}'\, d\mathbf{r} \tag{8.10}$$

The inner and outer integrals in (8.10) are each performed over two triangles. However, as each triangle supports a maximum of three basis and testing functions, the computations involving one source and testing triangle can contribute to a maximum of nine matrix elements. Thus, the matrix fill becomes more expedient by performing a loop over source and testing triangle pairs, where all possible combinations of basis and testing functions are considered for each pair.

#### 8.3.1.1 Non-Near Terms

When the source and testing triangles are separated by a sufficient distance, (8.10) can be implemented as written. For a source and testing triangle, the contribution to (8.10) can be computed using an $M$-point numerical quadrature on each triangle, yielding

$$I = \frac{L_m L_n}{4\pi} \sum_{p=1}^{M} \sum_{q=1}^{M} w_p w_q \left[ \frac{j\omega\mu_l}{4} \boldsymbol{\rho}_m^{\pm}(\mathbf{r}_p) \cdot \boldsymbol{\rho}_n^{\pm}(\mathbf{r}_q') \pm \frac{j}{\omega\epsilon_l} \right] \frac{e^{-jk_l R_{pq}}}{R_{pq}} \tag{8.11}$$

where $p$ and $q$ refer to the testing and source coordinates, respectively, and the quadrature weights $w_p$ and $w_q$ have been normalized with respect to triangle area. The distance $R_{pq}$ is

$$R_{pq} = \sqrt{(x_p - x_q)^2 + (y_p - y_q)^2 + (z_p - z_q)^2} \tag{8.12}$$

The variable signs in (8.11) depend on the orientation of the basis and testing functions on each triangle.

#### 8.3.1.2 Near and Self Terms

When the source and testing triangles are close together or overlapping, the integrations must be treated carefully. One of the methods used in previous chapters was to replace the Green's function by a small argument approximation and compute the resulting integral analytically. This method yielded reasonable results because the size of the integration domain was small and inside the valid range of the approximation. On the other hand, triangular subdomains are sometimes large enough to invalidate such an approximation, so we will approach the problem differently in this case. A common approach is to rewrite the Green's function as

$$\frac{e^{-jk_l r}}{r} = \left[ \frac{e^{-jk_l r}}{r} - \frac{1}{r} \right] + \frac{1}{r} \tag{8.13}$$

which is referred to as a *singularity extraction*. The first term on the right-hand side is well behaved, having the limit

$$\lim_{r \to 0} \left[ \frac{e^{-jk_l r}}{r} - \frac{1}{r} \right] = -jk_l \tag{8.14}$$

which can be used as-is in (8.11). Inserting the second term on the right-hand side of (8.13) into (8.10) yields singular potential integrals of the form

$$I_1 = \int_T \int_{T'} \boldsymbol{\rho}_m^{\pm}(\mathbf{r}) \cdot \boldsymbol{\rho}_n^{\pm}(\mathbf{r}') \frac{1}{r} \, d\mathbf{r}' \, d\mathbf{r} \tag{8.15}$$

and

$$I_2 = \int_T \int_{T'} \frac{1}{r} \, d\mathbf{r}' \, d\mathbf{r} \tag{8.16}$$

## Self Terms

When $T$ and $T'$ overlap, we can compute the inner and outer integrations in (8.15) and (8.16) analytically. For brevity, we will consider the integrals in terms of the basis vectors $\boldsymbol{\rho}_{m,n}(\mathbf{r}) = \boldsymbol{\rho}_{m,n}^-(\mathbf{r})$ on $T^-$, and results for $\boldsymbol{\rho}_{m,n}^+(\mathbf{r})$ can be obtained by a change of sign. To begin, we convert $\boldsymbol{\rho}_{m,n}(\mathbf{r})$ to simplex coordinates (see Section 10.2.1), yielding

$$\boldsymbol{\rho}_{m,n}(\mathbf{r}) = (1 - \lambda_1 - \lambda_2)\mathbf{v}_1 + \lambda_1\mathbf{v}_2 + \lambda_2\mathbf{v}_3 - \mathbf{v}_{m,n} \tag{8.17}$$

where $\mathbf{v}_1$, $\mathbf{v}_2$ and $\mathbf{v}_3$ are the triangle nodes and $\mathbf{v}_{m,n}$ is the node opposite edge $m$ or $n$ on the triangle. Substitution of (8.17) into (8.15) yields

$$I_1 = \int_T \int_{T'} \Big[ (1 - \lambda_1 - \lambda_2)\mathbf{v}_1 + \lambda_1\mathbf{v}_2 + \lambda_2\mathbf{v}_3 - \mathbf{v}_m \Big]$$
$$\cdot \Big[ (1 - \lambda_1' - \lambda_2')\mathbf{v}_1 + \lambda_1'\mathbf{v}_2 + \lambda_2'\mathbf{v}_3 - \mathbf{v}_n \Big] \frac{1}{r} \, d\lambda_1' \, d\lambda_2' \, d\lambda_1 \, d\lambda_2 \tag{8.18}$$

Multiplication of all the terms and performing some lengthy term collection yields an integrand given by

$$\Big[ \lambda_1\lambda_1'(a_{11} - 2a_{12} + a_{22}) + \lambda_1\lambda_2'(a_{11} - a_{13} - a_{12} + a_{23})$$
$$+ \lambda_2\lambda_2'(a_{11} - 2a_{13} + a_{33}) + \lambda_1'\lambda_2(a_{11} - a_{12} - a_{13} + a_{23})$$
$$+ \lambda_1(-a_{11} + a_{1n} + a_{12} - a_{2n}) + \lambda_2(-a_{11} + a_{1n} + a_{13} - a_{3n})$$
$$+ \lambda_1'(-a_{11} + a_{1m} + a_{12} - a_{2m}) + \lambda_2'(-a_{11} + a_{1m} + a_{13} - a_{3m})$$
$$+ a_{11} - a_{1n} - a_{1m} + a_{mn} \Big] \frac{1}{r} \tag{8.19}$$

where

$$\begin{array}{lll}
a_{11} = \mathbf{v}_1 \cdot \mathbf{v}_1 & a_{12} = \mathbf{v}_1 \cdot \mathbf{v}_2 & a_{13} = \mathbf{v}_1 \cdot \mathbf{v}_3 \\
a_{22} = \mathbf{v}_2 \cdot \mathbf{v}_2 & a_{23} = \mathbf{v}_2 \cdot \mathbf{v}_3 & a_{33} = \mathbf{v}_3 \cdot \mathbf{v}_3 \\
a_{1n} = \mathbf{v}_1 \cdot \mathbf{v}_n & a_{1m} = \mathbf{v}_1 \cdot \mathbf{v}_m & a_{2n} = \mathbf{v}_2 \cdot \mathbf{v}_n \\
a_{2m} = \mathbf{v}_2 \cdot \mathbf{v}_m & a_{3n} = \mathbf{v}_3 \cdot \mathbf{v}_n & a_{3m} = \mathbf{v}_3 \cdot \mathbf{v}_m \\
& a_{mn} = \mathbf{v}_n \cdot \mathbf{v}_n &
\end{array} \tag{8.20}$$

This results in a set of integrals of the form

$$\int_T \int_{T'} \lambda_i \lambda_j' \frac{1}{|\mathbf{r} - \mathbf{r}'|} \, d\lambda_1' \, d\lambda_2' \, d\lambda_1 d\lambda_2 \tag{8.21}$$

The various permutations of this integral were evaluated analytically by Eibert and Hansen in [8], and are

$$
\frac{1}{4A^2} \int_T \int_{T'} \lambda_1 \lambda_1' \frac{1}{|\mathbf{r}-\mathbf{r}'|} \, dT \, dT' = \frac{\log\left[\frac{b+\sqrt{a}\sqrt{c}}{b-c+\sqrt{c}\sqrt{a-2b+c}}\right]}{40\sqrt{c}}
$$

$$
+ \frac{\log\left[\frac{-b+c+\sqrt{c}\sqrt{a-2b+c}}{-b+\sqrt{a}\sqrt{c}}\right]}{40\sqrt{c}} + \frac{\sqrt{a}\sqrt{a-2b+c}-\sqrt{c}\sqrt{a-2b+c}}{120(a-2b+c)^{\frac{3}{2}}}
$$

$$
+ \frac{(2a-5b+3c)\log\left[\frac{(a-b+\sqrt{a}\sqrt{a-2b+c})(c-b+\sqrt{c}\sqrt{a-2b+c})}{(b-a+\sqrt{a}\sqrt{a-2b+c})(b-c+\sqrt{c}\sqrt{a-2b+c})}\right]}{120(a-2b+c)^{\frac{3}{2}}}
$$

$$
+ \frac{(2a+b)\log\left[\frac{(b+\sqrt{a}\sqrt{c})(a-b+\sqrt{a}\sqrt{a-2b+c})}{(-b+\sqrt{a}\sqrt{c})(-a+b+\sqrt{a}\sqrt{a-2b+c})}\right]}{120a^{\frac{3}{2}}}
$$

$$
+ \frac{-\sqrt{a}\sqrt{c}+\sqrt{a}\sqrt{a-2b+c}}{120a^{\frac{3}{2}}} \tag{8.22}
$$

$$
\frac{1}{4A^2} \int_T \int_{T'} \lambda_1 \lambda_2' \frac{1}{|\mathbf{r}-\mathbf{r}'|} \, dT \, dT' = \frac{\log\left[\frac{b+\sqrt{a}\sqrt{c}}{b-c+\sqrt{c}\sqrt{a-2b+c}}\right]}{120\sqrt{c}}
$$

$$
+ \frac{\log\left[\frac{a-b+\sqrt{a}\sqrt{a-2b+c}}{-b+\sqrt{a}\sqrt{c}}\right]}{120\sqrt{a}} + \frac{-\sqrt{a}\sqrt{a-2b+c}+\sqrt{c}\sqrt{a-2b+c}}{120(a-2b+c)^{\frac{3}{2}}}
$$

$$
+ \frac{(2a+3b)\log\left[\frac{b+\sqrt{a}\sqrt{c}}{-a+b+\sqrt{a}\sqrt{a-2b+c}}\right]}{120a^{\frac{3}{2}}} + \frac{\sqrt{a}\sqrt{a-2b+c}-\sqrt{c}\sqrt{a-2b+c}}{120(a-2b+c)^{\frac{3}{2}}}
$$

$$
+ \frac{(a-3b+2c)\log\left[\frac{-b+c+\sqrt{c}\sqrt{a-2b+c}}{-a+b+\sqrt{a}\sqrt{a-2b+c}}\right]}{120(a-2b+c)^{\frac{3}{2}}} + \frac{-3\sqrt{a}\sqrt{c}+3\sqrt{c}\sqrt{a-2b+c}}{120c^{\frac{3}{2}}}
$$

$$
+ \frac{(3b+2c)\log\left[\frac{-b+c+\sqrt{c}\sqrt{a-2b+c}}{-b+\sqrt{a}\sqrt{c}}\right]}{120c^{\frac{3}{2}}} + \frac{-3\sqrt{a}\sqrt{c}+3\sqrt{a}\sqrt{a-2b+c}}{120a^{\frac{3}{2}}}
$$

$$
+ \frac{(2a-3b+c)\log\left[\frac{a-b+\sqrt{a}\sqrt{a-2b+c}}{b-c+\sqrt{c}\sqrt{a-2b+c}}\right]}{120(a-2b+c)^{\frac{3}{2}}} \tag{8.23}
$$

$$\frac{1}{4A^2} \int_T \int_{T'} \lambda_1' \frac{1}{|\mathbf{r} - \mathbf{r}'|} \, dT \, dT' = \frac{-\log\left[\frac{-b+\sqrt{a}\sqrt{c}}{a-b+\sqrt{a}\sqrt{a-2b+c}}\right]}{24\sqrt{a}}$$

$$+ \frac{\log\left[\frac{b+\sqrt{a}\sqrt{c}}{b-c+\sqrt{c}\sqrt{a-2b+c}}\right]}{24\sqrt{c}} + \frac{-\sqrt{a}\sqrt{c} + \sqrt{a}\sqrt{a-2b+c}}{24a^{\frac{3}{2}}}$$

$$+ \frac{(a+b)\log\left[\frac{b+\sqrt{a}\sqrt{c}}{-a+b+\sqrt{a}\sqrt{a-2b+c}}\right]}{24a^{\frac{3}{2}}} + \frac{\log\left(\frac{a-b+\sqrt{a}\sqrt{a-2b+c}}{b-c+\sqrt{c}\sqrt{a-2b+c}}\right)}{24\sqrt{a-2b+c}}$$

$$- \frac{\log\left[\frac{-b+\sqrt{a}\sqrt{c}}{-b+c+\sqrt{c}\sqrt{a-2b+c}}\right]}{12\sqrt{c}} + \frac{\sqrt{a}\sqrt{a-2b+c} - \sqrt{c}\sqrt{a-2b+c}}{24(a-2b+c)^{\frac{3}{2}}}$$

$$+ \frac{(a-3b+2c)\log\left[\frac{-b+c+\sqrt{c}\sqrt{a-2b+c}}{-a+b+\sqrt{a}\sqrt{a-2b+c}}\right]}{24(a-2b+c)^{\frac{3}{2}}} \tag{8.24}$$

where

$$\begin{aligned} a &= (\mathbf{v}_1 - \mathbf{v}_3) \cdot (\mathbf{v}_1 - \mathbf{v}_3) \\ b &= (\mathbf{v}_1 - \mathbf{v}_3) \cdot (\mathbf{v}_1 - \mathbf{v}_2) \\ c &= (\mathbf{v}_1 - \mathbf{v}_2) \cdot (\mathbf{v}_1 - \mathbf{v}_2) \end{aligned} \tag{8.25}$$

All integrals involving terms in the integrand of (8.19) can be obtained via (8.22), (8.23), and (8.24) by permuting the vertex indices appropriately. The corresponding solution to (8.16) is

$$I_2 = \frac{1}{4A^2} \int_T \int_{T'} \frac{1}{|\mathbf{r} - \mathbf{r}'|} \, dT \, dT' =$$

$$\frac{\log\left[\frac{(a-b+\sqrt{a}\sqrt{a-2b+c})(b+\sqrt{a}\sqrt{c})}{(-b+\sqrt{a}\sqrt{c})(-a+b+\sqrt{a}\sqrt{a-2b+c})}\right]}{6\sqrt{a}} + \frac{\log\left[\frac{(b+\sqrt{a}\sqrt{c})(-b+c+\sqrt{c}\sqrt{a-2b+c})}{(b-c+\sqrt{c}\sqrt{a-2b+c})(-b+\sqrt{a}\sqrt{c})}\right]}{6\sqrt{c}}$$

$$+ \frac{\log\left[\frac{(a-b+\sqrt{a}\sqrt{a-2b+c})(-b+c+\sqrt{c}\sqrt{a-2b+c})}{(b-c+\sqrt{c}\sqrt{a-2b+c})(-a+b+\sqrt{a}\sqrt{a-2b+c})}\right]}{6\sqrt{a-2b+c}} \tag{8.26}$$

Note that (8.26) does not depend on any basis function, so it only needs to be calculated once per triangle.

*Near Terms*

When $T$ and $T'$ are close together, (8.15) and (8.16) must be treated carefully to achieve maximum accuracy. Following our approach in computing the near terms in the thin wire EFIE, we will compute the outermost integral numerically and the innermost analytically. For the innermost integrals, we will use the method outlined in [5], which contains analytic expressions for potential

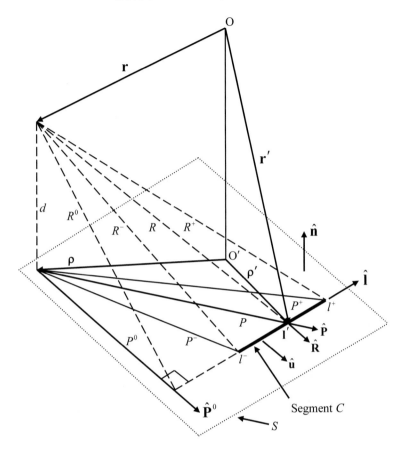

**FIGURE 8.5:** Geometric Quantities for Polygon Segment $C$

integrals over $N$ sided planar polygons. As this method involves the triangle edges, consider the line segment $C$ with endpoints $\mathbf{r}^-$ and $\mathbf{r}^+$ located in $S$ (the plane of the polygon) as illustrated in Figure 8.5. The projections of source and observation vectors $\mathbf{r}'$ and $\mathbf{r}$ onto $S$ are $\rho'$ and $\rho$, respectively. The quantities in the figure are computed as follows:

$$\rho^{\pm} = \mathbf{r}^{\pm} - \hat{\mathbf{n}}(\hat{\mathbf{n}} \cdot \mathbf{r}^{\pm}) \tag{8.27}$$

$$\hat{\mathbf{l}} = \frac{\rho^+ - \rho^-}{|\rho^+ - \rho^-|} \tag{8.28}$$

$$\hat{\mathbf{u}} = \hat{\mathbf{l}} \times \hat{\mathbf{n}} \tag{8.29}$$

$$l^\pm = (\rho^\pm - \rho) \cdot \hat{\mathbf{l}}$$ (8.30)

$$P^0 = |(\rho^\pm - \rho) \cdot \hat{\mathbf{u}}|$$ (8.31)

$$P^\pm = |(\rho^\pm - \rho)| = \sqrt{(P^0)^2 + (l^\pm)^2}$$ (8.32)

$$\hat{\mathbf{P}}^0 = \frac{(\rho^\pm - \rho) - l^\pm \hat{\mathbf{l}}}{P^0}$$ (8.33)

$$R^0 = \sqrt{(P^0)^2 + d^2}$$ (8.34)

$$R^\pm = \sqrt{(P^\pm)^2 + d^2}$$ (8.35)

The distance from the observation point to $S$ is

$$d = \hat{\mathbf{n}} \cdot (\mathbf{r} - \mathbf{r}^\pm)$$ (8.36)

Using the above, integrals of the form

$$I_1 = \int_{T'} \frac{\rho' - \rho}{r} \, d\mathbf{r}'$$ (8.37)

are evaluated via the following sum over the $N$ polygon edges

$$I_1 = \frac{1}{2} \sum_{i=1}^{N} \hat{\mathbf{u}}_i \left[ (R_i^0)^2 \log \frac{R_i^+ + l_i^+}{R_i^- + l_i^-} + l_i^+ R_i^+ - l_i^- R_i^- \right]$$ (8.38)

The integral

$$I_2 = \int_{T'} \frac{1}{r} \, d\mathbf{r}'$$ (8.39)

is evaluated similarly, yielding

$$I_2 = \sum_{i=1}^{N} \hat{\mathbf{P}}_i^0 \cdot \hat{\mathbf{u}}_i \left[ P_i^0 \log \frac{R_i^+ + l_i^+}{R_i^- + l_i^-} - |d| \left( \tan^{-1} \frac{P_i^0 l_i^+}{(R_i^0)^2 + |d| R_i^+} \right. \right.$$
$$\left. \left. - \tan^{-1} \frac{P_i^0 l_i^-}{(R_i^0)^2 + |d| R_i^-} \right) \right]$$ (8.40)

Note that if the observation point $\rho$ lies anywhere along an edge, the contribution from that edge to (8.38) is zero. In addition, (8.38) and (8.3.1.2) suffer from numerical issues when $\mathbf{r}$ lies in $S$ along the extension of an edge

$(R_i^0 = 0)$. In such cases, the observation should be moved a very small distance away from the edge, i.e.,

$$\mathbf{r} = \mathbf{r} + \epsilon \hat{\mathbf{u}} \qquad (8.41)$$

To evaluate the innermost integral of (8.15), we write the basis function $\rho_n(\mathbf{r}')$ as

$$\rho_n(\mathbf{r}') = \rho' - \rho_n \qquad (8.42)$$

where $\rho_n$ is the projection of $\mathbf{v}_n$ on $S$. We can then write

$$\rho' - \rho_n = (\rho' - \rho) + (\rho - \rho_n) \qquad (8.43)$$

which leads to

$$\int_{T'} \frac{\rho_n(\mathbf{r}')}{r} \, d\mathbf{r}' = \int_{T'} \frac{\rho' - \rho}{r} \, d\mathbf{r}' + (\rho - \rho_n) \int_{T'} \frac{1}{r} \, d\mathbf{r}' \qquad (8.44)$$

The two integrals on the right can now be evaluated via (8.38) and (8.3.1.2), and the innermost integral of (8.16) can be computed directly via (8.3.1.2). Integrals involving $\rho_n^+(\mathbf{r})$ are obtained similarly, with a change of sign.

*Duffy Transform*

The Duffy Transform [12, 13, 14, 15] is a method of regularizing integrals over a triangle with a $1/r$ singularity at a vertex, allowing a numerical quadrature to be used. The region of integration is converted from a triangle to a rectangle following the transformation

$$\int_T \frac{f(\mathbf{r}')}{r(\mathbf{r}')} \, d\mathbf{r}' = \int_0^1 \int_0^1 \frac{f(u, \gamma)}{r(u, \gamma)} |J(u, \gamma)| \, du \, d\gamma \qquad (8.45)$$

where $J(u, \gamma)$ is the Jacobian. To illustrate, let us apply the transform to the integral

$$\int_T \frac{1}{r} \, d\mathbf{r}' = \int_0^1 \int_0^1 \frac{|J(u, \gamma)|}{r(u, \gamma)} \, du \, d\gamma \qquad (8.46)$$

where the triangle is defined by the points $\mathbf{v}_1 = (x_1, y_1)$, $\mathbf{v}_2 = (x_2, y_2)$, and $\mathbf{v}_3 = (x_3, y_3)$, and the vector $\mathbf{r}'$ on $T'$ is

$$\mathbf{r}' = (1 - u)(1 - \gamma)\mathbf{v}_1 + u\mathbf{v}_2 + \gamma(1 - u)\mathbf{v}_3 \qquad (8.47)$$

The Jacobian is

$$\begin{aligned} J(u, \gamma) = (1 - u) \Big[ &\big(x_2 - x_1 + \gamma[x_1 - x_3]\big)\big(y_3 - y_1\big) \\ &- \big(y_2 - y_1 + \gamma[y_1 - y_3]\big)\big(x_3 - x_1\big) \Big] \end{aligned} \qquad (8.48)$$

and fixing the observation (singular) point $\mathbf{r}$ at $(x_1, y_1)$, $r$ can be written as

$$r(x,y) = \sqrt{(x_1 - x)^2 + (y_1 - y)^2} = \sqrt{a(u,\gamma) + b(u,\gamma)} = r(u,\gamma) \quad (8.49)$$

where

$$a(u,\gamma) = x_1^2 + ux_1x_2 + u^2x_2^2 + 2(1-u)(ux_2 - x_1)\left[(1-\gamma)x_1 + \gamma x_3\right]$$
$$+ (1-u)^2\left[(1-\gamma)x_1 + \gamma x_3\right]^2 \quad (8.50)$$

and

$$b(u,\gamma) = y_1^2 + uy_1y_2 + u^2y_2^2 + 2(1-u)(uy_2 - y_1)\left[(1-\gamma)y_1 + \gamma y_3\right]$$
$$+ (1-u)^2\left[(1-\gamma)y_1 + \gamma y_3\right]^2 \quad (8.51)$$

The resulting $(1-u)$ term in the numerator acts to cancel the singularity in the denominator, allowing the integral to be performed using an $M$-point numerical quadrature rule on a rectangular subdomain. When the observation point is located in the interior of the triangle, it can be divided into three sub-triangles with the new vertex placed at the singular point.

This method does suffer from some disadvantages. Since it is still a purely numerical method, compute time will be higher due to the number of numerical computations required at each quadrature point. The transformation is accurate only for triangles with reasonably good aspect ratios, with thinner triangles requiring increasing numbers of quadrature points to retain the same level of accuracy [15]. In spite of these drawbacks, the method remains attractive in cases where the function $f(\mathbf{r})$ is not integrable analytically.

### Other Singularity Extractions

It was suggested in [15] that because the first term in (8.13) has a discontinuous derivative at $r = 0$, using standard numerical quadrature will produce inaccurate results. They suggest using a different singularity extraction given by

$$\frac{e^{-jkr}}{r} = \left[\frac{e^{-jkr}}{r} - \frac{1}{r} + \frac{k^2r}{2}\right] + \frac{1}{r} - \frac{k^2r}{2} \quad (8.52)$$

where the first term now has two continuous derivatives. They conclude that using (8.52) results in more accurate results than (8.13) for the same number of quadrature points. It is this author's experience that while this may be true, the singularity extraction in (8.13) is still accurate enough for most three-dimensional problems. It is worth noting that the appendix of [15] also contains expressions for the evaluation of integrals of the form

$$\int_T r^n \, d\mathbf{r}' \quad (8.53)$$

$$\int_T r^n \left[ \mathbf{r}' - \mathbf{v} \right] d\mathbf{r}' \tag{8.54}$$

$$\int_T \nabla r^n \, d\mathbf{r}' \qquad \mathbf{r} \notin T \tag{8.55}$$

$$\int_T \left[ \nabla r^n \right] \left[ \mathbf{r}' \times \mathbf{v} \right] d\mathbf{r}' \qquad \mathbf{r} \notin T \tag{8.56}$$

where $\mathbf{v}$ is any vertex of $T$.

*Example Singular Integral Evaluation*

Let us compute (8.16) using (8.26) and (8.3.1.2). For the latter case, we apply a Gaussian quadrature rule of order $N$ (Section 10.2.4) to compute the outermost integral. The integration domain is an equilateral triangle with sides of length 1. The results are summarized in Table 8.2. The comparison matches to two significant digits at higher quadrature orders, however the results demonstrate that for overlapping triangles (8.26) remains the superior choice. We next calculate (8.16) over a pair of co-planar equilateral triangles of side length 1 that share a common edge. For this task we use Gaussian quadrature in computing both the inner and outer integrals, and then (8.3.1.2) to compute the innermost integral analytically, leaving the outer integration unchanged. The results are summarized in Table 8.3. The comparison is fairly good, suggesting that either method may be sufficient in this case. We now make the internal angle between triangle faces much smaller. In this case we choose an angle of 10 degrees, and the results are summarized in Table 8.4. The results from quadrature alone compare poorly to those using the analytic inner integration. It is obvious from this example that analytic inner integrations should be used to calculate all near matrix elements. To determine when to apply the near-integration formulas, the programmer can compare the distances between triangle centroids. A reasonable threshold might be some fraction of a wavelength, such as $0.1 - 0.2\lambda$.

**TABLE 8.2:** Overlapping Equilateral Triangle Double Integration

| Order 1 | Order 2 | Order 3 | Order 4 | Order 5 | Eqn. (8.26) |
|---------|---------|---------|---------|---------|-------------|
| 0.98767 | 0.85005 | 0.84709 | 0.83107 | 0.82830 | 0.82392 |

**TABLE 8.3:** Near Equilateral Triangle Double Integration (Co-planar)

| Integration Method | Order 1 | Order 2 | Order 3 | Order 4 | Order 5 |
|---|---|---|---|---|---|
| All Quadrature | 0.32475 | 0.34364 | 0.34724 | 0.35330 | 0.35794 |
| Quadrature + (8.3.1.2) | 0.32922 | 0.34331 | 0.34553 | 0.34903 | 0.35009 |

**TABLE 8.4:** Near Equilateral Triangle Double Integration (Non Co-planar)

| Integration Method | Order 1 | Order 2 | Order 3 | Order 4 | Order 5 |
|---|---|---|---|---|---|
| All Quadrature | 3.7357 | 2.1169 | 4.4860 | 1.5794 | 1.4878 |
| Quadrature + (8.3.1.2) | 0.86129 | 0.73681 | 0.73302 | 0.72283 | 0.72166 |

### 8.3.2 $\mathcal{K}$ Operator

Substitution of the magnetic basis and electric testing functions into (3.186) yields in Region $R_l$ matrix elements given by

$$Z_{mn}^{EM(l)} = K(\mathbf{f}_m, \mathbf{g}_n) \tag{8.57}$$

where

$$
K(\mathbf{f}_m, \mathbf{g}_n) = \frac{1}{2} \int_{\mathbf{f}_m, \mathbf{g}_n} \mathbf{f}_m(\mathbf{r}) \cdot \left[ \hat{\mathbf{n}}_l(\mathbf{r}) \times \mathbf{g}_n(\mathbf{r}) \right] d\mathbf{r}
$$
$$
- \frac{1}{4\pi} \int_{\mathbf{f}_m} \mathbf{f}_m(\mathbf{r}) \cdot \int_{\mathbf{g}_n} \left[ (\mathbf{r} - \mathbf{r}') \times \mathbf{g}_n(\mathbf{r}') \right] \frac{1 + jk_l r}{r^3} e^{-jk_l r} \, d\mathbf{r}' \, d\mathbf{r} \tag{8.58}
$$

The first term on the right-hand side of (8.58) is used when the source and testing triangles overlap, and the second term when they do not.

#### 8.3.2.1 Non-Near Terms

When the source and testing triangles are separated by a sufficient distance, (8.58) can be implemented as written. For a source and testing triangle, the contribution to (8.58) can be obtained using an $M$-point numerical quadrature, yielding

$$\eta_0 \frac{L_m L_n}{8 A_m} \sum_{p=1}^{M} w_p \, \rho_m^{\pm}(\mathbf{r}_p) \cdot \left[ \hat{\mathbf{n}}_l(\mathbf{r}_p) \times \rho_n^{\pm}(\mathbf{r}_p) \right] \tag{8.59}$$

when the triangles overlap, and

$$-\eta_0 \frac{L_m L_n}{16\pi} \sum_{p=1}^{M} \sum_{q=1}^{M} w_p w_q \, \boldsymbol{\rho}_m^{\pm}(\mathbf{r}_p) \cdot \left[ (\mathbf{r}_p - \mathbf{r}_q') \times \boldsymbol{\rho}_n^{\pm}(\mathbf{r}_q') \right] \frac{1 + jk_l R_{pq}}{R_{pq}^3} e^{-jk_l R_{pq}}$$

(8.60)

when they do not. The quadrature weights are again normalized with respect to triangle area, and the variable signs depend on the orientation of the basis and testing functions on each triangle. The $\eta_0$ scale factor of the magnetic basis function has also been included in both terms.

### 8.3.2.2 Near Terms

When the source and testing triangles are close together, the second term on the right-hand side of (8.58) can exhibit strongly singular behavior, and special care should be taken when computing that integration. In this case, we will use an approach similar to what was used with the $\mathcal{L}$ operator. Following [16], we rewrite the innermost integral, perform a singularity extraction, and evaluate the singular terms analytically. In addition to the quantities in Figure 8.5, we define the additional quantities shown in Figure 8.6, where $\boldsymbol{\rho}_n(\mathbf{r}') = \boldsymbol{\rho}_n^-(\mathbf{r}')$ on $T_n^-$. Making the definition

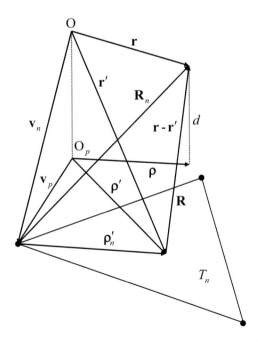

**FIGURE 8.6:** Geometric Quantities For $\mathcal{K}$ Operator Singularity Extraction

$$\mathbf{R}_n = \mathbf{r} - \mathbf{v}_n \tag{8.61}$$

where $\mathbf{v}_n$ is the vertex opposite edge $n$ on $T_n$, and noting that

$$\mathbf{r} - \mathbf{r}' = \mathbf{R}_n - \boldsymbol{\rho}_n(\mathbf{r}') \tag{8.62}$$

we can write

$$(\mathbf{r} - \mathbf{r}') \times \boldsymbol{\rho}_n(\mathbf{r}') = \mathbf{R}_n \times \boldsymbol{\rho}_n(\mathbf{r}') \tag{8.63}$$

Since $\mathbf{R}_n$ is a constant vector, we can write the innermost integral in the second part of (8.58) as

$$\int_{T_n} \left[ (\mathbf{r} - \mathbf{r}') \times \boldsymbol{\rho}_n(\mathbf{r}') \right] \frac{1 + jk_l r}{r^3} e^{-jk_l r} \, d\mathbf{r}' = \mathbf{R}_n \times \int_{T_n} \boldsymbol{\rho}_n(\mathbf{r}') \frac{1 + jk_l r}{r^3} e^{-jk_l r} \, d\mathbf{r}' \tag{8.64}$$

We next perform the following singularity extraction [16]

$$\mathbf{R}_n \times \int_{T_n} \boldsymbol{\rho}_n(\mathbf{r}') \frac{1 + jk_l r}{r^3} e^{-jk_l r} \, d\mathbf{r}' = \mathbf{R}_n \times \left[ \int_{T_n} \boldsymbol{\rho}_n(\mathbf{r}') \cdot \right.$$

$$\left. \frac{(1 + jk_l r)e^{-jk_l r} - (1 + \frac{1}{2}k_l^2 r^2)}{r^3} + \mathbf{a}_n(\mathbf{r}) + \frac{k_l^2}{2} \mathbf{b}_n(\mathbf{r}) \right] \tag{8.65}$$

where

$$\mathbf{a}_n(\mathbf{r}) = \int_{T_n} \frac{\boldsymbol{\rho}_n(\mathbf{r}')}{r^3} \, d\mathbf{r}' \tag{8.66}$$

and

$$\mathbf{b}_n(\mathbf{r}) = \int_{T_n} \frac{\boldsymbol{\rho}_n(\mathbf{r}')}{r} \, d\mathbf{r}' \tag{8.67}$$

The first term on the right-hand side in (8.65) is well behaved and can be evaluated using numerical quadrature. The remaining two integrals $\mathbf{a}_n(\mathbf{r})$ and $\mathbf{b}_n(\mathbf{r})$ are the singular contributions and will be evaluated analytically. The integral $\mathbf{b}_n(\mathbf{r})$ can be computed via (8.44). The integral $\mathbf{a}_n(\mathbf{r})$ can be computed by rewriting it following (8.43), which yields

$$\int_{T_n} \frac{\boldsymbol{\rho}_n(\mathbf{r}')}{r^3} \, d\mathbf{r}' = \int_{T_n} \frac{\boldsymbol{\rho}' - \boldsymbol{\rho}}{r^3} \, d\mathbf{r}' + (\boldsymbol{\rho} - \boldsymbol{\rho}_n) \int_{T_n} \frac{1}{r^3} \, d\mathbf{r}' \tag{8.68}$$

The solutions to these integrals are [16]

$$\int_{T_n} \frac{\boldsymbol{\rho}' - \boldsymbol{\rho}}{r^3} \, d\mathbf{r}' = -\sum_{i=1}^{N} \hat{\mathbf{u}}_i \log \frac{R_i^+ + l_i^+}{R_i^- + l_i^-} \tag{8.69}$$

and

$$
\int_{T_n} \frac{1}{r^3} \, d\mathbf{r}' = - \sum_{i=1}^{N} \hat{\mathbf{P}}_i^0 \cdot \hat{\mathbf{u}}_i \frac{1}{|d|} \left[ \tan^{-1} \frac{|d| l_i^+}{P_i^0 R_i^+} - \tan^{-1} \frac{|d| l_i^-}{P_i^0 R_i^-} \right.
$$
$$
\left. + \tan^{-1} \frac{l_i^-}{P_i^0} - \tan^{-1} \frac{l_i^+}{P_i^0} \right] \tag{8.70}
$$

for $d \neq 0$, and

$$
\int_{T_n} \frac{1}{r^3} \, d\mathbf{r}' = - \sum_{i=1}^{N} \hat{\mathbf{P}}_i^0 \cdot \hat{\mathbf{u}}_i \left[ \frac{l_i^+}{P_i^0 R_i^+} - \frac{l_i^-}{P_i^0 R_i^-} \right] \tag{8.71}
$$

for $d = 0$. The sums in (8.69) and (8.71) may also have numerical problems when $\mathbf{r}$ lies in $S$ along the extension of an edge. In these cases, the observation point can again be slightly modified following (8.41). Results for $\rho_n^+(\mathbf{r}')$ on $T_n^+$ are also obtained using (8.65) with a change of sign.

*Example Singular Integral Evaluation*

Let us compare a straightforward Gaussian quadrature to (8.69–8.71) in computing the singular portion of (8.65). For this comparison, we will compute the scalar quantity

$$
\left| \mathbf{a}_n(\mathbf{r}) + \frac{k^2}{2} \mathbf{b}_n(\mathbf{r}) \right| \tag{8.72}
$$

The integrations are performed over an equilateral triangle with sides of length $1\lambda$, and quadratures of orders 3, 4, and 5 are used. We first place the observation point above the center of the triangle and the separation distance is varied, with results plotted in Figure 8.7a. The observation point is then placed above

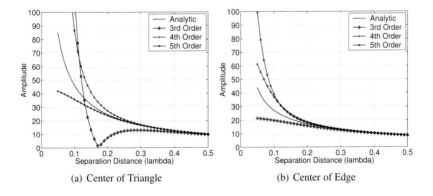

(a) Center of Triangle      (b) Center of Edge

**FIGURE 8.7**: K Operator Near Integration

the midpoint of an edge, and the distance again varied, with results shown in Figure 8.7b. In both cases the comparison is good at larger distances but grows increasingly poor at shorter ones, as expected. These results suggest that the singularity extraction method should be applied for triangle pairs separated by 0.2–0.3$\lambda$ or less.

### 8.3.3  Excitation

The right-hand side vector elements for the EFIE are given by (3.191). Using an $N$-point numerical quadrature, the contribution to (3.191) from a single testing triangle can be written as

$$\frac{L_m}{2} \sum_{p=1}^{M} w_p \boldsymbol{\rho}_m^\pm(\mathbf{r}_p) \cdot \mathbf{E}^i(\mathbf{r}_p) \tag{8.73}$$

where the quadrature weight is normalized with respect to triangle area, and the variable sign depends on the orientation of the testing function on the triangle.

#### 8.3.3.1  Plane Wave Excitation

For incident plane waves of $\hat{\boldsymbol{\theta}}^i$- or $\hat{\boldsymbol{\phi}}^i$-polarization having an electric field of unit amplitude, (3.191) in the exterior region $R_0$ can be written as

$$V_m^{E(\theta,\phi)} = \frac{L_m}{2A_m}(\hat{\boldsymbol{\theta}}^i, \hat{\boldsymbol{\phi}}^i) \cdot \int_{\mathbf{f}_m} \boldsymbol{\rho}_m^\pm(\mathbf{r}) e^{jk_0\mathbf{r}\cdot\hat{\mathbf{r}}^i}\, d\mathbf{r} \tag{8.74}$$

and following (8.73), the contribution from a single testing triangle is

$$\frac{L_m}{2}(\hat{\boldsymbol{\theta}}^i, \hat{\boldsymbol{\phi}}^i) \cdot \sum_{p=1}^{M} w_p\, \boldsymbol{\rho}_m^\pm(\mathbf{r}_p) \cdot e^{jk_0\mathbf{r}_p\cdot\hat{\mathbf{r}}^i} \tag{8.75}$$

#### 8.3.3.2  Planar Antenna Excitation

Because planar antennas have many practical applications they are often simulated, making a feedpoint model of the antenna terminals necessary. In some applications, a high-fidelity feed model may be needed to obtain accurate results. Though such specialized models are beyond the scope of this text, the delta-gap model remains useful in computing impedance and radiation patterns in many cases. To develop a delta-gap model, consider the planar bowtie antenna illustrated in Figure 8.8a. We divide the antenna into two halves along edge $m$ and connect a voltage generator of amplitude $V_{in}$ to the triangles sharing the edge, as shown in Figure 8.8b. If we assume that a very small gap of width $d$ exists between the two triangles, the electric field in the gap is

$$\mathbf{E}^i = \frac{V_{in}}{d}\hat{\mathbf{t}}_m \tag{8.76}$$

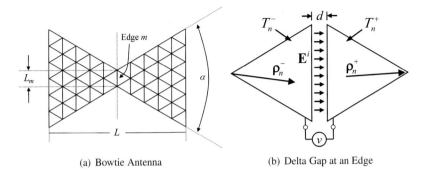

(a) Bowtie Antenna　　　　　　　(b) Delta Gap at an Edge

**FIGURE 8.8:** Delta-Gap Feed Model

where $\hat{\mathbf{t}}_m$ is the vector normal to edge $m$ in the plane. Inserting (8.76) into (3.191), and using the fact that the normal component of the RWG function is unity along the edge, (3.191) evaluates to

$$V_m^{E(l)} = L_m V_{in} \tag{8.77}$$

Once the MOM linear system has been solved, the input impedance may be desired. As the electric current vector element $I_m^{J(l)}$ comprises the linear current density flowing across edge $m$, the total electric current flowing across the edge is then

$$I_{in} = L_m I_m^{J(l)} \tag{8.78}$$

and the input impedance is

$$Z_{in} = \frac{V_{in}}{I_{in}} = \frac{V_{in}}{L_m I_m^{J(l)}} \tag{8.79}$$

## 8.4　MFIE

The MFIE (3.179) also has $\mathcal{K}$ and $\mathcal{L}$ operators, which were discretized and tested via the Moment Method in (3.187) and (3.188). Substitution of the electric basis and magnetic testing functions into (3.187) yields in Region $R_l$ matrix elements given by

$$Z_{mn}^{HJ(l)} = -K(\mathbf{g}_m, \mathbf{f}_n) \tag{8.80}$$

and substitution of the magnetic basis and testing functions into (3.188) yields

$$Z_{mn}^{HM(l)} = \frac{\epsilon_l}{\mu_l} L(\mathbf{g}_m, \mathbf{g}_n) \tag{8.81}$$

where the expressions for L and K were developed in Sections 8.3.1 and 8.3.2, respectively. Note that the L elements must be scaled by $\eta_0^2$ to account for the scale factors of the magnetic basis and testing functions.

### 8.4.1   Excitation

The right-hand side vector elements for the MFIE are given by (3.192). Using an $N$-point numerical quadrature, the contribution to (3.192) from a single testing triangle can be written as

$$\frac{L_m}{2} \sum_{p=1}^{M} w_p \boldsymbol{\rho}_m^{\pm}(\mathbf{r}_p) \cdot \mathbf{H}^i(\mathbf{r}_p) \tag{8.82}$$

where the quadrature weight is normalized with respect to triangle area, and the variable sign depends on the orientation of the testing function on the triangle.

#### 8.4.1.1   Plane Wave Excitation

Given $\hat{\boldsymbol{\theta}}^i$ and $\hat{\boldsymbol{\phi}}^i$-polarized incident electric fields, the corresponding magnetic fields are $-\hat{\boldsymbol{\phi}}^i$ and $\hat{\boldsymbol{\theta}}^i$-polarized, respectively. Thus in the exterior region $R_0$ we can re-use the elements of the EFIE right-hand side, yielding MFIE right-hand side elements given by

$$V_m^{H,\theta} = -V_m^{E,\phi} \tag{8.83}$$

and

$$V_m^{H,\phi} = V_m^{E,\theta} \tag{8.84}$$

---

## 8.5   nMFIE

The nMFIE (3.180) contains the $\hat{\mathbf{n}} \times \mathcal{L}$ and $\hat{\mathbf{n}} \times \mathcal{K}$ operators, which were discretized and tested via the Moment Method in (3.189) and (3.190).

## 8.5.1 $\hat{n} \times \mathcal{K}$ Operator

Substituting the electric basis and testing functions into (3.189) yields in Region $R_l$ matrix elements given by

$$Z_{mn}^{nHJ(l)} = \text{nK}(\mathbf{f}_m, \mathbf{f}_n) \tag{8.85}$$

where

$$\text{nK}(\mathbf{f}_m, \mathbf{f}_n) = \frac{1}{2} \int_{\mathbf{f}_m, \mathbf{f}_n} \mathbf{f}_m(\mathbf{r}) \cdot \mathbf{f}_n(\mathbf{r}) \, d\mathbf{r}$$

$$+ \frac{1}{4\pi} \int_{\mathbf{f}_m} \mathbf{f}_m(\mathbf{r}) \cdot \left[ \hat{\mathbf{n}}_l(\mathbf{r}) \times \int_{\mathbf{f}_n} \left[ (\mathbf{r} - \mathbf{r}') \times \mathbf{f}_n(\mathbf{r}') \right] \frac{1 + jk_l r}{r^3} e^{-jk_l r} \, d\mathbf{r}' \right] d\mathbf{r} \tag{8.86}$$

The first term on the right-hand side of (8.86) is used when the source and testing triangles overlap, and the second term when they do not. We note that (8.86) is nearly identical to (8.58), except for the sign change on the second term and the re-location of $\hat{\mathbf{n}}_l(\mathbf{r})$.

### 8.5.1.1 Non-Near Terms

When the source and testing triangles are separated by a sufficient distance, (8.86) can be implemented as written. For a source and testing triangle, the contribution to (8.86) can be obtained using an $M$-point numerical quadrature, yielding

$$\frac{L_m L_n}{8 A_m} \sum_{p=1}^{M} w_p \boldsymbol{\rho}_m^{\pm}(\mathbf{r}_p) \cdot \boldsymbol{\rho}_n^{\pm}(\mathbf{r}_p) \tag{8.87}$$

when the triangles overlap, and

$$\frac{L_m L_n}{16\pi} \sum_{p=1}^{M} \sum_{q=1}^{M} w_p w_q \boldsymbol{\rho}_m^{\pm}(\mathbf{r}_p) \cdot \left[ \hat{\mathbf{n}}_l(\mathbf{r}_p) \times (\mathbf{r}_p - \mathbf{r}_q') \times \boldsymbol{\rho}_n^{\pm}(\mathbf{r}_q') \right] \frac{1 + jk_l R_{pq}}{R_{pq}^3} e^{-jk_l R_{pq}}$$

when they do not.

### 8.5.1.2 Near Terms

The innermost integral in the second part of (8.86) is identical to that of (8.58). To save time, the matrix fill process can use this fact to compute the innermost integral of the K and nK terms simultaneously.

## 8.5.2 $\hat{n} \times \mathcal{L}$ Operator

Substituting the magnetic basis and electric testing functions into (3.190) yields in Region $R_l$ matrix elements given by

$$Z_{mn}^{nHM(l)} = \text{nL}(\mathbf{f}_m, \mathbf{g}_n) \tag{8.88}$$

where

$$
\begin{aligned}
\mathrm{nL}(\mathbf{f}_m, \mathbf{g}_n) =& jw\epsilon_l \int_{\mathbf{f}_m} \mathbf{f}_m(\mathbf{r}) \cdot \left[ \hat{\mathbf{n}}_l(\mathbf{r}) \times \int_{\mathbf{g}_n} \mathbf{g}_n(\mathbf{r}') \frac{e^{-jk_l r}}{4\pi r} \, d\mathbf{r}' \right] d\mathbf{r} \\
&- \frac{j}{w\mu_l} \oint_{\delta\mathbf{f}_m} \hat{\mathbf{c}}_m(\mathbf{r}) \cdot \left[ \mathbf{f}_m(\mathbf{r}) \times \hat{\mathbf{n}}_l(\mathbf{r}) \right] \int_{\mathbf{g}_n} \nabla' \cdot \mathbf{g}_n(\mathbf{r}') \frac{e^{-jk_l r}}{4\pi r} \, d\mathbf{r}' \, d\mathbf{r}
\end{aligned}
$$

(8.89)

and (3.206) has been used to re-write the second part of (3.190). Note that since $\hat{\mathbf{c}}_m(\mathbf{r})$ is normal to the boundary in the plane of each triangle, it may have a different value along edge $m$ in $T_m^-$ than it does in $T_m^+$. Thus, the contour integral in the second part (8.89) must be performed over the boundaries of $T_m^-$ and $T_m^+$ separately (a total of six edges).

### 8.5.2.1   Non-Near Terms

When the source and testing triangles are separated by a sufficient distance, (8.89) can be implemented as written. For a source and testing triangle, the contribution to (8.89) can be broken into two parts. The first term comprises a $N$-point numerical quadrature over the source and testing triangles, and is

$$
I_1 = jw\epsilon_l \eta_0 \frac{L_m L_n}{16\pi} \sum_{p=1}^{M} \sum_{q=1}^{M} w_p w_q \boldsymbol{\rho}_m^{\pm}(\mathbf{r}_p) \cdot \left[ \hat{\mathbf{n}}_l(\mathbf{r}_p) \times \boldsymbol{\rho}_n^{\pm}(\mathbf{r}_q') \right] \frac{e^{-jk_l R_{pq}}}{R_{pq}}
$$

(8.90)

The second term comprises a two-dimensional quadrature over the source triangle, and a one-dimensional quadrature over the three edges of the testing triangle. This term can be written as

$$
I_2 = \pm \frac{j\eta_0}{w\mu_l} \frac{L_m L_n}{8\pi} \sum_{e=1}^{3} \sum_{p=1}^{M} \sum_{q=1}^{M} w_{e,p} w_q \mathbf{c}_m(\mathbf{r}_{e,p}) \cdot \left[ \boldsymbol{\rho}_m^{\pm}(\mathbf{r}_{e,p}) \times \hat{\mathbf{n}}_l(\mathbf{r}_{e,p}) \right] \frac{e^{-jk_l R_{pq}}}{R_{pq}}
$$

(8.91)

where the quadrature weights are again normalized, and the variable signs depend on the orientation of the basis and testing functions on each triangle. The $\eta_0$ scale factor of the magnetic basis function has also been included in both terms. Experience has shown that for a mesh having at least 10 edges per wavelength, 3-5 quadrature points per edge is sufficient in computing (8.91).

### 8.5.2.2   Near and Self Terms

For the near and self terms, we will compute the innermost integrals analytically, and the outermost numerically. As the innermost integrals of both terms in (8.89) have the same form as (8.37) and (8.39), these expressions will support computation of the L and nL near terms simultaneously.

### 8.5.3 Excitation

The right-hand side vector elements for the nMFIE are given by (3.193). Using an $N$-point numerical quadrature, the contribution to (3.193) from a single testing triangle can be written as

$$\frac{L_m}{2} \sum_{p=1}^{M} w_p \boldsymbol{\rho}_m^{\pm}(\mathbf{r}_p) \cdot \left[\hat{\mathbf{n}}_l(\mathbf{r}_p) \times \mathbf{H}^i(\mathbf{r}_p)\right] \tag{8.92}$$

where the quadrature weight is normalized with respect to triangle area, and the variable sign depends on the orientation of the testing function on the triangle.

#### 8.5.3.1 Plane Wave Excitation

Given $\hat{\boldsymbol{\theta}}^i$ and $\hat{\boldsymbol{\phi}}^i$-polarized incident electric fields, the corresponding magnetic fields are $-\hat{\boldsymbol{\phi}}^i$ and $\hat{\boldsymbol{\theta}}^i$-polarized, respectively. Thus, for an electric field of unit amplitude, the nMFIE right-hand side vectors in the exterior region $R_0$ can be written as

$$V_m^{nH(\theta,\phi)} = \frac{L_m}{2\eta_0 A_m} \int_{\mathbf{f}_m} \boldsymbol{\rho}_m^{\pm}(\mathbf{r} \cdot \left[\hat{\mathbf{n}}_0(\mathbf{r}) \times (-\hat{\boldsymbol{\phi}}^i, \hat{\boldsymbol{\theta}}^i)\right] e^{jk_0 \mathbf{r} \cdot \hat{\mathbf{r}}^i} \, d\mathbf{r} \tag{8.93}$$

and following (8.92), the contribution to (8.93) from a single testing triangle is

$$\frac{L_m}{2\eta_0} \sum_{p=1}^{M} w_p \, \boldsymbol{\rho}_m^{\pm}(\mathbf{r}_p) \cdot \left[\hat{\mathbf{n}}_0(\mathbf{r}_p) \times (-\hat{\boldsymbol{\phi}}^i, \hat{\boldsymbol{\theta}}^i)\right] e^{jk_0 \mathbf{r}_p \cdot \hat{\mathbf{r}}^i} \tag{8.94}$$

---

## 8.6 Enforcement of Boundary Conditions

The matrix system (3.194) and basis functions have been defined region-wise, as discussed in Sections 3.6.3.1 and 8.2.2. However, the boundary conditions stipulate that there are linearly dependent unknowns that must be combined before the system can be solved. In this section, we will present a generalized procedure for treating the boundary conditions along interfaces and junctions of general type, following very closely the method outlined in [1].

### 8.6.1 Classification of Edges and Junctions

In treating the boundary conditions and imposing the desired integral equation formulation, we classify each edge $e_n$ into one of three types:

1. **Dielectric edge or junction**: $e_n$ lies on the intersection of two or more dielectric surfaces but does not touch any conducting surfaces.

2. **Conducting edge or junction**: $e_n$ lies on the intersection of open or closed conducting surfaces but does not touch any dielectric surfaces.

3. **Composite conducting-dielectric junction**: $e_n$ lies on the intersection of at least one open or closed conducting surface and at least one dielectric surface.

We will consider the rules for treating boundary conditions at each type of edge or junction in this section.

### 8.6.1.1 Dielectric Edges and Junctions

The boundary conditions along dielectric interfaces requires that the oriented basis functions assigned to regular edges or junctions have the same coefficient. Therefore, following Section 7.10.1.1, all basis functions of the same type (electric and magnetic) assigned to $e_n$ are combined. Thus, the first rule is as follows:

- **Rule 1**: Because the coefficients of the oriented basis functions of the same type assigned to a dielectric edge or junction have the same value, these unknowns must be combined into a single unknown. This is accomplished by combining the corresponding columns of the system matrix.

### 8.6.1.2 Conducting Edges and Junctions

We next consider the boundary conditions at conducting edges or junctions, which may contain a mixture of open and closed surfaces. Since the magnetic field is not necessarily continuous across a conducting surface, the electric current $\mathbf{J}$ must have independent values on the opposite sides of the surface. Thus, the electric basis functions assigned to a conducting edge or junction cannot be combined. There is one exception to this rule, however: when all conducting surfaces connected to $e_n$ are open, and $e_n$ is completely in the interior of a dielectric region. Consider an open conducting surface $S$ lying completely inside a homogeneous dielectric region, where $e_n$ comprises a regular edge on $S$ that has been assigned the oriented electric basis functions $\mathbf{f}_1$ and $\mathbf{f}_2$. In a homogeneous region, these basis functions produce the same fields with opposite signs due to their orientation. Therefore, the fields can be generated by only one basis function, and one of the basis functions can be removed. Generalizing this to a junction of $N$ conducting surfaces, we get our second rule:

- **Rule 2**: As the magnetic field is not necessarily continuous across conducting surfaces, the electric basis functions assigned to a conducting

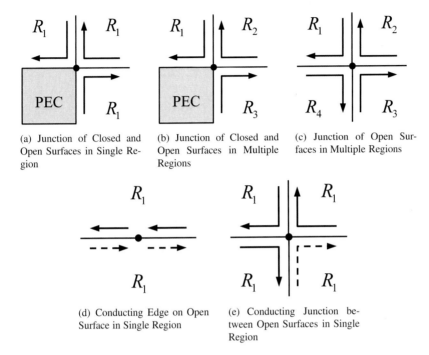

(a) Junction of Closed and Open Surfaces in Single Region

(b) Junction of Closed and Open Surfaces in Multiple Regions

(c) Junction of Open Surfaces in Multiple Regions

(d) Conducting Edge on Open Surface in Single Region

(e) Conducting Junction between Open Surfaces in Single Region

**FIGURE 8.9:** Basis Functions at Conducting Edges and Junctions

edge or junction are independent. In addition, if all the surfaces meeting at $e_n$ are open conducting surfaces, and $e_n$ lies completely in the interior of a homogeneous dielectric region, one of the electric basis functions assigned to $e_n$ is removed.

Several types of conducting edges and junctions that can be encountered are illustrated in Figures 8.9a–8.9e. In each figure, solid lines indicate open conducting surfaces, and the shaded areas are closed conducting regions. The basis functions drawn with dashed lines in Figures 8.9d and 8.9e indicate those that can be removed.

### 8.6.1.3 Composite Conducting-Dielectric Junctions

Now let $e_n$ comprise a composite conducting-dielectric junction. Again, because the magnetic field is not continuous across open metallic surfaces, the electric current must have independent values on the opposite sides of any open conducting surface meeting at $e_n$. Combining this with the previously stated rules, we obtain our third rule:

- **Rule 3**: Unknowns on opposite sides of open conducting surfaces meet-

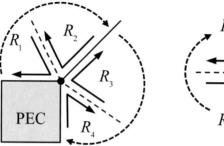
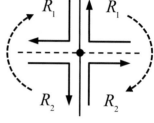

(a) Closed and Open Conductors, Multiple Dielectric Regions

(b) Open Conductors, Two Dielectric Regions

**FIGURE 8.10:** Basis Functions at Composite Junctions

ing at a conducting-dielectric junction cannot be combined. In addition, unknowns on opposite sides of dielectric surfaces that lie in between open or closed conductors must be combined into a single unknown.

Two examples of composite junctions are illustrated in Figures 8.10a and 8.10b. Solid lines indicate open conducting surfaces, dashed lines indicate dielectric interfaces, and shaded areas are closed conducting regions. The curved, dashed arrows indicate where basis functions are combined.

## 8.6.2   Reducing the Overdetermined System

Having combined all linearly dependent unknowns at edges and junctions by combining the columns of the system matrix, the linear system is now over-determined. To obtain a well-defined system, the number of equations must be reduced to that of the remaining unknowns. The way this is done depends on the integral equation formulation. Our approach here is the EFIE-CFIE-PMCHWT approach of Section 3.6.3.1, where the EFIE and CFIE are applied at conducting edges and junctions, and PMCHWT is applied on dielectric edges and junctions. We describe the approach in detail in this section.

### 8.6.2.1   PMCHWT at Dielectric Edges and Junctions

For a regular edge on the interface between two dielectric regions, the PM-CHWT formulation is simply a summation of the EFIEs and MFIEs on each side of the interface. This can be generalized for dielectric junctions as follows. Let $e_n$ be a dielectric edge or junction. We then sum the adjacent EFIEs and

MFIEs, respectively, as follows

$$\sum_{l=1}^{M} \text{EFIE}_n^{(l)} \tag{8.95}$$

and

$$\sum_{l=1}^{M} \text{MFIE}_n^{(l)} \tag{8.96}$$

where $M$ is the number of dielectric regions meeting at $e_n$, and $\text{EFIE}_n^{(l)}$ and $\text{MFIE}_n^{(l)}$ comprise the EFIE and MFIE tested with the RWG function assigned to $e_n$ in Region $R_l$. This is accomplished by combining the corresponding rows of the system matrix.

### 8.6.2.2 EFIE and CFIE at Conducting Edges and Junctions

Let $e_n$ be an edge on an open conducting surface, or a junction between several open conducting surfaces. If $e_n$ is completely inside one dielectric region, by Rule 2 we have also removed one of the unknowns associated with the electric basis functions assigned to $e_n$. In addition, we now remove the EFIE from the system equations, which was tested with the corresponding electric basis function.

Next, let $e_n$ be an edge on a closed conducting surface, or at the junction of a closed conduction surface and at least one open conducting surface. For each remaining electric basis function assigned to $e_n$, we combine the EFIE tested by the corresponding testing function on the open surfaces, with the CFIE (3.181) tested on the closed conducting surfaces.

To illustrate, consider the junction in Figure 8.11a, where an open conducting surface (solid line) connects to a closed conducting region (shaded area). There are two dielectric regions $R_1$ and $R_2$ in addition to the closed conducting region $R_3$. As $S_{12}$ (or $S_{21}$) is an open surface, the fields are not necessarily continuous across it, and following Rule 2, the electric basis functions $\mathbf{f}_1$ and $\mathbf{f}_2$ are considered to be independent. The choice of integral equations are as follows: The testing function associated with $\mathbf{f}_1$ will test in $R_1$ the EFIE on $S_{12}$, and the CFIE on $S_{13}$, which are added to a common matrix element. Likewise, the testing function associated with $\mathbf{f}_2$ will test in $R_2$ the EFIE on $S_{21}$ and the CFIE on $S_{23}$.

### 8.6.2.3 EFIE and CFIE at Composite Conducting-Dielectric Junctions

Finally, we consider general conducting-dielectric junctions. If no closed conducting surfaces meet at $e_n$, then by Rule 2, we combine the EFIEs between

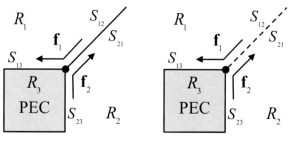

(a) Conducting Junction With Open and Closed Conducting Surfaces

(b) Composite Junction With Closed Conducting Surface

**FIGURE 8.11:** Integral Equation Treatment at Junctions

any two metallic surfaces using the sum

$$\sum_{l=1}^{M} \mathrm{EFIE}_n^{(l)} \tag{8.97}$$

where $M$ is the number of dielectric regions between the surfaces. If a closed conducting surface meets at $e_n$, then the CFIEs are combined between any two metallic surfaces in a similar manner. To illustrate, consider the composite junction of Figure 8.11b, which is similar to that of 8.11a, except that $S_{12}$ and $S_{21}$ now comprise a dielectric interface. The choice of integral equations are as follows: The testing function associated with $\mathbf{f}_1$ will test in $R_1$ the CFIE on $S_{12}$ and $S_{13}$. Likewise, the testing function associated with $\mathbf{f}_2$ will test in $R_2$ the CFIE on $S_{21}$ and $S_{23}$. As the boundary conditions on $S_{12}$ stipulate that $\mathbf{f}_1$ and $\mathbf{f}_2$ must have the same coefficient, the tested CFIEs that lie on and between $S_{13}$ and $S_{23}$ will contribute to the same matrix element.

## 8.7   Notes on Software Implementation

In this section we discuss some elements of a practical software implementation of the techniques presented in this chapter, as well as important issues and considerations.

### 8.7.1 Pre-Processing and Bookkeeping

In practice, the procedures outlined in Sections 8.2.2 and 8.6 condense into a set of pre-processing steps, resulting in a bookkeeping database used in the construction of the system matrix. Because the implementation details will differ between software codes, we will address these steps at a high level.

#### 8.7.1.1 Region and Interface Assignments

The first step involves the construction of several lists. The first is a list of all unique dielectric regions and their dielectric parameters $\epsilon$ and $\mu$. Each region is assigned a unique ID, with the regions comprising free space and conductor assigned a special value (such as 0 or $-1$) used internally by the software. The second list describes all interfaces, where each interface is assigned a unique ID as well as the IDs of the regions that are interior and exterior to that interface (as determined by the surface normal). Each triangle is then assigned to an interface by appending the corresponding interface ID to each line of the connectivity list in the facet file. This is typically done when the facet file is being constructed.

#### 8.7.1.2 Geometry Processing

Applying the edge-finding algorithm (Section 8.1.2) to the facet file, we next identify all boundary, regular, and junction edges in the triangle mesh, as well as the triangles (and interfaces) that connect to each edge. With the connectivity between neighboring triangles established, we can perform basic error checking. Triangles connected at a regular edge that are assigned to the incorrect interfaces can be found, and the presence of boundary (naked) edges on supposed closed interfaces can be reported.

#### 8.7.1.3 Assignment and Orientation of Basis Functions

Each triangle is now assigned two data structures: one stores the indexes and orientations of the basis functions on the *interior* side of the triangle, the other contains similar data for the *exterior* side. These structures are now populated by performing a loop over all non-boundary edges, where for each edge, we work in a local two-dimensional coordinate system. To illustrate, consider edge $e_n$ with endpoints $\mathbf{e}_1$ and $\mathbf{e}_2$, which lies at the junction of $M$ triangles. On triangle $T_m$, we denote the node opposite $e_n$ as $\mathbf{v}_m$. We then define the references axes in the local coordinate system as

$$\hat{\mathbf{z}} = \frac{\mathbf{e}_2 - \mathbf{e}_1}{|\mathbf{e}_2 - \mathbf{e}_1|} \tag{8.98}$$

$$\hat{\mathbf{x}} = (\mathbf{v}_1 - \mathbf{e}_1) - \left[(\mathbf{v}_1 - \mathbf{e}_1) \cdot \hat{\mathbf{z}}\right]\hat{\mathbf{z}} \tag{8.99}$$

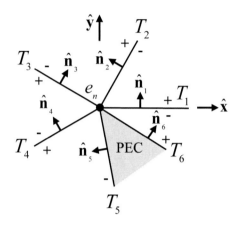

**FIGURE 8.12:** Local Reference Coordinates at Junction

$$\hat{\mathbf{y}} = \hat{\mathbf{z}} \times \hat{\mathbf{x}} \tag{8.100}$$

where the plane of $T_1$ lies along the local $\hat{\mathbf{x}}$ axis, and all other triangles at the junction are sorted in a counterclockwise manner as shown in Figure 8.12. A walk through all triangles in the counterclockwise direction is now all that is required. The normal vector on each triangle is used to determine whether we populate the data structure for its interior or exterior side. The vector orientations of the basis functions at $e_n$ are assigned in back-to-front manner on each triangle, as shown with the plus and minus signs. The rules outlined in Section 8.6.1 are enforced at each step. The indexes assigned to each basis function is taken from a running count of the number of electric or magnetic basis functions assigned so far, where the counts are either incremented or left alone at each step during the walk on $e_n$, depending on how the rules are applied. Once all edges have been processed, we add to each region a pointer to each triangle on its boundary, and to the data structure assigned to the side of the triangle facing into the region.

## 8.7.2　Matrix and Right-Hand Side Fill

Matrix filling is performed on a per-region basis, as the integral operators depend on the wavenumber in each region, and only those triangles assigned to each region are considered. Similarly, calculations for the right-hand side are performed only in those regions where an impressed field exists. For scattering problems, this is the free space region $R_0$. The bookkeeping information generated during the pre-processing steps is utilized to determine the orientation and index of the basis and testing functions on the source and testing triangles, as well as the direction of the normal vector on each interface.

### 8.7.3 Parallelization

There are many areas in Moment Method codes that can be parallelized effectively. The routines that compute the system matrix and right-hand side vectors are obvious candidates. Multiple right-hand sides can also be solved in parallel, as well as the matrix-vector product in an iterative solver. We now discuss some approaches to these problems for shared memory and distributed memory systems.

#### 8.7.3.1 Shared Memory Systems

On a shared memory system, all geometry and book-keeping information as well as the system matrix are addressable by all threads. As system memory is typically at a premium, the obvious choice is to allocate space for a single system matrix, and to then do the matrix fill in parallel. The matrix can be computed efficiently using a loop over triangle pairs, where the integral operators are computed on separate threads. This requires thread synchronization every time a value is added to the system matrix, though as the time needed is small compared to that spent computing the operators, a mutex lock will suffice in resolving thread contention. For matrix factorization and right-hand side solving, third-party software libraries such as LAPACK are typically used. Many of these are often single-threaded, though some vendor-specific implementations have built-in parallelism. As the storage needed for the right-hand side vectors is small compared to the system matrix, multiple right-hand sides can be solved in parallel. This is ideal for monostatic scattering problems, where the number of incidence angles may exceed the number of available processors. If the number of right-hand sides is small, individual right-hand side vectors can be computed by performing the required quadratures and assignments in parallel.

#### 8.7.3.2 Distributed Memory Systems

The design and implementation of a solver on distributed memory systems depends on several factors. The most important considerations are the maximum size of the problem to be attempted, the type of processors and total memory installed on each node, and the speed and topology of the network connections. If the problem is small enough that it can be solved on a single node, one can simply utilize all the available processors to decrease the matrix fill time. In this case, each node can load the geometry into local memory and create the needed edge list and any other information, then compute and transmit its system matrix and right-hand side vector contributions to the master node.

If the geometry and system matrix are too large to be stored on a single node, they must be efficiently divided among the available nodes. A software mechanism must be devised that allows a node to obtain any geometry that is not stored locally. Though it may not be difficult to store parts of the system matrix on different nodes, a distributed matrix factorization is more difficult and is beyond the scope of this text. A distributed iterative solver may be a more attractive choice in this case, as each node only needs to compute the portion of the matrix-vector product involving what it stores locally. The results from each node can then be transmitted to a master node and added to the global result vector at each iteration.

### 8.7.4 Triangle Mesh Considerations

One of the most important elements in any Moment Method simulation is the quality of the triangle mesh that is used. We have already discussed at length the importance of obtaining accurate integrations for the matrix elements, however the model itself is also key to obtaining a good solution. Generating a good model of a realistic object is not a trivial task. Significant time and effort may be invested generating a triangular mesh that accurately describes the object and its various dielectric interfaces and junctions. Such a model must also have a sufficient number of unknowns at the frequency of interest, and must also be watertight and free of any T-junctions.

#### 8.7.4.1 Aspect Ratio

Most computer-aided design (CAD) tools have the capability to generate a triangle mesh from a curved surface, however the quality of these meshes can vary. It is well known that meshes having triangles with poor aspect ratios yield a more poorly conditioned system matrix. Thus, when building a mesh to be used in a Moment Method simulation, care should be taken to ensure that each triangle has a reasonable aspect ratio and that it does not have small internal angles. An example of such a mesh is shown in Figure 8.13a, where we have meshed a circular plate using a very simple algorithm. The triangles in this mesh grow progressively smaller and thinner the closer they are to the center. Though it is not generally possible to create a mesh comprising perfectly equilateral triangles, many meshing tools will allow the user to analyze and adjust the distribution of aspect ratios in a mesh. In areas where the triangles are too thin, these tools often allow for restructuring or rebuilding the triangles to improve their shape. A better mesh is illustrated in Figure 8.13b, where the aspect ratios have been greatly improved.

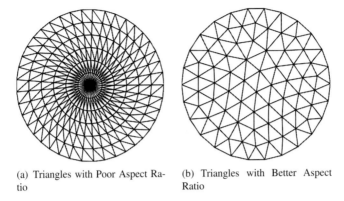

(a) Triangles with Poor Aspect Ratio

(b) Triangles with Better Aspect Ratio

**FIGURE 8.13:** Examples of Triangle Mesh Aspect Ratio

### 8.7.4.2 T-Junctions

In Section 8.1.2.1 we discussed the importance of having shared nodes between triangles. If nodes are not joined at an edge, the surface will not have logical connectivity at the edge, and the edge-finding algorithm will not locate it. On interfaces that bound a closed region, the mesh must be free of any boundary or naked edges. Such a mesh is referred to as "watertight." Some CAD programs may generate meshes where the nodes of one triangle are not co-located with those of adjacent triangles. Consider the meshes shown in Figure 8.14. The mesh in Figure 8.14a is properly constructed, however the mesh in Figure 8.14b has several disjoint intersections, which are called *T-junctions*. These connections cannot be logically resolved through the connectivity list, and will result in boundary edges where there should be none. When the number of triangles in the model grows to be quite high, T-junctions can become difficult to find and eliminate. Though it is sometimes necessary to remove t-junctions manually, most CAD programs can automatically locate and highlight these edges, and in some cases repair the problem with little or no user intervention.

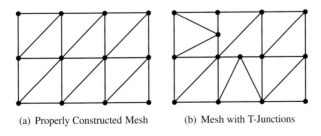

(a) Properly Constructed Mesh

(b) Mesh with T-Junctions

**FIGURE 8.14:** Triangle Mesh Connectivities

## 8.8   Examples

In this section we will use the formulations and methods outlined in this chapter to compute the radar cross section of several test articles. We will first consider a series of conducting, dielectric, and coated spheres, and then a set of benchmark targets that were constructed and measured by the Electromagnetic Code Consortium (EMCC) [17]. We will then consider the input impedance of several antennas including the well-known bowtie and Archimedean spiral. We will then consider the monoconic RV discussed previously in Section 7.10.2.4. For each case, the surface meshes are constructed such that there are at least ten edges (unknowns) per wavelength in each dielectric region, and the CFIE is applied on closed conductors with $\alpha = 0.5$.

### 8.8.1   Serenity

The examples in this chapter are computing using the Tripoint Industries code *Serenity*, which is part of the *lucernhammer* suite of radar cross section codes. *Serenity* implements the methods outlined in this chapter, and supports multiple dielectric regions and junctions of general configuration. It has an automatic edge-finding algorithm, adjustable Gaussian Quadrature integration, and carefully written near and self term routines for all matrix elements. It supports a full-matrix approach with direct factorization and iterative solvers, as well as an MLFMA-accelerated iterative solver, which is discussed in more detail in Chapter 9. It is written using a C++ framework, with double-precision complex variables used in computing the integrations, and single-precision complex variables used for storing and factoring the system matrix. It is parallelized on shared memory systems using POSIX threads (pthreads), with fully parallel matrix fill and right-hand side fill and solve.

### 8.8.2   Test System

The examples in this chapter were run on a computer with a quad-core Intel Xeon (X3320) processor @ 2.5 GHz and 8 GB of RAM, running Ubuntu Linux Version 12. *Serenity* was compiled using the GNU C++ Compiler (Version 4.6.3), with optimizations (-O2 and -march=native). It uses the LAPACK implementation called ATLAS [18, 19], which has a multi-threaded LU factorization. *Serenity* used a maximum of four simultaneous threads on this system when possible.

### 8.8.3  Spheres

In this section we will consider a series of conducting, fully dielectric, and coated spheres.

#### 8.8.3.1  Conducting Sphere

In this section, we consider a conducting sphere with a radius of 0.5 meters. The exact expressions for the scattered field of a conducting sphere were previously summarized in Section 7.9.1.1. For the MOM simulation, a facet model comprising 5120 near-equilateral triangles is used.

*Monostatic RCS versus Frequency*

We first compute the monostatic RCS for frequencies between 10 and 750 MHz. Figures 8.15a and 8.15b compare the Mie series RCS to that of the EFIE and nMFIE, respectively. The comparisons are good at lower frequencies, though at higher ones the nMFIE results are corrupted by internal resonances. In Figure 8.16 is plotted the results from the CFIE, where the resonance problems have been eliminated. In Figure 8.17 are plotted the condition numbers obtained using the LAPACK function cgecon for each case. We note the periodic internal resonances in the EFIE and nMFIE results, which correspond with the corruptions seen in Figures 8.15a and 8.15b. The condition number for the CFIE remains smooth, as expected. We also note that the EFIE solution becomes singular at low frequencies, which is due to low-frequency breakdown of the discretized matrix elements resulting from the $\mathcal{L}$ operator. As $\omega \to 0$, the surface currents decouple into solenoidal (divergence-free) and irrotational (curl-free) components, and the contribution from the vector part of (8.10) becomes lost due to the frequency scaling [20]. This problem can be mitigated through the use of basis functions that model the components separately, such as the Loop-Star basis functions discussed in [21].

*Bistatic RCS*

Using the same simulation parameters as before, we compute the bistatic RCS at 250, 500, and 750 MHz. The CFIE results are compared to the Mie series in Figures 8.18a–8.18f. The comparison is very good at all frequencies.

*Near Field*

Using (3.108), we next compute the near field of the conducting sphere illuminated by an $\hat{\mathbf{x}}$-polarized plane wave incident from the $+\hat{\mathbf{z}}$ direction at 750 MHz. In Figures 8.19a and 8.19b is shown the real part of the scattered and total electric field, respectively, in the $yz$ plane.

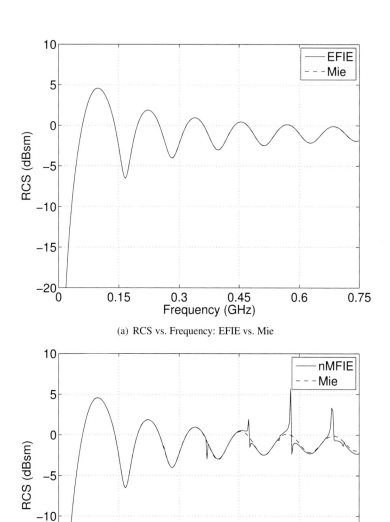

(a) RCS vs. Frequency: EFIE vs. Mie

(b) RCS vs. Frequency: nMFIE vs. Mie

**FIGURE 8.15:** Conducting Sphere: EFIE and nMFIE vs. Mie

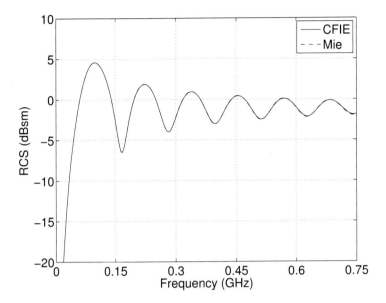

**FIGURE 8.16:** Conducting Sphere: RCS vs. Frequency, CFIE vs. Mie

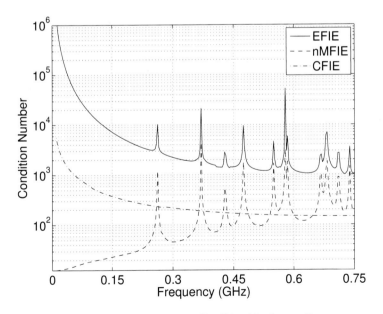

**FIGURE 8.17:** Conducting Sphere: Condition Number vs. Frequency

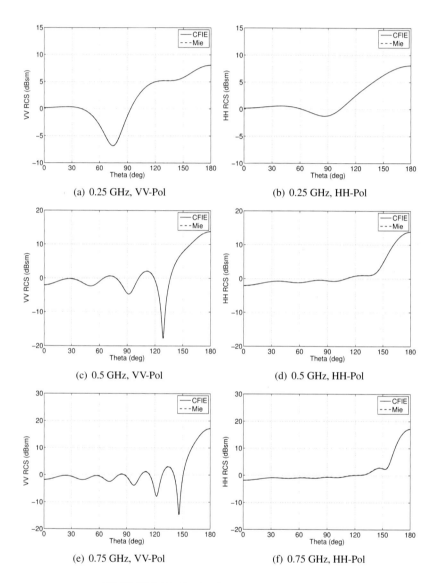

(a) 0.25 GHz, VV-Pol

(b) 0.25 GHz, HH-Pol

(c) 0.5 GHz, VV-Pol

(d) 0.5 GHz, HH-Pol

(e) 0.75 GHz, VV-Pol

(f) 0.75 GHz, HH-Pol

**FIGURE 8.18:** Conducting Sphere: Bistatic RCS

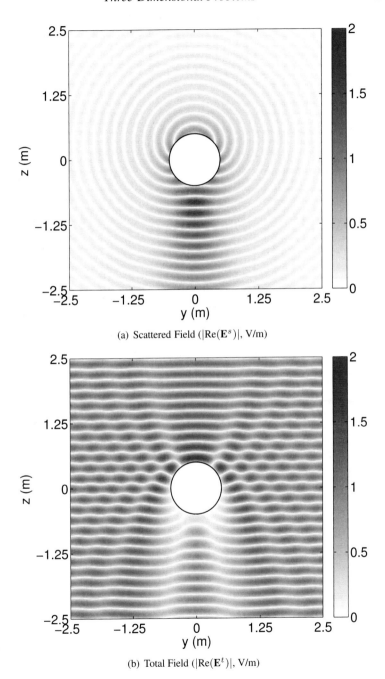

(a) Scattered Field ($|\mathrm{Re}(\mathbf{E}^s)|$, V/m)

(b) Total Field ($|\mathrm{Re}(\mathbf{E}^t)|$, V/m)

**FIGURE 8.19:** Conducting Sphere: Near Field at 750 MHz

### 8.8.3.2 Dielectric Sphere

In this section, we consider a dielectric sphere with a radius of 0.5 meters. The exact expressions for the scattered field of a stratified sphere were previously summarized in Section 7.9.1.2. For the MOM simulation, we re-use the facet model from Section 8.8.3.1.

*Monostatic RCS versus Frequency*

We first compute the RCS for frequencies between 10 and 750 MHz. In Figure 8.20a we compare the Mie series RCS to that of the MOM in backscattering ($\theta^i = \theta^s = 0$) for a lossless sphere ($\epsilon_r = 2.56$). The comparison is very good. In Figure 8.20b is a similar comparison for a lossy sphere ($\epsilon_r = 2.56 - j.102$), where the comparison is again very good. The forward-scattering RCS ($\theta^i = 0, \theta^s = \pi$) for each case is summarized in Figures 8.21a and 8.21b, respectively, where the comparison is again quite good.

*Bistatic RCS*

Using the same simulation parameters as before, we consider the bistatic RCS at 250, 500, and 750 MHz. The MOM results are compared to the Mie series for the lossless sphere in Figures 8.22a–8.22f. The comparison is very good. The results for the lossy sphere are compared in Figures 8.23a–8.23f, where the comparison is again very good.

*Near Field*

Using (3.108) and (3.112), we compute the near field of the lossless dielectric sphere illuminated by an $\hat{x}$-polarized plane wave incident from the $+\hat{z}$ direction at 750 MHz. Figures 8.24a and 8.24b show the real part of the scattered and total electric field, respectively, in the $yz$ plane.

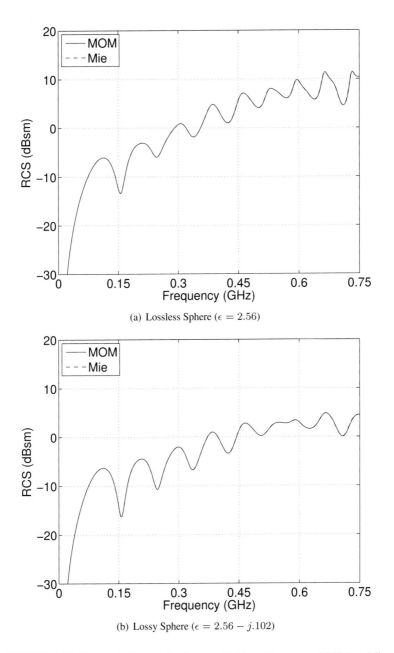

(a) Lossless Sphere ($\epsilon = 2.56$)

(b) Lossy Sphere ($\epsilon = 2.56 - j.102$)

**FIGURE 8.20:** Dielectric Sphere: Backscatter RCS vs. Frequency, MOM vs. Mie

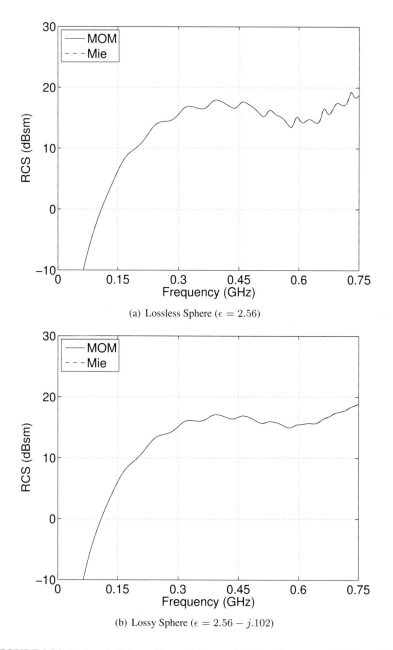

(a) Lossless Sphere ($\epsilon = 2.56$)

(b) Lossy Sphere ($\epsilon = 2.56 - j.102$)

**FIGURE 8.21:** Dielectric Sphere: Forward Scatter RCS vs. Frequency, MOM vs. Mie

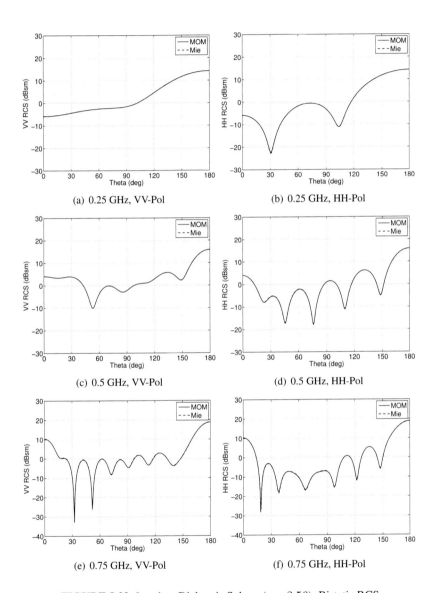

(a) 0.25 GHz, VV-Pol

(b) 0.25 GHz, HH-Pol

(c) 0.5 GHz, VV-Pol

(d) 0.5 GHz, HH-Pol

(e) 0.75 GHz, VV-Pol

(f) 0.75 GHz, HH-Pol

**FIGURE 8.22:** Lossless Dielectric Sphere ($\epsilon = 2.56$): Bistatic RCS

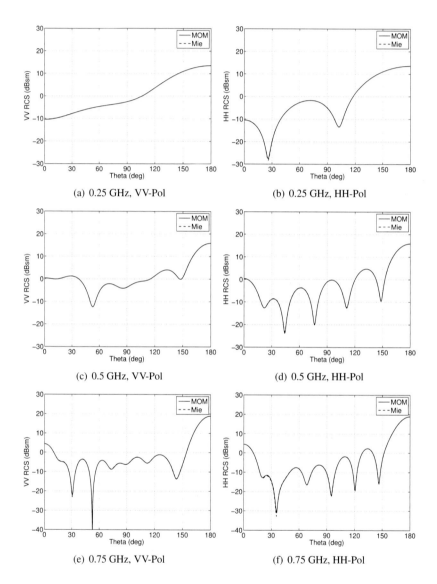

(a) 0.25 GHz, VV-Pol

(b) 0.25 GHz, HH-Pol

(c) 0.5 GHz, VV-Pol

(d) 0.5 GHz, HH-Pol

(e) 0.75 GHz, VV-Pol

(f) 0.75 GHz, HH-Pol

**FIGURE 8.23:** Lossy Dielectric Sphere ($\epsilon = 2.56 - j.102$): Bistatic RCS

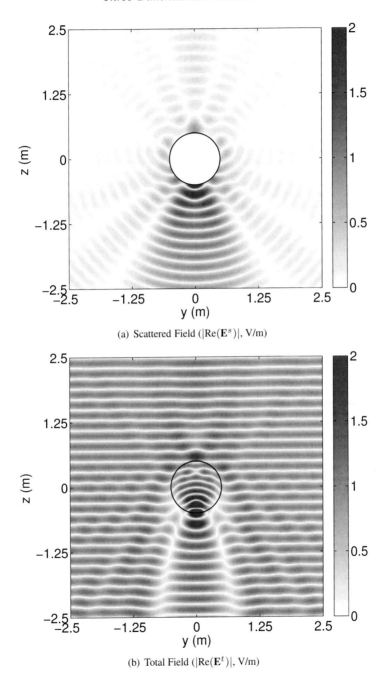

(a) Scattered Field ($|\text{Re}(\mathbf{E}^s)|$, V/m)

(b) Total Field ($|\text{Re}(\mathbf{E}^t)|$, V/m)

**FIGURE 8.24:** Dielectric Sphere: Near Field at 750 MHz

### 8.8.3.3   Coated Sphere

We next consider a conducting sphere with a thick dielectric coating. The radius of the conducting core is 0.25 meters, which we represent using 1280 near-equilateral triangles. The thickness of the dielectric coating is 0.25 meters, yielding an outermost radius of 0.5 meters. In describing the exterior surface, we again use the facet model from Section 8.8.3.1. This yields a total of 6400 triangles for the inner and outer surfaces.

*Monostatic RCS versus Frequency*

We first compute the RCS of the coated sphere for frequencies between 10 and 750 MHz. In Figure 8.25a we compare the Mie series to that of the MOM in backscattering ($\theta^i = \theta^s = 0$) for a lossless coating ($\epsilon_r = 2.56$). The comparison is very good. In Figure 8.25b is a similar comparison for a lossy coating ($\epsilon_r = 2.56 - j.102$). The forward-scattering RCS ($\theta^i = 0, \theta^s = \pi$) for each case is summarized in 8.26a and 8.26b, respectively, where the comparison is again quite good.

*Bistatic RCS*

Using the same parameters as before, we next consider the bistatic RCS at 0.25, 0.5, and 0.75 GHz. The MOM results are compared to the Mie series for the lossless coating in Figures 7.14a–7.14f. The comparison is very good. The results for the lossy coating are compared in Figures 7.15a–7.15f, where the comparison is again very good.

*Near Field*

Using (3.108) and (3.112), we compute the near field of the lossless coated sphere illuminated by an $\hat{\mathbf{x}}$-polarized plane wave incident from the $+\hat{\mathbf{z}}$ direction at 750 MHz. Figures 8.29a and 8.29b show the real part of the scattered and total electric field, respectively, in the $yz$ plane.

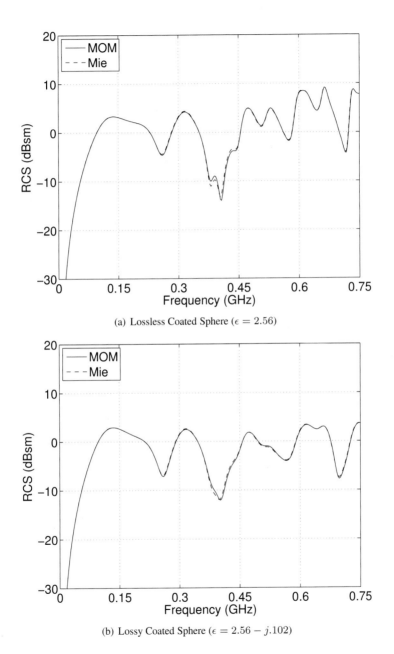

(a) Lossless Coated Sphere ($\epsilon = 2.56$)

(b) Lossy Coated Sphere ($\epsilon = 2.56 - j.102$)

**FIGURE 8.25:** Coated Sphere: Backscatter RCS vs. Frequency, MOM vs. Mie

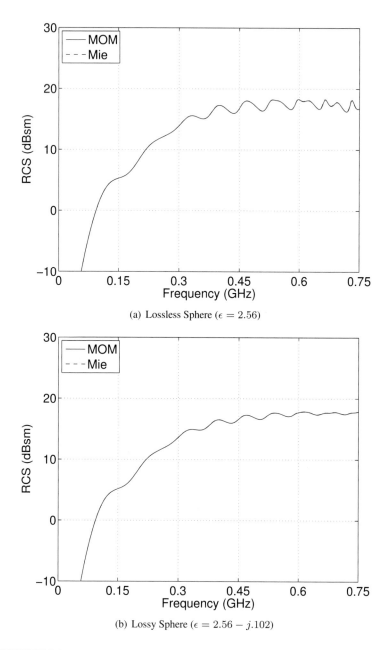

(a) Lossless Sphere ($\epsilon = 2.56$)

(b) Lossy Sphere ($\epsilon = 2.56 - j.102$)

**FIGURE 8.26:** Coated Sphere: Forward Scatter RCS vs. Frequency, MOM vs. Mie

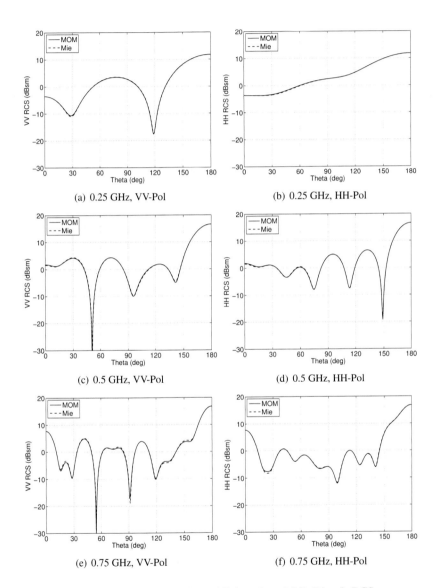

(a) 0.25 GHz, VV-Pol

(b) 0.25 GHz, HH-Pol

(c) 0.5 GHz, VV-Pol

(d) 0.5 GHz, HH-Pol

(e) 0.75 GHz, VV-Pol

(f) 0.75 GHz, HH-Pol

**FIGURE 8.27:** Lossless Coated Sphere ($\epsilon = 2.56$): Bistatic RCS

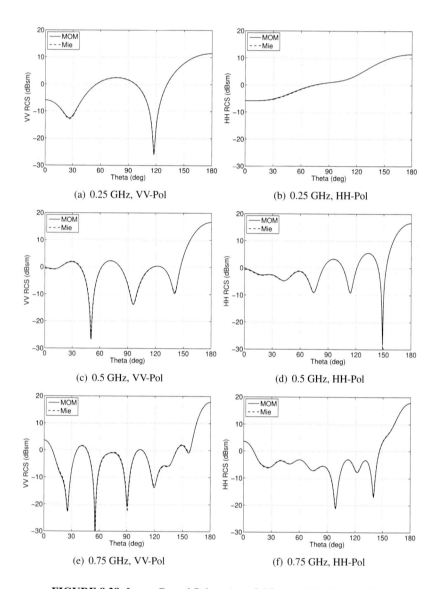

**FIGURE 8.28:** Lossy Coated Sphere ($\epsilon = 2.56 - j.102$): Bistatic RCS

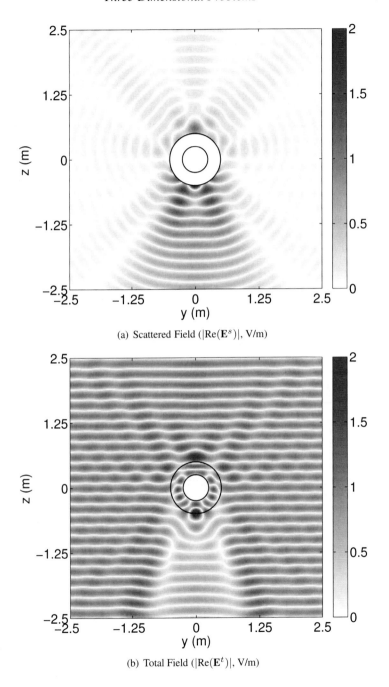

(a) Scattered Field ($|\mathrm{Re}(\mathbf{E}^s)|$, V/m)

(b) Total Field ($|\mathrm{Re}(\mathbf{E}^t)|$, V/m)

**FIGURE 8.29:** Coated Sphere: Near Field at 750 MHz

### 8.8.4   EMCC Plate Benchmark Targets

We next consider the EMCC benchmark plate radar targets described and measured in [22]. Because tip and edge diffractions were of primary interest in this article, the measurements were performed using a conical cut 10 degrees from the plane of each target. The test articles were fabricated from aluminum foil or thin pieces of high-density foam coated with conducting paint. Therefore, our surface models comprise a thin sheet of planar facets, and the EFIE is used. In the plots that follow, the measurements have been shifted in angle slightly to better align them with the numerical results.

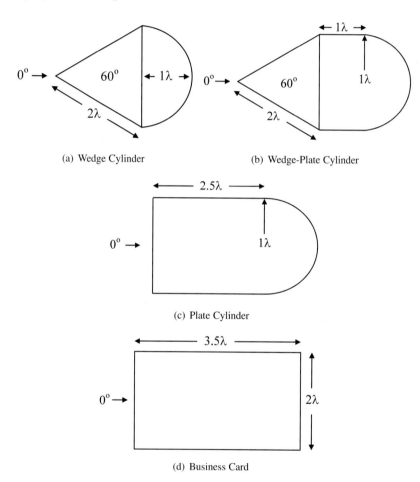

(a) Wedge Cylinder

(b) Wedge-Plate Cylinder

(c) Plate Cylinder

(d) Business Card

**FIGURE 8.30:** EMCC Plate Benchmark Targets

### 8.8.4.1 Wedge Cylinder

The wedge cylinder is shown in Figure 8.30a. The corresponding facet model comprises 3072 triangles, and is shown in Figure 8.31. The monostatic RCS from the EFIE is compared to the EMCC measurements in Figures 8.35a and 8.35b for vertical and horizontal polarizations, respectively. The comparisons are very good for both polarizations, but are noticeably better in the horizontal case. In [22] the measurements were adjusted by 2dB before plotting, however no such adjustment has been made here.

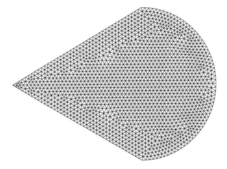

**FIGURE 8.31:** Triangular Surface Model of the Wedge Cylinder

### 8.8.4.2 Wedge-Plate Cylinder

The wedge-plate cylinder is shown in Figure 8.30b. The corresponding facet model comprises 4656 triangles, and is shown in Figure 8.32. The monostatic RCS from the EFIE is compared to the EMCC measurements in Figures 8.36a and 8.36b for vertical and horizontal polarizations, respectively. The comparisons are very good for both polarizations, but again slightly better in the horizontal case.

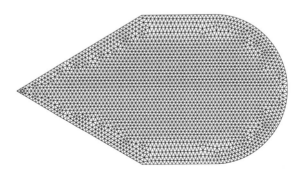

**FIGURE 8.32:** Triangular Surface Model of the Wedge-Plate Cylinder

### 8.8.4.3  Plate Cylinder

The plate cylinder is shown in Figure 8.30c. The corresponding facet model comprises 5058 triangles, and is shown in Figure 8.33. The monostatic RCS from the EFIE is compared to the EMCC measurements in Figures 8.37a and 8.37b for vertical and horizontal polarizations, respectively. The comparison is quite good, however the measurement has a higher amplitude than expected between $-45° \leq \theta \leq 45°$ and $100° \leq \theta \leq 160°$ in Figure 8.37a, the cause of which is unknown.

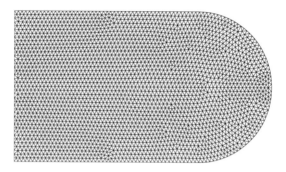

**FIGURE 8.33:** Triangular Surface Model of the Plate Cylinder

### 8.8.4.4  Business Card

The business card is shown in Figure 8.30d. The corresponding facet model comprises 5662 triangles, and is shown in Figure 8.34. The monostatic RCS from the EFIE is compared to the EMCC measurements in Figures 8.38a and 8.38b for vertical and horizontal polarizations, respectively. The comparison is very good for horizontal polarization, though the computed results are lower than the measurement for vertical polarization, which was also noted in [22].

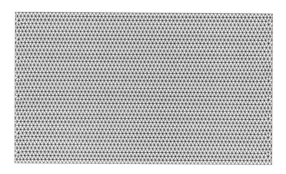

**FIGURE 8.34:** Triangular Surface Model of the Business Card

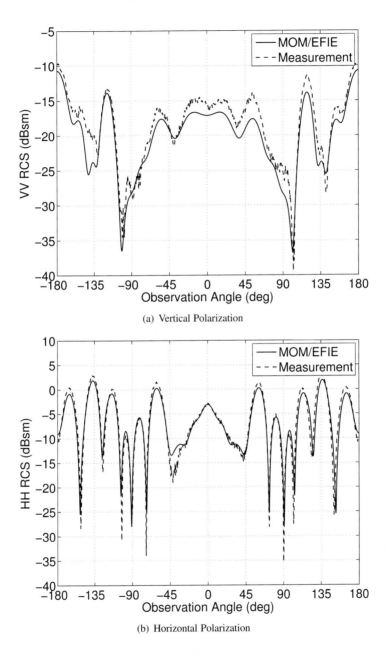

(a) Vertical Polarization

(b) Horizontal Polarization

**FIGURE 8.35:** EMCC Wedge Cylinder: EFIE vs. Measurement

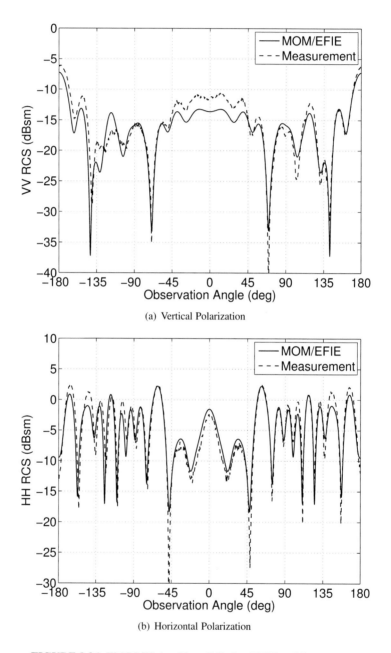

(a) Vertical Polarization

(b) Horizontal Polarization

**FIGURE 8.36:** EMCC Wedge-Plate Cylinder: EFIE vs. Measurement

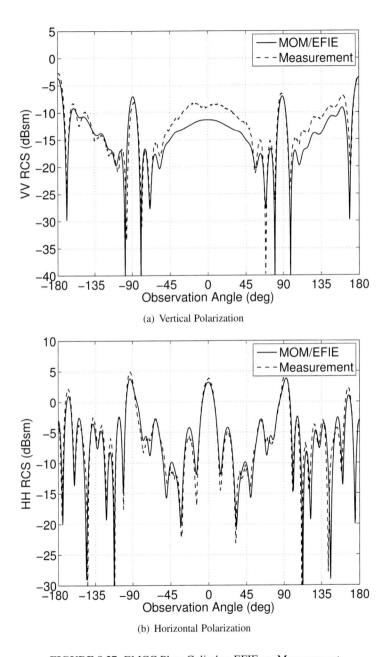

(a) Vertical Polarization

(b) Horizontal Polarization

**FIGURE 8.37:** EMCC Plate Cylinder: EFIE vs. Measurement

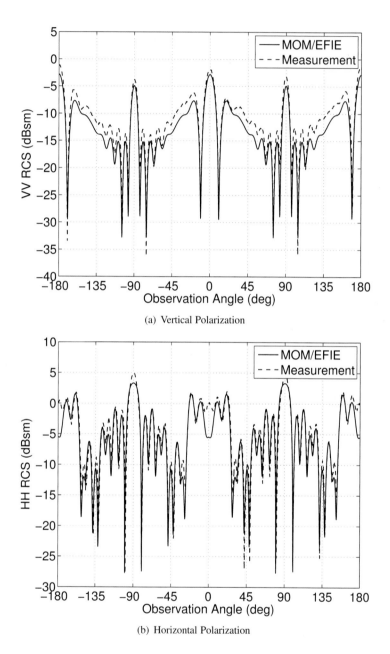

(a) Vertical Polarization

(b) Horizontal Polarization

**FIGURE 8.38:** EMCC Business Card: EFIE vs. Measurement

### 8.8.5 Strip Dipole Antenna

We next consider the input impedance of a thin strip dipole antenna whose dimensions are illustrated in Figure 8.39. This type of geometry is representative of thin foil chaff used in radar countermeasures. The strip considered has a length $L = 1$ meter and width $w = 3$ mm, so Figure 8.39 is not to scale. The facet model comprises 400 triangles, and it is fed at the centermost edge ($e_m$) using the delta-gap model outlined in Section 8.3.3.2. The strip is considered to be infinitely thin, and the EFIE is used. The computed input resistance and reactance are shown in Figure 8.40 for frequencies between 10 MHz and 1 GHz. The first resonance is at approximately 150 MHz, where the dipole is $\lambda/2$ in length.

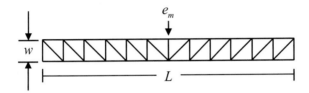

**FIGURE 8.39:** Strip Dipole Dimensions

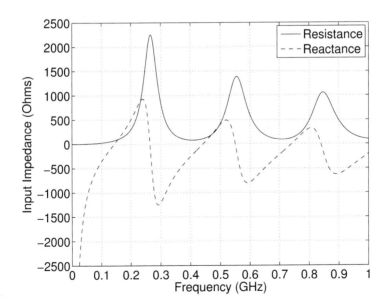

**FIGURE 8.40:** Strip Dipole Input Impedance

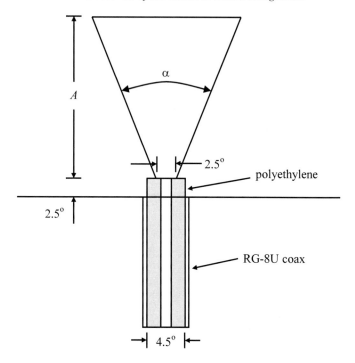

**FIGURE 8.41:** Bowtie Monopole Measurement Setup

## 8.8.6   Bowtie Antenna

The planar bowtie antenna depicted in Figure 8.8a is used extensively in applications where a wide operating bandwidth is necessary. A study of the input impedance and radiation characteristics of the bowtie antenna was presented previously in [23] for various flare angles and lengths. Their measurement setup is illustrated in Figure 8.41, where a half-bowtie was fed and measured against a ground plane that comprised a conducting disk 8 feet in diameter. The feed utilized a RG-8U coaxial cable that was trimmed and inserted through a hole in the disk. The fed end of the bowtie was truncated to facilitate an electrical connection to the center conductor of the cable. Measurements were obtained at a fixed frequency of 500 MHz using a slotted line with sliding probe and transmission line charts. Each antenna was fabricated to the maximum electrical length and physically cut down as the measurements were made. Therefore, the electrical dimensions of the feedpoint and ground plane remained constant.

For the MOM simulation, we will model a full bowtie in free space, and take into account that the input impedance obtained will be approximately twice that of a half-bowtie over a finite conducting ground plane. Where the

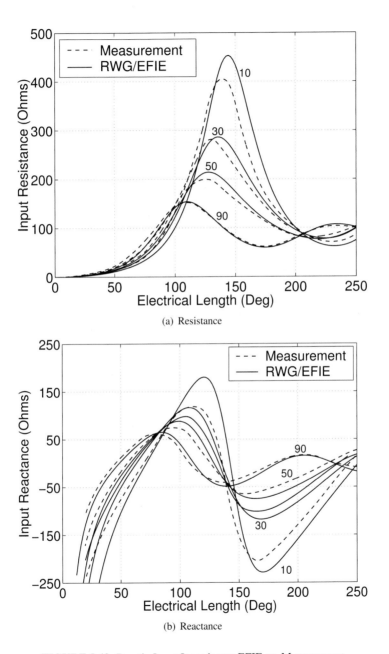

(a) Resistance

(b) Reactance

**FIGURE 8.42:** Bowtie Input Impedance: EFIE vs. Measurement

physical item was cut down to size at each step, in our case it makes more sense to fix the dimensions of the antenna and vary the frequency. We consider flare angles of 10, 30, 50, and 90 degrees, and the corresponding facet models comprise 826, 2010, 3462, and 7472 facets, respectively. Each facet model has a half-length of 1 meter and an edge length of 2 cm at the feedpoint (Figure 8.8a). In Figures 8.42a and 8.42b we compare the input resistance and reactance obtained from MOM with a delta-gap feed to the measurements in [23]. The comparison is fair, considering the differences between the physical and computer models. The disparity between the delta-gap feed model and the physical setup is an obvious difference, and likely contributes to the observed offset between the resonant frequencies as well as the impedance amplitudes. A similar observation was made in [24]. If we had instead used the magnetic frill in Section 5.2.2 and reduced the physical length of the model rather than changing the frequency, the comparison might be better, though it would be more time intensive in terms of setup and execution.

### 8.8.7   Archimedean Spiral Antenna

We next consider the characteristics of the Archimedean spiral antenna [25, 26], which has desirable broadband performance in terms of its input impedance as well as its circularly polarized radiation pattern. The geometry of this antenna is described in Figure 8.43a. The radius of each arm is linearly proportional to the angle and can be written as

$$r = b\phi + r_1 \tag{8.101}$$

and

$$r = b(\phi + \pi) + r_1 \tag{8.102}$$

where $r_1$ is a starting radius and $b$ is a constant that depends on the width $w$ and spacing $s$ of the arms of the antenna. If we choose a self-complimentary spiral with $s = w$, then

$$b = \frac{2w}{\pi} \tag{8.103}$$

If the spiral has an infinite number of turns, then by Babinet's Principle, the input impedance of the antenna should be approximately half that of the surrounding medium. If that medium is free space, then

$$Z_{in} = \frac{\eta_0}{2} \approx 188.4 \, \Omega \tag{8.104}$$

When a voltage is applied to the center terminals of this antenna, it behaves in a manner similar to a two-wire transmission line and gradually transitions into a radiating structure where the circumference of the spiral is one wavelength [26]. In theory, an antenna with the largest number of turns should have the

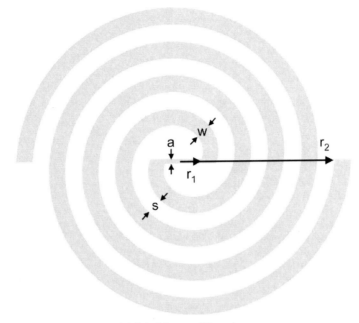

(a) Spiral Antenna Dimensions

(b) Example Spiral Antenna Mesh

**FIGURE 8.43:** Archimedean Spiral Antenna

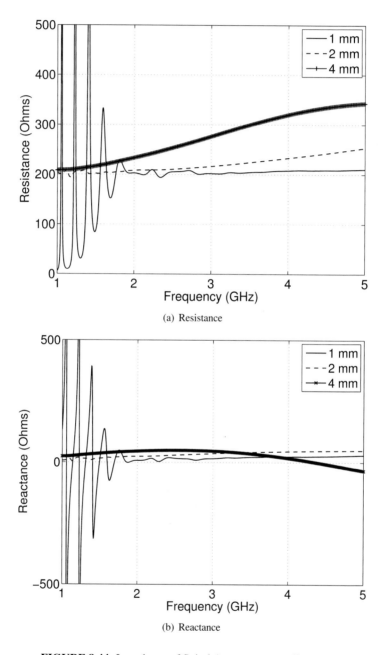

(a) Resistance

(b) Reactance

**FIGURE 8.44:** Impedance of Spiral Antennas versus Frequency

broadest bandwidth, however available space and size will limit its maximum size. A method of connecting the antenna to the feedline must also be devised. All of these considerations will influence the final input impedance, which we do not expect will agree perfectly with the theoretical value of $Z_{in}$. For the numerical model, we choose a starting radius $r_1 = w$ and bridge the arms of the spiral at the center using a flat strip of width $a = w/2$. A heavier triangulation is used near the center, and a lower density is applied to the arms. This tessellation scheme is illustrated in Figure 8.43b, where the arms each comprise two and a half turns $(0 \le \phi \le \frac{5}{2}\pi)$.

We will now attempt to find the dimensions of a spiral antenna having the flattest impedance characteristics between 1 and 5 GHz. To do so, we construct three spiral antennas, each with 8 turns per arm $(0 \le \phi \le 16\pi)$, with widths $w$ of 1, 2, and 4 mm. A delta-gap source is used to excite the edge at the center of the strip bridging the arms. The input resistance and reactance of each antenna are shown in Figures 8.44a and 8.44b, respectively. The antenna of 1 mm width has a nearly constant input resistance of approximately 210 Ohms between 2 and 5 GHz, and an input reactance of around 20 Ohms. The other two antennas have a wider variation throughout this range, though their impedances are somewhat smoother at lower frequencies, which is expected given their larger overall dimensions. We expect that by adding more turns to the 1-mm antenna, its usable range could be extended down to 1 GHz with little impact on its higher-frequency performance. Using the 1-mm design as a starting point, a laboratory model could now be fabricated and a more accurate estimate of the impedance measured using a vector network analyzer (VNA).

### 8.8.8 Monoconic Reentry Vehicle with Dielectric Nose

Finally, we examine the monoconic reentry vehicle with dielectric nose previously described in Section 7.10.2.4. The facet model of this object comprises three separate parts: the nose, septum, and frustum, which are discretized into 1944, 472, and 5824 triangles, respectively, which are shown in an exploded view in Figure 8.45. Because multiple bounces occur inside the dielectric nose, giving rise to more complex fields in that region, the level of facetization on the nose and septum is higher than on the frustum.

*Results*

We compare the monostatic RCS obtained from *Serenity* to that obtained by *Galaxy* at 1.1 GHz. The MOM/BOR geometry and other run settings remain the same as in Section 7.10.2.4. The results for a lossless nose tip $(\epsilon = 5)$ are compared in vertical and horizontal polarizations in Figures 8.46a and 8.46b, respectively. The comparison is very good. The results for a lossy nose tip $(\epsilon = 5 - j0.25)$ are compared in vertical and horizontal polarizations in Figures

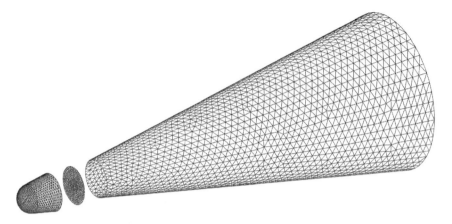

**FIGURE 8.45:** Monoconic RV with Dielectric Nose (Exploded View)

8.47a and 8.47b, respectively, where the comparison is again good. We observe that with the lossy nose tip, the RCS has been reduced slightly at forward aspect angles, as expected.

**TABLE 8.5:** Run Metrics

| Object | $N_J$ | $N_M$ | RAM (MB) | $T_{fill}(s)$ | $T_{LU}(s)$ |
|---|---|---|---|---|---|
| PEC Sphere (EFIE) | 7680 | 0 | 450 | 43 | 170 |
| PEC Sphere (nMFIE) | 7680 | 0 | 450 | 95 | 170 |
| PEC Sphere (CFIE) | 7680 | 0 | 450 | 105 | 170 |
| Dielectric Sphere | 7680 | 7680 | 1800 | 160 | 1427 |
| Coated Sphere | 9600 | 7680 | 2278 | 240 | 2035 |
| Wedge Cylinder | 4537 | 0 | 157 | 16 | 37 |
| Wedge-Plate Cylinder | 6893 | 0 | 362 | 35 | 131 |
| Plate Cylinder | 7486 | 0 | 427 | 42 | 163 |
| Business Card | 8383 | 0 | 536 | 53 | 225 |
| Strip Dipole | 399 | 0 | 1.2 | 0.1 | 0.1 |
| Bowtie (10°) | 1151 | 0 | 10 | 4 | 1 |
| Bowtie (30°) | 2911 | 0 | 65 | 30 | 10 |
| Bowtie (50°) | 5067 | 0 | 196.9 | 86 | 51 |
| Bowtie (90°) | 11013 | 0 | 925 | 425 | 527 |
| Spiral | 4333 | 0 | 143 | 27 | 31 |
| RV | 12340 | 2896 | 1171 | 309 | 1398 |

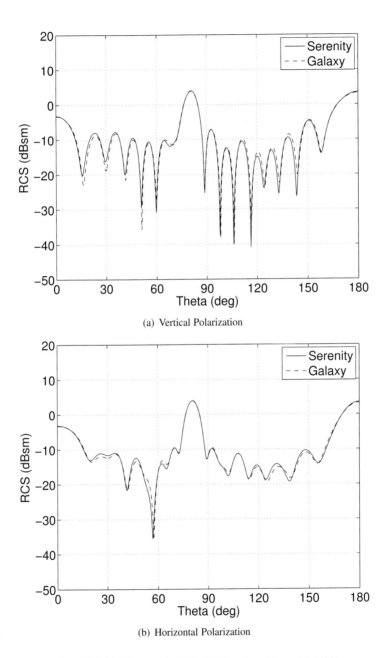

(a) Vertical Polarization

(b) Horizontal Polarization

**FIGURE 8.46:** Monoconic RV RCS: Lossless Nose at 1.1 GHz

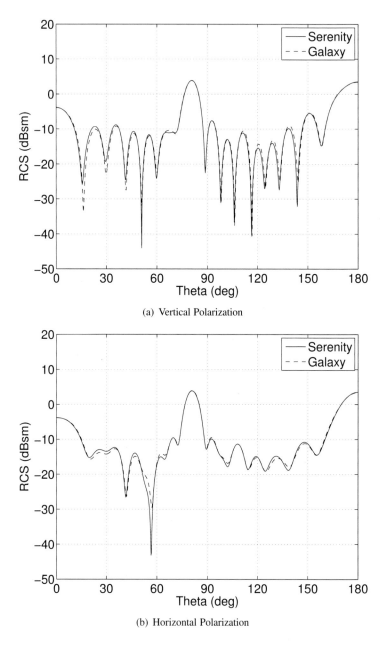

(a) Vertical Polarization

(b) Horizontal Polarization

**FIGURE 8.47:** Monoconic RV RCS: Lossy Nose at 1.1 GHz

### 8.8.9 Summary of Examples

Run metrics for the examples in this section are summarized in Table 8.5. Listed are the total number of electric and magnetic basis functions, system memory (RAM) required to store the system matrix, matrix fill time ($T_{fill}$), and LU factorization time ($T_{LU}$) using the LAPACK function `cgetrf`. As the number of matrix fills, factorizations and right-hand sides vary between examples, that timing information is not shown. We note that in 8 GB of RAM and using 8 bytes for each single-precision complex value, a matrix with just under 32768 unknowns could be handled before all memory would be exhausted. The actual number is probably even less, since a fair amount of memory is required to store the geometry and other data. If the number of unknowns was to now increase to hundreds of thousands or even millions, there is no way the matrix could be stored, even if we installed the maximum amount of memory allowed by the host system's motherboard (32 GB). We will consider an approach to handling problems of this size in the next chapter.

## References

[1] P. Ylä-Oijala, M. Taskinen, and J. Sarvas, "Surface integral equation method for general composite metallic and dielectric structures with junctions," *Progress in Electromagnetics Research*, vol. 52, pp. 81–108, 2005.

[2] Rhinoceros 3D, Robert McNeel and Associates, http://www.rhino3d.com.

[3] L. Piegl and W. Tiller, *The NURBS Book*. Springer, 1995.

[4] J. Foley, A. van Dam, S. Feiner, and J. Hughes, *Computer Graphics: Principles and Practice*. Addison-Wesley, 1996.

[5] D. Wilton, S. Rao, A. Glisson, D. Schaubert, M. Al-Bundak, and C. M. Butler, "Potential integrals for uniform and linear source distributions on polygonal and polyhedral domains," *IEEE Trans. Antennas Propagat.*, vol. 32, pp. 276–281, March 1984.

[6] S. Caorsi, D. Moreno, and F. Sidoti, "Theoretical and numerical treatment of surface integrals involving the free-space Green's function," *IEEE Trans. Antennas Propagat.*, vol. 41, pp. 1296–1301, September 1993.

[7] R. D. Graglia, "On the numerical integration of the linear shape functions times the 3-D Green's function or its gradient on a plane triangle," *IEEE Trans. Antennas Propagat.*, vol. 41, pp. 1448–1454, October 1993.

[8] T. F. Eibert and V. Hansen, "On the calculation of potential integrals for linear source distributions on triangular domains," *IEEE Trans. Antennas Propagat.*, vol. 43, pp. 1499–1502, December 1995.

[9] J. Jin, *The Finite Element Method in Electromagnetics*. John Wiley and Sons, 1993.

[10] S. Rao, D. Wilton, and A. Glisson, "Electromagnetic scattering by surfaces of arbitrary shape," *IEEE Trans. Antennas Propagat.*, vol. 30, pp. 409–418, May 1982.

[11] R. D. Graglia, D. R. Wilton, and A. F. Peterson, "Higher order interpolatory vector bases for computational electromagnetics," *IEEE Trans. Antennas Propagat.*, vol. 45, pp. 329–342, March 1997.

[12] M. G. Duffy, "Quadrature over a pyramid or cube of integrands with a singularity at a vertex," *SIAM J. Numer. Anal*, vol. 19, pp. 1260–1262, December 1982.

[13] A. Tzoulis and T. Eibert, "Review of singular potential integrals for method of moments solutions of surface integral equations," *Advances in Radio Science*, vol. 2, pp. 97–99, 2004.

[14] W. C. Chew, J. M. Jin, E. Michielssen, and J. Song, *Fast and Efficient Algorithms in Computational Electromagnetics*. Artech House, 2001.

[15] P. Ylä-Oijala and M. Taskinen, "Calculation of CFIE impedance matrix elements with RWG and n x RWG functions," *IEEE Trans. Antennas Propagat.*, vol. 51, pp. 1837–1846, August 2003.

[16] R. E. Hodges and Y. Rahmat-Samii, "The evaluation of MFIE integrals with the use of vector triangle basis functions," *Microw. Opt. Tech. Lett.*, vol. 14, pp. 9–14, January 1997.

[17] K. Faison, "The Electromagnetics Code Consortium," *IEEE Antennas Propagat. Magazine*, vol. 30, pp. 19–23, February 1990.

[18] R. C. Whaley and A. Petitet, "Minimizing development and maintenance costs in supporting persistently optimized BLAS," *Software: Practice and Experience*, vol. 35, pp. 101–121, February 2005.

[19] R. C. Whaley, A. Petitet, and J. J. Dongarra, "Automated empirical optimization of software and the ATLAS project," *Parallel Computing*, vol. 27, no. 1–2, pp. 3–35, 2001.

[20] J. S. Zhao and W. C. Chew, "Integral equation solution of Maxwell's equations from zero frequency to microwave frequencies," *IEEE Trans. Antennas Propagat.*, vol. 48, pp. 1635–1645, October 2000.

[21] J. F. Lee, R. Lee, and R. J. Burkholder, "Loop star basis functions and a robust preconditioner for EFIE scattering problems," *IEEE Trans. Antennas Propagat.*, vol. 51, pp. 1855–1863, August 2003.

[22] A. C. Woo, H. T. G. Wang, M. J. Schuh, and M. L. Sanders, "Benchmark plate radar targets for the validation of computational electromagnetics programs," *IEEE Antennas Propagat. Magazine*, vol. 34, pp. 52–56, December 1992.

[23] G. H. Brown and J. O. M. Woodward, "Experimentally determined radiation characteristics of conical and triangular antennas," *RCA Review*, pp. 425–452, 1952.

[24] C. Leat, N. Shuley, and G. Stickley, "Triangular-patch model of bowtie antennas: Validation against Brown and Woodward," *IEE Proc.-Microw. Antennas Propag.*, vol. 145, pp. 465–470, December 1998.

[25] W. L. Curtis, "Spiral antennas," *IRE Trans. Antennas Propagat.*, vol. 8, pp. 298–306, May 1960.

[26] J. A. Kaiser, "The Archimedean two-wire spiral antenna," *IRE Trans. Antennas Propagat.*, vol. 8, pp. 312–323, May 1960.

# Chapter 9

## *The Fast Multipole Method*

In previous chapters we observed and discussed the issues and complexities encountered in applying the Method of Moments to larger problems. Foremost among these are the limits in storing the MOM system matrix in memory, and the CPU requirements of the LU factorization, which increases exponentially with the number of unknowns. Though one of the iterative solvers discussed in Section 4.2 may yield a shorter compute time for a small number of right-hand sides, the run time may still be quite high due to the number of matrix-vector products per iteration. If the system matrix will not fit into memory, there are few options besides changing the surface geometry to reduce the number of unknowns, or using virtual (disk) memory and an out-of-core solver, though this may make the problem intractable due to the far slower read/write speed of the hard disk. This motivates the search for a technique that will reduce or eliminate many of these difficulties.

Let us consider the $N$-body problem, often encountered in many areas of applied physics. The most classic example of the $N$-body problem is the motion of celestial objects interacting with each other gravitationally. For example, if we wish to know the position and velocity of a star at some point in time, the differential equations of motion must be numerically integrated in time. This requires a calculation of the gravitational forces from all other stars at each time step in the integration. Similarly, as the other stars are also moving, the gravitational forces on them and their position and velocity must also be computed, resulting in an $N \times N$ problem at each point in time, where $N$ is the number of stars. When $N$ is small, a brute-force pairwise calculation of the forces can be carried out at each time step. However, when $N$ becomes very large, the number of calculations in the brute-force approach may grow so large as to be impractical or impossible.

A successful approach in reducing the complexity of the $N$-body problem was introduced by Greengard and Rokhlin in [1], and is called the Fast Multipole Method (FMM). In this method, the gravitational potential functions of a group of nearby particles are re-written in terms of their multipole expansions, which are combined into a single function that represents the potential of that group. The interaction between that group and any particles far away from that group can then be computed using the multipole expansion. Thus, the FMM allows for grouping together many nearby particles and treating them as if they are a single particle.

We can also consider vector electromagnetic problems and the Moment Method in this context. The assembly of the system matrix can be viewed as another type of $N$-body problem, where each matrix element is the field radiated by one basis function and received by another. Though the matrix is full and of size $N \times N$, the amplitude of the fields drops off quickly with increasing distance between basis functions. Thus, it makes sense to consider whether the interaction between groups of basis functions sufficiently far away from one another can be treated using the FMM. As we will see in this chapter, the FMM can be applied to vector Helmholtz problems, allowing for a fast computation of the matrix-vector product in an iterative solver, and eliminating the need to store many of the MOM matrix elements explicitly. This results in a speed increase and a reduction in memory requirement, allowing us to solve existing problems faster, as well as larger problems that could not be attempted before.

## 9.1   Matrix-Vector Product

Let us first consider the matrix-vector product in the context of the Moment Method. The system matrix $\mathbf{Z}$ represents the interactions (fields) between basis and testing functions, and the right-hand side vector $\mathbf{y}$ comprises the excitation of each basis function. When we compute the matrix-vector product

$$\mathbf{x} = \mathbf{Z}\mathbf{y} \tag{9.1}$$

the rows of the column vector $\mathbf{x}$ are the coherent sum of those fields "received" by each testing function due to the fields "radiated" by all basis functions. Noting that the amplitude of the interaction between basis and testing functions diminishes quickly with distance, let us then divide the space around a basis function into two regions: a *near*-field region with stronger interactions, and a *far*-field region with weaker interactions. This is illustrated in Figure 9.1, where the points inside the shaded circular region of diameter $d$ around point $A$ are in the *near* region, and those outside are in the *far* region. Note that these are *not* the near and far field regions described in Section 3.5.

Consider the next row $m$ of $\mathbf{Z}$, which we denote as $\mathbf{z}_{m*}$. The inner product of this row with the column vector $\mathbf{b}$ yields the column vector element $a_m$, which comprises the fields received by testing function $m$ due to all basis functions. This can be written as

$$x_m = \mathbf{z}_{m*} \cdot \mathbf{y} = [Z_{m1}, Z_{m2}, Z_{m3}, \ldots, Z_{mN}] \cdot [y_1, y_2, y_3, \ldots, y_N] \tag{9.2}$$

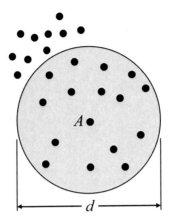

**FIGURE 9.1:** Near and Far Points

As some basis functions are in the near region of testing function $m$ and the others are in the far region, the vectors can be re-ordered so that (9.2) reads as

$$x_m = \mathbf{z}_{m*} \cdot \mathbf{y} = \left[ \mathbf{z}_{m*}^{near} \; \mathbf{z}_{m*}^{far} \right] \cdot \left[ \mathbf{y}^{near} \mathbf{y}^{far} \right] = \mathbf{z}_{m*}^{near} \cdot \mathbf{y}^{near} + \mathbf{z}_{m*}^{far} \cdot \mathbf{y}^{far} \quad (9.3)$$

where the elements in the sub-vectors $\mathbf{z}_{m*}^{near}$ and $\mathbf{y}^{near}$ are due to the basis functions in the region near to testing function $m$, and those in $\mathbf{z}_{m*}^{far}$ and $\mathbf{y}^{far}$ are due to those in the far region. If we now write the system matrix as

$$\mathbf{Z} = \mathbf{Z}^{near} + \mathbf{Z}^{far} \quad (9.4)$$

where $\mathbf{Z}^{near}$ and $\mathbf{Z}^{far}$ are sparse, the matrix-vector product can be written as the sum of a near product and a far product, yielding

$$\mathbf{x} = \mathbf{x}^{near} + \mathbf{x}^{far} = \mathbf{Z}^{near}\mathbf{y} + \mathbf{Z}^{far}\mathbf{y} \quad (9.5)$$

In the fast multipole method, the matrix-vector product is performed in two steps: a matrix-vector product involving $\mathbf{Z}^{near}$, where the elements are stored explicitly and computed the usual way, and a matrix-vector product involving $\mathbf{Z}^{far}$, where individual elements are not computed or available explicitly. We will discuss the practical implementation of these steps throughout the rest of this chapter. As we will see, one the key benefits of this method is that a large number of the matrix elements are located in $\mathbf{Z}^{far}$, yielding a very sparse $\mathbf{Z}^{near}$ and a tremendous savings in memory.

## 9.2   Addition Theorem

Consider again the three dimensional Green's function (3.39), which can be written as

$$G(\mathbf{r}, \mathbf{r}') = \frac{e^{-jk|\mathbf{r}-\mathbf{r}'|}}{4\pi|\mathbf{r} - \mathbf{r}'|} \tag{9.6}$$

where $\mathbf{r}'$ and $\mathbf{r}$ are the source and field points, respectively. If we now add to the field point a small vector offset $\mathbf{x}$, we can write (9.6) as

$$\frac{e^{-jk|\mathbf{r}-\mathbf{r}'+\mathbf{x}|}}{4\pi|\mathbf{r} - \mathbf{r}' + \mathbf{x}|} = \frac{e^{-jk|\mathbf{R}+\mathbf{x}|}}{4\pi|\mathbf{R} + \mathbf{x}|} \tag{9.7}$$

where $\mathbf{R} = \mathbf{r} - \mathbf{r}'$. Provided that $|\mathbf{x}| < |\mathbf{R}|$, we now define the *addition theorem* for spherical waves, which is [2, 3]

$$\frac{e^{-jk|\mathbf{R}+\mathbf{x}|}}{4\pi|\mathbf{R} + \mathbf{x}|} = -\frac{jk}{4\pi} \sum_{l=0}^{\infty} (-1)^l (2l + 1) j_l\left(k|\mathbf{x}|\right) h_l^{(2)}\left(k|\mathbf{R}|\right) P_l\left(\hat{\mathbf{x}} \cdot \hat{\mathbf{R}}\right) \tag{9.8}$$

where $j_1(x)$ is a spherical Bessel function of the first kind, and $P_l(x)$ a Legendre polynomial of order $l$. $h_l^{(2)}(x)$ is a spherical Hankel function of the second kind, i.e.,

$$h_l^{(2)}(x) = \sqrt{\frac{\pi}{2x}} H_{l+1/2}^{(2)}(x) = j_l(x) - jn_l(x) \tag{9.9}$$

where $j_l(x)$ and $n_l(x)$ are spherical Bessel functions of the first and second kinds, respectively. Looking closely at (9.8), we see that it is a way of writing a wave radiating from a point as if it were radiating from a different point nearby. Using Stratton [4], (9.8) can be converted to a surface integral on the unit sphere via the relationship

$$4\pi(-j^l) j_l\left(k|\mathbf{x}|\right) P_l\left(\hat{\mathbf{x}} \cdot \hat{\mathbf{R}}\right) = \int_1 e^{-jk\hat{\mathbf{k}} \cdot \mathbf{x}} P_l\left(\hat{\mathbf{k}} \cdot \hat{\mathbf{R}}\right) dS \tag{9.10}$$

where $\hat{\mathbf{k}}$ is the radially oriented unit vector on the sphere. Using this expression allows us to then write

$$\frac{e^{-jk|\mathbf{R}+\mathbf{x}|}}{4\pi|\mathbf{R} + \mathbf{x}|} = \frac{k}{(4\pi)^2} \int_1 e^{-jk\hat{\mathbf{k}} \cdot \mathbf{x}} \sum_{l=0}^{\infty} (-j)^{l+1} (2l + 1) h_l^{(2)}\left(k|\mathbf{R}|\right) P_l(\hat{\mathbf{k}} \cdot \hat{\mathbf{R}}) dS$$

$$\tag{9.11}$$

where the integration and summation have been exchanged. This exchange will be legitimate provided that we truncate the summation at some finite order $L$,

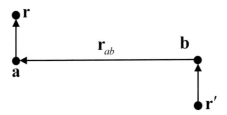

**FIGURE 9.2:** Wave Translation

which limits the permissible values for $\mathbf{R}$ and $\mathbf{d}$. Making this truncation, we can now write (9.11) as

$$\frac{e^{-jk|\mathbf{R}+\mathbf{x}|}}{4\pi|\mathbf{R}+\mathbf{x}|} = \int_1 e^{-jk\hat{\mathbf{k}}\cdot\mathbf{x}} T_L(k,\hat{\mathbf{k}},\mathbf{R})\, dS \qquad (9.12)$$

where

$$T_L(k,\hat{\mathbf{k}},\mathbf{R}) = \frac{k}{(4\pi)^2} \sum_{l=0}^{L}(-j)^{l+1}(2l+1)h_l^{(2)}(k|\mathbf{R}|)P_l(\hat{\mathbf{k}}\cdot\hat{\mathbf{R}}) \qquad (9.13)$$

We refer to $T_L(k,\hat{\mathbf{k}},\mathbf{R})$ as the *transfer function*. This function allows us to convert outgoing spherical waves at a source point to a set of incoming spherical waves at a field point. In practice, the integration on the unit sphere in (9.12) is carried out numerically with quadratures in the $\theta$ and $\phi$ dimensions. We will discuss the types of quadratures and the sampling rates in more detail in later sections.

## 9.2.1 Wave Translation

Let us illustrate the addition theorem by calculating the Green's function between a source point $\mathbf{r}'$ and observation point $\mathbf{r}$. We first we define the vector

$$\mathbf{v} = \mathbf{r} - \mathbf{r}' \qquad (9.14)$$

and then introduce two points $\mathbf{a}$ and $\mathbf{b}$ that are very close to $\mathbf{r}$ and $\mathbf{r}'$, respectively, as shown in Figure 9.2. We can now write (9.14) as

$$\mathbf{v} = (\mathbf{r} - \mathbf{a}) + (\mathbf{a} - \mathbf{b}) - (\mathbf{r}' - \mathbf{b}) \qquad (9.15)$$

or

$$\mathbf{v} = \mathbf{r}_{ra} + \mathbf{r}_{ab} - \mathbf{r}_{r'b} \qquad (9.16)$$

Using the above, (9.12) can be written as

$$\frac{e^{-jk|\mathbf{r}-\mathbf{r}'|}}{4\pi|\mathbf{r}-\mathbf{r}'|} = \int_1 e^{-jk\hat{\mathbf{k}}\cdot(\mathbf{r}_{ra}-\mathbf{r}_{r'b})}T_L(k,\hat{\mathbf{k}},\mathbf{r}_{ab})\, dS \qquad (9.17)$$

where

$$T_L(k, \hat{\mathbf{k}}, \mathbf{r}_{ab}) = \frac{k}{(4\pi)^2} \sum_{l=0}^{L} (-j)^{l+1} (2l+1) h_l^{(2)} (k|\mathbf{r}_{ab}|) P_l (\hat{\mathbf{k}} \cdot \hat{\mathbf{r}}_{ab}) \quad (9.18)$$

This is an important result, as the transfer function depends only on $\mathbf{r}_{ab}$. If we were to move $\mathbf{r}$ and $\mathbf{r}'$ a small distance from their previous location, the transfer function remains unchanged. Therefore, we can compute the interaction between *any* two points near **a** and **b** using the *same* transfer function. Let us now write (9.17) as

$$\frac{e^{-jk|\mathbf{r}-\mathbf{r}'|}}{4\pi|\mathbf{r}-\mathbf{r}'|} = \int_1 R(\mathbf{r}, \hat{\mathbf{k}}) \, T_L(k, \hat{\mathbf{k}}, \mathbf{r}_{ab}) T(\mathbf{r}', \hat{\mathbf{k}}) \, dS \quad (9.19)$$

where at the source point $\mathbf{r}'$, we define the *radiation function*

$$T(\mathbf{r}', \hat{\mathbf{k}}) = e^{jk\hat{\mathbf{k}} \cdot \mathbf{r}_{r'b}} \quad (9.20)$$

and at the field point $\mathbf{r}$, we define the *receive function*

$$R(\mathbf{r}, \hat{\mathbf{k}}) = e^{-jk\hat{\mathbf{k}} \cdot \mathbf{r}_{ra}} \quad (9.21)$$

If we now move $\mathbf{r}'$ or $\mathbf{r}$ to another location, only the radiation or receive function changes. As a result, we can now compute a sum of the Green's functions evaluated at the field point $\mathbf{r}$ due to *many* source points $\mathbf{r}'_n$ located close to **b**. Using (9.19), this can be written as

$$\sum_{n=1}^{N} \frac{e^{-jk|\mathbf{r}-\mathbf{r}'_n|}}{4\pi|\mathbf{r}-\mathbf{r}'_n|} = \int_1 R(\mathbf{r}, \hat{\mathbf{k}}) \, T_L(k, \hat{\mathbf{k}}, \mathbf{r}_{ab}) \sum_{n=1}^{N} T_n(\mathbf{r}', \hat{\mathbf{k}}) \, dS \quad (9.22)$$

where the radiation functions are

$$T_n(\mathbf{r}', \hat{\mathbf{k}}) = e^{jk\hat{\mathbf{k}} \cdot \mathbf{r}_{r'_n b}} \quad (9.23)$$

The arrangement in (9.22) is what allows us to quickly compute a matrix-vector product. The radiation functions for all source points are coherently added (*aggregated*) to create a local field at **b**. This field is then *transmitted* via the transfer function yielding a local field at **a**. The local field is then multiplied with the receive function at the field point and integrated on the unit sphere (*disaggregated*) yielding the desired sum. In a practical FMM implementation, the basis and testing functions are sorted into groups. The radiation and receive functions for all basis and testing functions, as well as the transfer functions linking together all groups, are precomputed. The far part of the matrix-vector product then comprises aggregations, transfers, and disaggregations between groups.

#### 9.2.1.1 Complex Wavenumbers

In lossy dielectric regions having a complex-valued wavenumber, the transfer function (9.18) remains valid, however special numerical routines are needed to compute the spherical Hankel function for non-integer orders and complex arguments. The subroutines by Amos [5] are ideal for this purpose, in particular the routines `cbesh` and `zbesh` for single- and double-precision complex, respectively. These routines are available via the Netlib repository at `http://www.netlib.org/amos`, and are often integrated in software libraries for computing special functions.

## 9.3 FMM Matrix Elements

Having discussed the addition theorem and explained the concept of wave translation and its place in the matrix-vector product, we will now derive expressions for the radiation and receive functions used to compute the MOM matrix elements (3.185–3.190). Though the examples in this chapter use a three-dimensional approach with the RWG basis and testing functions from Chapter 8, the results in this section are independent of basis function. We remind the reader that though the radiation and receive functions to follow are derived in the context of a single source-field pair, the FMM matrix-vector product operates on aggregated (summed) radiation functions, so the individual matrix elements will not be available explicitly. We also allow the points **a** and **b** to remain undefined for now. We will specify them later when the basis functions are sorted into groups.

### 9.3.1 EFIE

The EFIE (3.178) contains $\mathcal{L}$ and $\mathcal{K}$ operators, which were discretized and tested via the Moment Method in (3.185) and (3.186). We will derive the corresponding radiation and receive functions in this section.

#### 9.3.1.1 $\mathcal{L}$ Operator

Let us focus our attention on the second term on the right in (3.185). As the FMM interactions are not in the near zone, we do not need to re-distribute the differential operators as in Section 3.6.2.3. Instead, we will follow Section 3.4.4 and move them so that they operate only on the Green's function. This yields in Region $R_l$ matrix elements given by

$$Z_{mn}^{EJ(l)} = \mathbf{L}(\mathbf{f}_m, \mathbf{f}_n) \tag{9.24}$$

where

$$L(\mathbf{f}_m, \mathbf{f}_n) = j\omega\mu_l \int_{\mathbf{f}_m} \mathbf{f}_m(\mathbf{r}) \cdot \int_{\mathbf{f}_n} \mathbf{f}_n(\mathbf{r}')\left[1 - \frac{1}{k_l^2}\nabla\nabla'\right]G(\mathbf{r}, \mathbf{r}') \, d\mathbf{r}' \, d\mathbf{r} \quad (9.25)$$

We next apply the differential operations to the Green's function in (9.17). This is accomplished by noting that the $T_L(k, \hat{\mathbf{k}}, \mathbf{r}_{ab})$ is not a function of $r$ or $r'$, and the vectors $\mathbf{r}_{ra}$ and $\mathbf{r}_{r'b}$ are radially directed vectors in field and source coordinates, respectively, with $\mathbf{a}$ and $\mathbf{b}$ as the origin. Inserting the result into (9.25) yields an expression of the form

$$L(\mathbf{f}_m, \mathbf{f}_n) = \int_1 \mathbf{R}^{\mathcal{L}}(\mathbf{f}_m, \hat{\mathbf{k}}) \cdot T_L(k_l, \hat{\mathbf{k}}, \mathbf{r}_{ab})\mathbf{T}^{\mathcal{L}}(\mathbf{f}_n, \hat{\mathbf{k}}) \, dS \quad (9.26)$$

where the radiation function for basis function $\mathbf{f}_n(\mathbf{r}')$ is

$$\mathbf{T}^{\mathcal{L}}(\mathbf{f}_n, \hat{\mathbf{k}}) = \left[1 - \hat{\mathbf{k}}\hat{\mathbf{k}}\right] \cdot \int_{\mathbf{f}_n} \mathbf{f}_n(\mathbf{r}')e^{jk_l\hat{\mathbf{k}}\cdot\mathbf{r}_{r'b}} \, d\mathbf{r}' \quad (9.27)$$

and the receive function for testing function $\mathbf{f}_m(\mathbf{r})$ is

$$\mathbf{R}^{\mathcal{L}}(\mathbf{f}_m, \hat{\mathbf{k}}) = j\omega\mu_l \int_{\mathbf{f}_m} \mathbf{f}_m(\mathbf{r})e^{-jk_l\hat{\mathbf{k}}\cdot\mathbf{r}_{ra}} \, d\mathbf{r} \quad (9.28)$$

We note that the expression $\left[1 - \hat{\mathbf{k}}\hat{\mathbf{k}}\right]$ in (9.27) acts to remove the components of the integral along $\hat{\mathbf{k}}$, leaving only $\hat{\theta}$ and $\hat{\phi}$ components.

### 9.3.1.2  $\mathcal{K}$ Operator

As the basis and testing functions are well separated, the supports of $\mathbf{f}_m$ and $\mathbf{g}_n$ do not overlap and we retain only the first term on the right-hand side of (3.186). This yields in Region $R_l$ matrix elements given by

$$Z_{mn}^{EM(l)} = K(\mathbf{f}_m, \mathbf{g}_n) \quad (9.29)$$

where

$$K(\mathbf{f}_m, \mathbf{g}_n) = \int_{\mathbf{f}_m} \mathbf{f}_m(\mathbf{r}) \cdot \int_{\mathbf{g}_n} \mathbf{g}_n(\mathbf{r}) \times \nabla'G(\mathbf{r}, \mathbf{r}') \, d\mathbf{r}' \, d\mathbf{r} \quad (9.30)$$

Taking the gradient of (9.17) and re-arranging the cross product yields

$$K(\mathbf{f}_m, \mathbf{g}_n) = \int_1 \mathbf{R}^{\mathcal{K}}(\mathbf{f}_m, \hat{\mathbf{k}}) \cdot T_L(k_l, \hat{\mathbf{k}}, \mathbf{r}_{ab})\mathbf{T}^{\mathcal{K}}(\mathbf{g}_n, \hat{\mathbf{k}}) \, dS \quad (9.31)$$

where the radiation function for basis function $\mathbf{g}_n(\mathbf{r}')$ is

$$\mathbf{T}^{\mathcal{K}}(\mathbf{g}_n, \hat{\mathbf{k}}) = \int_{\mathbf{g}_n} \mathbf{g}_n(\mathbf{r}')e^{jk_l\hat{\mathbf{k}}\cdot\mathbf{r}_{r'b}} \, d\mathbf{r}' \quad (9.32)$$

and the receive function for testing function $\mathbf{f}_m(\mathbf{r})$ is

$$\mathbf{R}^{\mathcal{K}}(\mathbf{f}_m, \hat{\mathbf{k}}) = jk_l\, \hat{\mathbf{k}} \times \int_{\mathbf{f}_m} \mathbf{f}_m(\mathbf{r}) e^{-jk_l\hat{\mathbf{k}}\cdot\mathbf{r}_{ra}}\, d\mathbf{r} \qquad (9.33)$$

The cross product with $\hat{\mathbf{k}}$ in (9.33) results in the receive function having only $\hat{\boldsymbol{\theta}}$ and $\hat{\boldsymbol{\phi}}$ components. An $\eta_0$ scale factor must also be applied to the radiation function to account for the magnetic basis function.

### 9.3.2 MFIE

The MFIE (3.179) also has $\mathcal{K}$ and $\mathcal{L}$ operators, which were discretized and tested via the Moment Method in (3.187) and (3.188). Substitution of the electric basis and magnetic testing functions into (3.187) yields in Region $R_l$ matrix elements given by

$$Z_{mn}^{HJ(l)} = -\mathrm{K}(\mathbf{g}_m, \mathbf{f}_n) \qquad (9.34)$$

and substitution of the magnetic basis and testing functions into (3.188) yields

$$Z_{mn}^{HM(l)} = \frac{\epsilon_l}{\mu_l}\mathrm{L}(\mathbf{g}_m, \mathbf{g}_n) \qquad (9.35)$$

where the L and K elements were defined in Sections 9.3.1.1 and 9.3.1.2. An $\eta_0$ scale factor must also be applied to the radiation and receive functions to account for the presence of the magnetic basis functions, where appropriate.

### 9.3.3 nMFIE

The nMFIE (3.180) contains the $\hat{\mathbf{n}} \times \mathcal{L}$ and $\hat{\mathbf{n}} \times \mathcal{K}$ operators, which were discretized and tested via the Moment Method in (3.189) and (3.190). We derive their radiation and receive functions in this section.

#### 9.3.3.1 $\hat{\mathbf{n}} \times \mathcal{K}$ Operator

As the FMM treats only well-separated basis functions, the supports of $\mathbf{f}_m$ and $\mathbf{f}_n$ will not overlap and we retain only the first term on the right-hand side of (3.188). This yields in Region $R_l$ matrix elements given by

$$Z_{mn}^{nHJ(l)} = \mathrm{nK}(\mathbf{f}_m, \mathbf{f}_n) \qquad (9.36)$$

where

$$\mathrm{nK}(\mathbf{f}_m, \mathbf{f}_n) = -\int_{\mathbf{f}_m} \mathbf{f}_m(\mathbf{r}) \cdot \left[ \hat{\mathbf{n}}_l(\mathbf{r}) \times \int_{\mathbf{f}_n} \mathbf{f}_n(\mathbf{r}) \times \nabla' G(\mathbf{r}, \mathbf{r}')\, d\mathbf{r}' \right] d\mathbf{r} \quad (9.37)$$

Taking the gradient of (9.17) and re-arranging the cross products yields

$$nK(\mathbf{f}_m, \mathbf{f}_n) = \int_1 \mathbf{R}^{\hat{n} \times K}(\mathbf{f}_m, \hat{\mathbf{k}}) \cdot T_L(k_l, \hat{\mathbf{k}}, \mathbf{r}_{ab}) \mathbf{T}^{\hat{n} \times K}(\mathbf{f}_n, \hat{\mathbf{k}}) \, dS \qquad (9.38)$$

where the radiation function for basis function $\mathbf{f}_n(\mathbf{r}')$ is

$$\mathbf{T}^{\hat{n} \times K}(\mathbf{f}_n, \hat{\mathbf{k}}) = \int_{\mathbf{f}_n} \mathbf{f}_n(\mathbf{r}') e^{jk_l \hat{\mathbf{k}} \cdot \mathbf{r}_{r'b}} \, d\mathbf{r}' \qquad (9.39)$$

and the receive function for testing function $\mathbf{f}_m(\mathbf{r})$ is

$$\mathbf{R}^{\hat{n} \times K}(\mathbf{f}_m, \hat{\mathbf{k}}) = -jk_l \, \hat{\mathbf{k}} \times \int_{\mathbf{f}_m} \left[ \mathbf{f}_m(\mathbf{r}) \times \hat{\mathbf{n}}_l(\mathbf{r}) \right] e^{-jk_l \hat{\mathbf{k}} \cdot \mathbf{r}_{ra}} \, d\mathbf{r} \qquad (9.40)$$

We note that due to the cross product with $\hat{\mathbf{k}}$, the receive function has only $\hat{\theta}$ and $\hat{\phi}$ components.

### 9.3.3.2  $\hat{n} \times \mathcal{L}$ Operator

Substituting the magnetic basis and electric testing functions into (3.190) yields in Region $R_l$ matrix elements given by

$$Z_{mn}^{nHM(l)} = nL(\mathbf{f}_m, \mathbf{g}_n) \qquad (9.41)$$

where

$$nL(\mathbf{f}_m, \mathbf{g}_n) = j\omega\epsilon_l \int_{\mathbf{f}_m} \mathbf{f}_m(\mathbf{r}) \cdot \left[ \hat{\mathbf{n}}_l(\mathbf{r}) \times \int_{\mathbf{g}_n} \mathbf{g}_n(\mathbf{r}') \left[ 1 - \frac{1}{k_l^2} \nabla\nabla' \right] G(\mathbf{r}, \mathbf{r}') \, d\mathbf{r}' \right] d\mathbf{r}$$
$$(9.42)$$

Applying the differential operators yields a result similar to that in Section 9.3.1.1, which is

$$nL(\mathbf{f}_m, \mathbf{g}_n) = \int_1 \mathbf{R}^{\hat{n} \times \mathcal{L}}(\mathbf{f}_m, \hat{\mathbf{k}}) \cdot T_L(k_l, \hat{\mathbf{k}}, \mathbf{r}_{ab}) \mathbf{T}^{\hat{n} \times \mathcal{L}}(\mathbf{g}_n, \hat{\mathbf{k}}) \, dS \qquad (9.43)$$

where the radiation function for basis function $\mathbf{g}_n(\mathbf{r}')$ is

$$\mathbf{T}^{\hat{n} \times \mathcal{L}}(\mathbf{g}_n, \hat{\mathbf{k}}) = \left[ 1 - \hat{\mathbf{k}}\hat{\mathbf{k}} \right] \cdot \int_{\mathbf{g}_n} \mathbf{g}_n(\mathbf{r}') e^{jk_l \hat{\mathbf{k}} \cdot \mathbf{r}_{r'b}} \, d\mathbf{r}' \qquad (9.44)$$

and the receive function for testing function $\mathbf{f}_m(\mathbf{r})$ is

$$\mathbf{R}^{\hat{n} \times \mathcal{L}}(\mathbf{f}_m, \hat{\mathbf{k}}) = j\omega\epsilon_l \int_{\mathbf{f}_m} \left[ \mathbf{f}_m(\mathbf{r}) \times \hat{\mathbf{n}}(\mathbf{r}) \right] e^{-jk_l \hat{\mathbf{k}} \cdot \mathbf{r}_{ra}} \, d\mathbf{r} \qquad (9.45)$$

We note that the radiation function has only $\hat{\theta}$ and $\hat{\phi}$ components. A $\eta_0$ scale factor must also be applied to account for the presence of the magnetic basis function.

### 9.3.4 Unit Sphere Decomposition

In the previous sections we noted that for each matrix element, the radiation or receive function had only $\hat{\theta}$ and $\hat{\phi}$ components. Thus, the inner product of the radiation and receive functions operates on only those components, and any $\hat{\mathbf{k}}$-oriented components are disregarded. Therefore, when pre-computing these functions, only the scalar component in each dimension has to be stored. For example, when pre-computing the value of the receive function $\mathbf{R}^{\mathcal{L}}$ at each quadrature point on the unit sphere, one would calculate and store the scalar values $\mathrm{R}_{\theta}^{\mathcal{L}}$ and $\mathrm{R}_{\phi}^{\mathcal{L}}$, which are

$$\mathrm{R}_{\theta,\phi}^{\mathcal{L}} = (\hat{\theta}, \hat{\phi}) \cdot \mathbf{R}^{\mathcal{L}} \tag{9.46}$$

and likewise one would store the scalar values $\mathrm{T}_{\theta}^{\mathcal{L}}$ and $\mathrm{T}_{\phi}^{\mathcal{L}}$ for the radiation function $\mathbf{T}^{\mathcal{L}}$.

Though the receive functions for the various operators have different forms, a brief review of the radiation functions in Sections 9.3.1–9.3.3 shows them to be identical when decomposed this way. Thus, only one radiation function needs to be computed for each basis function. These have the form

$$\mathbf{T}^{J}(\mathbf{f}_n, \hat{\mathbf{k}}) = \left[ 1 - \hat{\mathbf{k}}\hat{\mathbf{k}} \right] \cdot \int_{\mathbf{f}_n} \mathbf{f}_n(\mathbf{r}')e^{jk_l\hat{\mathbf{k}}\cdot\mathbf{r}'_b} \, d\mathbf{r}' \tag{9.47}$$

for the electric basis functions, and

$$\mathbf{T}^{M}(\mathbf{g}_n, \hat{\mathbf{k}}) = \left[ 1 - \hat{\mathbf{k}}\hat{\mathbf{k}} \right] \cdot \int_{\mathbf{g}_n} \mathbf{g}_n(\mathbf{r}')e^{jk_l\hat{\mathbf{k}}\cdot\mathbf{r}'_b} \, d\mathbf{r}' \tag{9.48}$$

for the magnetic basis functions. Though $\mathbf{T}^{J}$ and $\mathbf{T}^{M}$ are not necessarily the same, as $\mathbf{f}_n$ and $\mathbf{g}_n$ may differ, if we use RWG basis functions they are identical except for the $\eta_0$ scale factor in the magnetic basis function. Therefore, only one radiation function has to be stored for each RWG function.

---

## 9.4 One-Level Fast Multipole Algorithm

We will now begin looking at the practical implementation details of the FMM in the context of the one-level Fast Multipole Algorithm (FMA). We will discuss grouping of basis functions into near and far groups, truncation of the transfer function, sampling rates on the unit sphere, and calculation and storage of radiation and receive functions and the near matrix. With this complete, we will then draw on the elements of the one-level algorithm in developing the Multi-Level Fast Multipole Algorithm (MLFMA) in the next section.

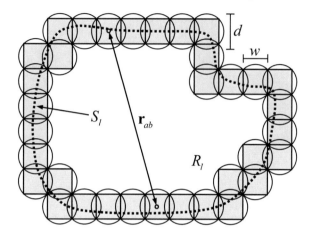

**FIGURE 9.3:** Spatial Subdivision and Grouping of Basis Functions

## 9.4.1  Grouping of Basis Functions

To determine which basis functions are well-separated, we will arrange
them into groups on a per-region basis. To do so, consider the dielectric region $R_l$ with bounding surface $S_l$. We first find the bounding box of minimum
size that encloses $S_l$. If using RWG functions, this can be done by testing the
bounding box against the endpoints of the supporting edges that are in $R_l$.
We then divide the bounding box into a number of small, equally sized cubes
having sides of length $w$ and a bounding sphere of diameter $d = \sqrt{3w^2}$. The
bounding box is then expanded such that there are a whole number of cubes in
each dimension. We set the length $w$ to some fraction of the wavelength in $R_l$,
such as $\lambda_l/2$ or $\lambda_l/4$. Each basis and testing function in $R_l$ is then assigned to
a cube. When using RWG functions, this is done by testing the center of the
supporting edge against the bounding box of each cube. Non-empty cubes we
refer to as *groups*, and all empty cubes are discarded. With this process complete, we are left with an arrangement similar to the one shown in Figure 9.3.
This process is repeated for each region, where the size, number, and layout
of the groups may be different due to the distribution of basis functions and
dielectric parameters in that region. We note that as a consequence of this per-
region grouping, it is possible that two basis functions may be well separated
in one region, but near in an adjacent region.

The points **a** and **b** were left undefined in Section 9.3. We now define these
to be the center of the cube in which the field and source points, respectively,
reside. The transfer function now acts to transmit the fields radiated by basis
functions in a source group centered at **a** to the testing functions in a receiving
group centered at **b**.

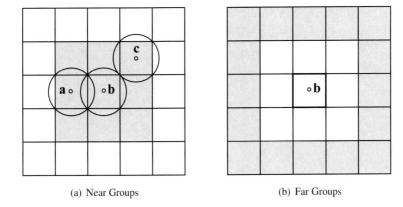

(a) Near Groups                 (b) Far Groups

**FIGURE 9.4:** Near and Far Groups

### 9.4.1.1   Classification of Near and Far Groups

We next defined what it means to be *well separated* in the context of the FMM. Consider two groups with bounding boxes centered at **a** and **b**. For the addition theorem of (9.8) to remain valid, the distance between $|\mathbf{r}_{ab}|$ between groups must be greater than or equal to at least one diameter $d$ [6]. Whether two groups are well separated is now a measure of the distance between their bounding spheres. Consider the groups as shown in Figure 9.4a, where the group at **b** lies in the middle. Groups sharing an edge with this one, such as the one at **a**, are separated by less than $d$. Those that share a corner, such as the one at **c**, are separated by exactly $d$. Thus, let us label as *near* all those groups that share an edge or a vertex. For the group at **b**, this includes itself and all the groups it touches, which comprises the shaded boxes in Figure 9.4a. The well-separated or *far groups* are those that are separated by at least one box, and comprise the shaded boxes in Figure 9.4b. The near-zone boxes that lie between the group at **b** and the far groups are often referred to as *buffer boxes*.

### 9.4.1.2   Near Matrix

The matrix elements between basis and testing functions in nearby groups are computed the usual way, and stored explicitly. For each basis function, this amounts to computing only the matrix elements that involve testing functions in nearby groups. Repeating this process for all testing functions and dielectric regions yields a system matrix $\mathbf{Z}^{near}$ that is very sparse. We will now discuss the storage and population of this sparse matrix.

*Compressed Sparse Row Format*

To store the sparse near matrix, we use a storage scheme called the Compressed Sparse Row (CSR) format [7]. To illustrate this format, consider the following sparse matrix $\mathbf{A}$

$$\mathbf{A} = \begin{bmatrix} 1 & 2 & 0 & 0 & 0 \\ 3 & 4 & 5 & 0 & 0 \\ 0 & 6 & 7 & 8 & 0 \\ 0 & 0 & 9 & 10 & 0 \\ 0 & 0 & 0 & 0 & 11 \end{bmatrix} \qquad (9.49)$$

which is of rank $N = 5$ and which has $N_z = 11$ nonzero entries. The CSR format comprises three vectors $\mathbf{a}$, $\mathbf{j}$, and $\mathbf{i}$, which in this case are

$$\mathbf{a} = \begin{bmatrix} 1 & 2 & 3 & 4 & 5 & 6 & 7 & 8 & 9 & 10 & 11 \end{bmatrix}$$
$$\mathbf{j} = \begin{bmatrix} 1 & 2 & 1 & 2 & 3 & 2 & 3 & 4 & 3 & 4 & 5 \end{bmatrix}$$
$$\mathbf{i} = \begin{bmatrix} 1 & 3 & 6 & 9 & 11 & 12 \end{bmatrix}$$

Vector $\mathbf{a}$ is of length $N_z$ and contains the nonzero elements of $\mathbf{A}$, stored row by row. Vector $\mathbf{j}$ is also of length $N_z$ and holds the column indexes of the matrix elements stored in $\mathbf{a}$. Vector $\mathbf{i}$ is of length $N + 1$, and stores the integer offset to each row in $\mathbf{a}$ and $\mathbf{j}$. Note that the last element of $\mathbf{i}$ contains an offset to the fictional row $m = N + 1$, which allows for quick indexing. The total storage requirements of the CSR format in bytes is therefore

$$N_{bytes} = N_z s_f + (N_z + N + 1)s_i \qquad (9.50)$$

where $s_f$ and $s_i$ are the size of the float and integer types,[1] respectively.

*Near Matrix Fill*

The construction of the near matrix comprises two steps. The first step involves identifying the nonzero rows and columns of $\mathbf{Z}^{near}$ and population of the CSR vectors $\mathbf{j}$ and $\mathbf{i}$, and is accomplished via Algorithm 6. Note that when basis and testing functions on opposite sides of a dielectric interface have the same coefficient index, the corresponding row and column are added to the list only once. Once completed, the linked list can be sorted by row and then by column. The vectors $\mathbf{j}$ and $\mathbf{i}$ are then easily filled. A side effect of the sort operations is that the columns of each row will be arranged in ascending order. This allows us to locate $Z_{mn}^{near}$ by performing a binary search of $\mathbf{j}$ in the range $\mathbf{i}(m)$ and $\mathbf{i}(m + 1)$ for column $n$. The second step involves calculating and storing the matrix elements, however the loop structure of Algorithm 6 is not efficient for this task. When using triangles and RWG functions, a loop over source and

---

[1] `float` and `int` are 4 bytes each on most systems.

---

**Algorithm 6** Near-Matrix Set-Up

---

Initialize empty linked list
**for all** Regions $R_l$ **do**
    **for all** Cubes $C_j$ in $R_l$ **do**
        **for all** Cubes $C_k$ near to $C_j$ **do**
            **for all** Testing functions in $C_j$ **do**
                **for all** Basis functions in $C_k$ **do**
                    Add row/column index to list if not already present
                **end for**
            **end for**
        **end for**
    **end for**
**end for**

---

testing triangles in each region is more efficient, in the manner prescribed by Chapter 8. For each triangle pair, one should then check the edges on which the basis and testing functions reside to see if they are well separated in that region.

### 9.4.2 Number of Multipoles

An accurate evaluation of the integral in (9.12) requires that we truncate the transfer function in (9.12) at some limit $L$, which we refer to as the number of *multipoles*. This limit depends on the size of the groups as well as the wavenumber, and expressions for it have been determined empirically. For real-valued $k$, Rokhlin prescribes in [2] the following formula for single precision calculations

$$L = kD + 5\log(kD + \pi) \tag{9.51}$$

and for double precision calculations,

$$L = kD + 10\log(kD + \pi) \tag{9.52}$$

where $D = |\mathbf{x}|$ from (9.12). The maximum possible value of $D$ to be accounted for occurs when $\mathbf{r} - \mathbf{r}'$ and $\mathbf{a} - \mathbf{b}$ are collinear. With the basis functions sorted by cubes, this value is equal to the bounding sphere diameter $d$, so we take $D = d$. The following expression was later suggested by Chew and Song in [8]

$$L = kd + \beta(kd)^{1/3} \tag{9.53}$$

where $\beta$ is the number of digits of accuracy required. They suggest that $\beta = 6$ is sufficient for most purposes. The same formula is also used in [9] for the optimal sampling of scattered fields on the sphere. Using (9.53), we will then assume the maximum bandwidth of the radiation and receive functions to be

$$L = ka + \beta(ka)^{1/3} \tag{9.54}$$

where $a = d/2$ is the bounding sphere radius.

### 9.4.2.1  Limiting $L$ for Transfer Functions

Though in each region there are many different distances between groups, using a single value of $L$ from (9.53) for all $|\mathbf{r}_{ab}|$ will lead to numerical problems. This is because the spherical Hankel function $h_l^{(2)}(x)$ becomes highly oscillatory for fixed $x$ and increasing $l$. Therefore, we cannot take $L$ to be much larger than the argument of the Hankel function, which is $|k_l \mathbf{r}_{ab}|$ in Region $R_l$. One remedy is to add additional near-zone buffer boxes around each group (see Section 9.4.1.1), but this will increase the size of the near matrix and is not often done. The approach taken in most practical implementations is to instead limit the value of $L$ to the minimum of either (9.53) or $|k_l \mathbf{r}_{ab}|$ in each region. Though this has the effect of reducing the accuracy of the expansion for groups having the smallest separations, in practice this does not adversely impact the results.

### 9.4.2.2  $L$ for Complex Wavenumbers

The expressions for $L$ in Section 9.4.2 were derived empirically for real-valued wavenumbers. However, in lossy dielectric regions having complex wavenumbers, the number of terms required in the addition theorem increases, and (9.53) is no longer valid. When the losses are small, $L$ increases only slightly, and the approximation

$$L = |kd| + \beta(|kd|)^{1/3} \tag{9.55}$$

is adequate in most cases. For large losses the increase in $L$ may be much larger, and it must be determined adaptively [10]. For the examples in Section 9.7 that involve lossy dielectric materials, we use (9.55) to determine $L$.

### 9.4.3  Integration on the Sphere

Numerical integration of (9.19) requires that the radiation, receive and transfer functions are computed and stored at discrete angles in the $\theta$ and $\phi$ dimensions on the unit sphere. We must choose the sampling rate in each dimension carefully, so that we utilize the system memory effectively while still meeting the Nyquist rate. As these functions are of finite (or *quasi-finite*) bandwidth [11], we need to determine the extent of the bandwidth, how it affects the sampling rate, and how the bandwidth changes when taking the product of these functions.

Consider a function $f_1(x)$ having bandwidth $A$, sampled at a rate $N_1$, and a similar function $f_2(x)$ of bandwidth $B$ sampled at a rate $N_2$. To then calculate the integral

$$\int_{x_1}^{x_2} f_1(x)f_2(x)\, dx \tag{9.56}$$

we can use a numerical quadrature of the form

$$\int_{x_1}^{x_2} f_1(x)f_2(x)\, dx = \sum_{i=1}^{N_3} w(x_i)f_1(x_i)f_2(x_i) \tag{9.57}$$

where we use a new sampling rate $N_3$. Because the bandwidth of the product $f_1(x)f_2(x)$ is the sum of the individual bandwidths, the sample rate $N_3$ must be sufficiently high to adequately capture the bandwidth of the product. This presents a dilemma, as the sampling rate required to evaluate (9.19) is greater than the rate needed to sample the radiation, receive, and transfer functions individually. Though a naive approach of simply storing the functions at the higher rate is passable for a one-level algorithm on smaller problems, it still wastes memory. It is more efficient to store the individual functions at a lower sampling rate and upsample them to the higher rate when needed. However, this interpolation has consequences in terms of the additional compute time needed to upsample the functions, as well as the interpolation error. To simplify the present discussion, we will assume that all functions are sampled at the rate required for integration. We will then return to the issue of interpolation when we discuss the multi-level algorithm in Section 9.5.

### 9.4.3.1 Spherical Harmonic Representation

The optimal sampling rates and numerical quadrature rules for integrating expressions like (9.19) on the unit sphere are related to their spherical harmonic representation. Therefore, let us first consider the decomposition of a bandlimited function in terms of spherical harmonics. A function $f(\theta, \phi)$ is said to be *bandlimited* with bandwidth $L$ if it can be written as [12]

$$f(\theta, \phi) = \sum_{l=0}^{L} \sum_{m=-l}^{l} a_l^m Y_l^m(\theta, \phi) \tag{9.58}$$

where $a_l^m$ are the spherical harmonic coefficients, $Y_l^m$ are the spherical harmonics given by

$$Y_l^m(\theta, \phi) = P_l^m(\cos \theta)e^{jm\phi} \tag{9.59}$$

and $P_l^m(\cos \theta)$ is the normalized associated Legendre polynomial of degree $l$ and order $m$, given by

$$P_l^m(x) = \sqrt{\frac{2l+1}{4\pi}\frac{(l-m)!}{(l+m)!}}(1-x^2)^{m/2}\frac{d^m}{dx^m}P_l(x) \tag{9.60}$$

The coefficients $a_l^m$ are obtained using the integral

$$a_l^m = \int_0^{2\pi} \int_0^\pi f(\theta, \phi) Y_l^{m*}(\theta, \phi) \sin \theta \, d\theta d\phi \qquad (9.61)$$

Substituting (9.59) allows us to write (9.61) as

$$a_l^m = \int_0^{2\pi} e^{-jm\phi} \int_0^\pi f(\theta, \phi) P_l^m(\cos \theta) \sin \theta \, d\theta d\phi \qquad (9.62)$$

It is shown in [12] that for a function $f(\theta, \phi)$ of bandwidth $L$, the integral in (9.62) can be evaluated optimally using an $L$ point Gauss-Legendre quadrature rule in $\theta$, and an equally spaced $(2L + 1)$ point quadrature (such as Simpson's Rule) in $\phi$.

### 9.4.3.2  Total Bandwidth

We can now determine the sample rate required to numerically integrate (9.19). Referring to Section 9.4.2, the transfer functions in region $R_l$ have a bandwidth of

$$L_T = k_l d + \beta(k_l d)^{1/3} \qquad (9.63)$$

and the radiation and receive functions each have a bandwidth equal to

$$L_f = k_l d/2 + \beta(k_l d/2)^{1/3} \qquad (9.64)$$

Thus, the total bandwidth of (9.19) is $L_{tot} = L_T + 2L_f$. In a naive one-level algorithm, all functions are stored and quadratures evaluated using $L_{tot}$ points in $\theta$ and $(2L_{tot} + 1)$ points in $\phi$, yielding a total of

$$N_k = L_{tot}(2L_{tot} + 1) \qquad (9.65)$$

discrete quadrature points (directions) on the unit sphere.

### 9.4.3.3  Transfer Functions

Transfer functions between far group pairs are precomputed and stored for all unique vectors $|\mathbf{r}_{ab}|$ in each region. The use of axis-aligned bounding cubes results in a set of vectors whose lengths are multiples of the box edge length $w$ in each cardinal direction. As a result, some of these vectors may be repeated among multiple group pairs, and in those cases only one transfer function has to be stored. We then pre-multiply the transfer functions by the quadrature weights $w_s(\hat{\mathbf{k}})$ used for integration. Though this is not necessary in the one-level algorithm, it is essential in the multi-level algorithm.

#### 9.4.3.4   Radiation and Receive Functions

The radiation and receive functions in Section 9.3 are pre-computed and stored for all basis and testing functions, respectively, in each region. We continue using the rules outlined in Section 8.6 to enforce the wanted integral equations on each interface. In general, there will be separate radiation functions for the electric and magnetic basis functions, however when using RWG functions, only one radiation function is needed. Each testing function will have two receive functions: one to receive the fields radiated by the electric basis functions, and one to receive the fields radiated by the magnetic basis functions.

### 9.4.4   Matrix-Vector Product

We now discuss the calculation of the matrix-vector product (9.1) using the one-level algorithm, where the result vector $\mathbf{x}$ comprises the sum of near and and far products in (9.5), which are computed separately.

#### 9.4.4.1   Near Product

The near product comprises a relatively straightforward matrix-vector product between the right-hand side vector $\mathbf{y}$ and the near matrix $\mathbf{Z}^{near}$, stored in the sparse CSR format. As the inner products of the matrix rows with the right-hand side vector are sparse, the time spent computing the near product is typically small compared to that spent on the far product.

#### 9.4.4.2   Far Product

In computing the far product, four complex arrays of length $N_k$ are allocated for each group. Two arrays, denoted as $t^\theta$ and $t^\phi$, are used to store the $\hat{\theta}$ and $\hat{\phi}$ components of the aggregated radiation functions in each group. The other two arrays, denoted as $s^\theta$ and $s^\phi$, store the $\hat{\theta}$ and $\hat{\phi}$ components of the far fields received by each group from all far groups. The one-level far-multiply then comprises two phases: an aggregation phase and a transfer and disaggregation phase, described in Algorithms 7 and 8, respectively. All operations in the aggregation phase must be completed before moving to the disaggregation phase, and in general, two passes must be executed to yield the far product. In the first pass, the fields due to only the electric basis functions are aggregated and transferred. In the second pass, the fields due to the magnetic basis functions are aggregated and transferred. During the disaggregation step, the appropriate electric or magnetic receive function, respectively, is used for each testing function. Note that only one pass is required in the purely conducting case, as there are no magnetic basis functions.

---

**Algorithm 7** One-Level FMA Aggregation Phase

---

**for all** Regions $R_l$ **do**
  **for all** Groups $C_j$ in $R_l$ **do**
    $t_{C_j}^{\theta}(\hat{\mathbf{k}}) = 0$
    $t_{C_j}^{\phi}(\hat{\mathbf{k}}) = 0$
    **for all** Basis functions ($\mathbf{f}_n$ or $\mathbf{g}_n$) in $C_j$ **do**
      $t_{C_j}^{\theta}(\hat{\mathbf{k}}) = t_{C_j}^{\theta}(\hat{\mathbf{k}}) + y_n T_{\theta,n}^{J,M}(\hat{\mathbf{k}})$
      $t_{C_j}^{\phi}(\hat{\mathbf{k}}) = t_{C_j}^{\phi}(\hat{\mathbf{k}}) + y_n T_{\phi,n}^{J,M}(\hat{\mathbf{k}})$
    **end for**
  **end for**
**end for**

---

---

**Algorithm 8** One-Level FMA Transfer/Disaggregation Phase

---

**for all** Regions $R_l$ **do**
  **for all** Groups $C_j$ in $R_l$ **do**
    $s_{C_j}^{\theta}(\hat{\mathbf{k}}) = 0$
    $s_{C_j}^{\phi}(\hat{\mathbf{k}}) = 0$
    **for all** Groups $C_k$ far from $C_j$ **do**
      $s_{C_j}^{\theta}(\hat{\mathbf{k}}) = s_{C_j}^{\theta}(\hat{\mathbf{k}}) + t_{C_k}^{\theta}(\hat{\mathbf{k}})T_L(k_l, \hat{\mathbf{k}}, \mathbf{r}_{ab})$
      $s_{C_j}^{\phi}(\hat{\mathbf{k}}) = s_{C_j}^{\phi}(\hat{\mathbf{k}}) + t_{C_k}^{\phi}(\hat{\mathbf{k}})T_L(k_l, \hat{\mathbf{k}}, \mathbf{r}_{ab})$
    **end for**
    $x_m = x_m + \sum_{p=1}^{N_k}\left[R_{\theta,n}^{J,M}(\hat{\mathbf{k}}_p)s_{C_j}^{\theta}(\hat{\mathbf{k}}_p) + R_{\phi,n}^{J,M}(\hat{\mathbf{k}}_p)s_{C_j}^{\phi}(\hat{\mathbf{k}}_p)\right]$
  **end for**
**end for**

---

### 9.4.5 Computational Complexity

If we assume there are $N$ basis functions divided among $G$ groups, each group will have about $M = N/G$ basis functions and $B$ near groups (including itself). The total cost of the near multiply is then

$$T_1 = c_1 BNM \tag{9.66}$$

where $c_1$ is dependent on the software and host platform. The cost of aggregating the radiation functions $\sum_{n \in b} \mathbf{T}_n(\hat{\mathbf{k}})$ for all basis functions is

$$T_2 = c_2 KN \tag{9.67}$$

The cost for computing all transfers $\sum_b T_L(k, \hat{\mathbf{k}}, \mathbf{r}_{ab})$ is

$$T_3 = c_3 KG(G - B) \tag{9.68}$$

and the cost of disaggregations is similar to the aggregations and is

$$T_4 = c_4 K N \tag{9.69}$$

Numerical analysis shows that the total computational cost of the matrix-vector multiply to be [3]

$$T = C_1 NG + C_2 N^2/G \tag{9.70}$$

which is minimized by choosing $G = \sqrt{C_2 N/C_1}$. This results in an algorithm whose complexity is $O(N^{3/2})$.

---

## 9.5 Multi-Level Fast Multipole Algorithm (MLFMA)

In the one-level algorithm, the number of transfers grows exponentially with the size of the problem and the number of groups. Though it offers a potential $O(N^{3/2})$ complexity and an improvement in memory utilization, an efficiency of $O(N \log N)$ can be obtained from a multi-level algorithm [3]. The Multi-Level Fast Multipole Algorithm (MLFMA) extends the one-level algorithm through recursive subdivision of the geometry into multiple levels with bounding cubes of progressively smaller size. As we will see, grouping the basis functions this way greatly reduces the number of transfers, further accelerating the far product. There are, however, some differences and subtleties that must be handled properly for the multi-level algorithm to be effective, and we will discuss those in detail in this section.

### 9.5.1 Spatial Subdivision via Octree

First let us consider what will happen if we increase the group size in the one-level algorithm. This will reduce the overall number of transfers at the expense of increasing the size of the near matrix, and shift more of the compute time to the near product. The sampling rates on the unit sphere also increase, increasing the storage requirements for the radiation, receive, and transfer functions. If we can devise a way to use large and small groups together, we can use a lower sample rate while simultaneously reducing the number of transfers. The answer to this problem is a recursive subdivision of the geometry into a hierarchy of progressively smaller groups. If we use a tree-based scheme, larger groups are divided into recursively smaller groups with a clear parent-child relationship. An excellent choice for this is the octree [13], a data structure commonly used in computer graphics. In generating an octree for region $R_l$, the bounding surface $S_l$ is first enclosed in a single large cube. This cube is then divided into eight smaller cubes, which are themselves divided into eight

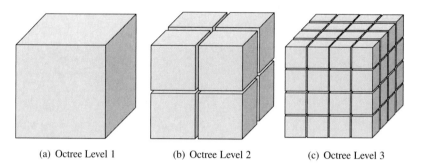

(a) Octree Level 1  (b) Octree Level 2  (c) Octree Level 3

**FIGURE 9.5:** Octree Levels 1–3

smaller cubes and so on, as illustrated in Figure 9.5. The approach is very similar to the one-level algorithm, where now the basis and testing functions are sorted and assigned to the cubes on each level. This process is very fast as only those functions assigned to a parent can be assigned to its children. The subdivision is stopped when the size of the cubes reaches some fraction of a wavelength in $R_l$, such as $\lambda_l/2$ or $\lambda_l/4$. The cubes on the finest level of the tree are called *leaf* cubes.

On levels 1 and 2, all groups are near groups. Thus, in regions where the recursion yields at most 2 levels, we do not use the FMM. Instead, we treat all interactions in those regions as near interactions and store them explicitly in the near matrix. Starting then at the coarsest level (level 3), all near and far groups are located as usual. On the higher levels, only the children of groups in the near zone of a parent group (see Section 9.4.1.1) are considered as candidate near and far groups for that parent's children. This greatly reduces the average number of far groups, and limits the number of unique transfer vectors to a maximum of 316 on each level. This is especially advantageous as the total number of transfers can be very high in larger problems.

### 9.5.2 Near Matrix

In the MLFMA, the near matrix is assembled using the groups on the finest level of the octree, following the steps outlined in Algorithm 6. The near matrix fill and near product are then performed in the same manner as the one-level algorithm.

### 9.5.3 Sampling Rates

Transfer functions are computed for all unique transfer vectors (a maximum of 316) on each octree level, where $L$ is given by (9.53) and limited for the closer groups as in Section 9.4.2.1. As the number of unique transfer functions on each level is limited, they can be precomputed and stored at the sampling rate required for integration (9.65), eliminating any need to interpolate them. Radiation and receive functions are computed and stored on the finest level of the octree, where the bandwidth (9.64) depends on the diameter of the leaf cubes. A consequence of this choice is that the radiation and receive functions must be interpolated when moving between levels, which is discussed in detail in the following sections.

### 9.5.4 Far Product

The far product in the MLFMA operates on levels 3 and higher in the octree. As in the one-level algorithm, the MLFMA comprises two phases: an aggregation phase and a transfer and disaggregation phase. However, we must extend the one-level algorithm, as the radiation and receive functions exist only at the finest level of the tree. In the aggregation phase, an upward pass is made through the octree, where radiation functions in each group are aggregated and passed up to their parents using interpolation and phase shifting. In the disaggregation phase, a downward pass is made through the octree, where transfers are made between all far groups and the received fields passed down to child groups through a combination of integration on the sphere, phase shifting, and "reverse interpolation" (anterpolation). The transfer scheme in the MLFMA is illustrated in Figures 9.6 - 9.8. Figure 9.6 depicts the transfers between groups on one of the coarser levels. On the next highest level in Figure 9.7, the far groups are children of groups that were considered near on the previous level. This is further emphasized on the finest level in Figure 9.8, where the number of transfers is now far less than it would have been in a one-level algorithm. We will now describe each phase in detail.

#### 9.5.4.1 Upward Pass (Aggregation)

The aggregation step comprises an upward pass through the tree. We begin on the finest level $l = l_{max}$, where the aggregated field $\mathbf{t}(\hat{\mathbf{k}}_{(l)})$ in each group is computed from the radiation functions and their corresponding excitation vector elements as

$$\mathbf{t}(\hat{\mathbf{k}}_{(l)}) = \sum_n y_n \left[ T_{\theta,n}^{J,M}(\hat{\mathbf{k}}_{(l)}) \hat{\theta} + T_{\phi,n}^{J,M}(\hat{\mathbf{k}}_{(l)}) \hat{\phi} \right] \quad (9.71)$$

Aggregated fields in groups on coarser levels are computed from phase-shifted versions of the aggregated fields of their children, as illustrated in Figure 9.9. For each parent-child pair, the phase shift can be written as

$$\mathbf{t}(\hat{\mathbf{k}}_{(l-1)}) = e^{jk\hat{\mathbf{k}}_{(l-1)} \cdot (\mathbf{r}_{b_{(l)}} - \mathbf{r}_{b_{(l-1)}})} \mathbf{t}(\hat{\mathbf{k}}_{(l)}) \quad (9.72)$$

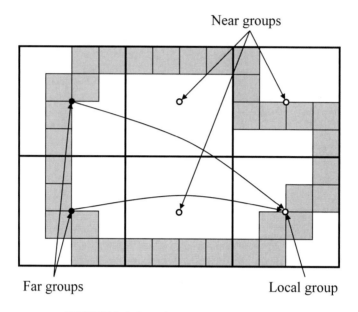

**FIGURE 9.6:** Transfers On MLFMA Level $M - 2$

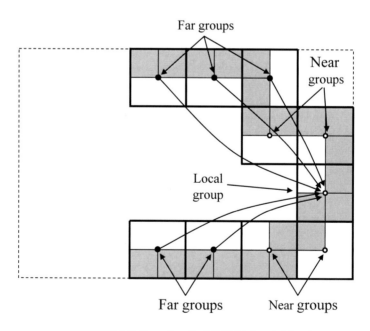

**FIGURE 9.7:** Transfers On MLFMA Level $M - 1$

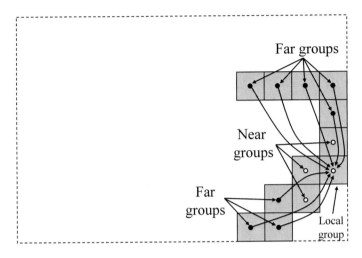

**FIGURE 9.8:** Transfers On MLFMA Level $M$

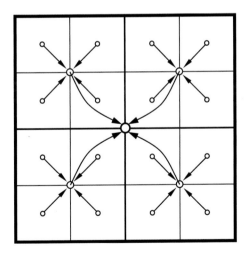

**FIGURE 9.9:** Upward Pass In MLFMA

where $\mathbf{r}_{b_{(l)}} - \mathbf{r}_{b_{(l-1)}}$ is the vector pointing from the center of the parent group to that of the child group. Note that after applying the phase shift, $\mathbf{t}(\hat{\mathbf{k}}_{(l-1)})$ will have a higher bandwidth, requiring a higher sampling rate $N_f^{(l-1)}$. Because $\mathbf{t}(\hat{\mathbf{k}}_{(l)})$ is sampled using $N_f^{(l)}$ points, it must be upsampled to $N_f^{(l-1)}$ points before applying the phase shift. Interpolation of the aggregated field at point $\hat{\mathbf{k}}_{p,(l-1)}$ on the coarser level can be written as [3]

$$\mathbf{t}(\hat{\mathbf{k}}_{p,(l-1)}) = e^{jk\hat{\mathbf{k}}_{p,(l-1)} \cdot (\mathbf{r}_{b_{(l)}} - \mathbf{r}_{b_{(l-1)}})} \sum_{q=1}^{N_f^{(l)}} W_{pq}\, \mathbf{t}(\hat{\mathbf{k}}_{q,(l)}) \qquad (9.73)$$

where $W_{pq}$ are elements of an interpolation matrix $\mathbf{W}_{(l-1,l)}$ of size $N_f^{(l-1)} \times N_f^{(l)}$. Interpolation of all sample points on the coarser level thus comprises a matrix-vector product between $\mathbf{W}_{(l-1,l)}$ and $\mathbf{t}(\hat{\mathbf{k}}_{(l)})$, allowing us to write (9.73) as

$$\mathbf{t}(\hat{\mathbf{k}}_{(l-1)}) = e^{jk\hat{\mathbf{k}}_{(l-1)} \cdot (\mathbf{r}_{b_{(l)}} - \mathbf{r}_{b_{(l-1)}})} \left[ \mathbf{W}_{(l-1,l)} \cdot \mathbf{t}(\hat{\mathbf{k}}_{(l)}) \right] \qquad (9.74)$$

If we now form the transpose matrix $\mathbf{W}_{(l-1,l)}^{\mathrm{T}}$, we can obtain the values of $\mathbf{t}(\hat{\mathbf{k}}_{(l)})$ from those in $\mathbf{t}(\hat{\mathbf{k}}_{(l-1)})$, which is required in the downward pass. Furthermore, is is critical that the interpolation matrix $\mathbf{W}_{(l-1,l)}$ be made sparse to yield an efficient MLFMA. The choice of interpolation algorithm is discussed in more detail in Section 9.5.5.

*Implementation Notes*

The radiation functions are stored using $N_f^{(l_{max})}$ samples on the finest level of the tree, however the transfer functions on each level are stored at $N_k^{(l)}$, the rate required for integration on the sphere. Thus, the radiation functions must at some point be upsampled to $N_k^{(l)}$ before performing the transfer step in the downward pass. This author has found that the software implementation is easier and leads to less interpolation error if the sample rate during the upward pass is simply fixed at $N_k^{(l)}$ on each level. To accomplish this, the aggregated radiation functions on the finest level $l_{max}$ are upsampled to $N_k^{(l_{max})}$ using an interpolation matrix of size $N_k^{(l_{max})} \times N_f^{(l_{max})}$, and all interpolations and phase shifts between levels performed at the higher rate, using an interpolation matrix of size $N_k^{(l-1)} \times N_k^{(l)}$.

### 9.5.4.2  Downward Pass (Disaggregation)

Let us first consider an octree that has only two levels that are used in the MLFMA: a coarse level (level $l - 1$) and a fine level (level $l$). We start with a single group on the coarse level, and following Algorithm 8, we transfer the aggregated fields in all far groups to this one. Within this group now exists a local field given by

$$\mathbf{s}\big(\hat{\mathbf{k}}_{(l-1)}\big) = s^\theta\big(\hat{\mathbf{k}}_{(l-1)}\big)\hat{\boldsymbol{\theta}} + s^\phi\big(\hat{\mathbf{k}}_{(l-1)}\big)\hat{\boldsymbol{\phi}} \qquad (9.75)$$

which comprises a sum of transfers from all far groups. Because the transfer functions have been pre-multiplied by the unit-sphere quadrature weights, dis-aggregation of the local field with a receive function $\mathbf{R}_m^{J,M}$ yields result vector contributions given by

$$x_m = x_m + \sum_{p=1}^{N_k^{(l-1)}} \mathbf{R}_m^{J,M}\big(\hat{\mathbf{k}}_{p,(l-1)}\big) \cdot \mathbf{s}\big(\hat{\mathbf{k}}_{p,(l-1)}\big) \qquad (9.76)$$

where

$$\mathbf{R}_m^{J,M}\big(\hat{\mathbf{k}}_{(l-1)}\big) = R_{\theta,m}^{J,M}\big(\hat{\mathbf{k}}_{(l-1)}\big)\hat{\boldsymbol{\theta}} + R_{\phi,m}^{J,M}\big(\hat{\mathbf{k}}_{(l-1)}\big)\hat{\boldsymbol{\phi}} \qquad (9.77)$$

The receive functions are stored in one of the child groups on level $l$, and must be interpolated and phase-shifted to obtain $\mathbf{R}_m^{J,M}\big(\hat{\mathbf{k}}_{(l-1)}\big)$ on level $l-1$. However, doing this for every receive function individually is not efficient. Therefore, consider again the upsampling operation of (9.73). Inserting this expression into (9.76) and interchanging the order of the summations allows us to write

$$x_m = x_m + \sum_{q=1}^{N_f^{(l)}} \mathbf{R}_m^{J,M}\big(\hat{\mathbf{k}}_{q,(l)}\big) \sum_{p=1}^{N_k^{(l-1)}} W_{pq}\, \mathbf{s}\big(\hat{\mathbf{k}}_{p,(l-1)}\big)e^{-jk\hat{\mathbf{k}}_{(l-1)p}\cdot(\mathbf{r}_{b_l}-\mathbf{r}_{b_{l-1}})}$$

$$(9.78)$$

This expression upsamples the receive function $\mathbf{R}_m^{J,M}\big(\hat{\mathbf{k}}_{(l)}\big)$ on level $l$ to $\mathbf{R}_m^{J,M}\big(\hat{\mathbf{k}}_{(l-1)}\big)$ on level $l-1$, adds the required phase shift from the parent to child group, and performs the unit sphere integration on the coarser level at $N_k^{(l-1)}$ sample points. As a result, the innermost sum on the right-hand side of (9.78) does not depend on the receive function, resulting in a "filtered" or "downsampled" field on the finer level. This allows us to work with the receive functions individually, and is often referred to as *adjoint interpolation* or *an-terpolation* [3], though this author prefers the term *reverse interpolation* in this context. Note that the sign of the exponent in the phase shift has been reversed, as we are working with the receive function.

The contributions in (9.78) involve only the far fields transmitted to the parent group on level $l-1$, however there are also far fields transmitted to the child group on level $l$ that must be disaggregated with $\mathbf{R}_m^{J,M}$. Adding these fields to (9.78) yields an expression of the form

$$x_m = x_m + \sum_{q=1}^{N_f^{(l)}} \mathbf{R}_m^{J,M}\left(\hat{\mathbf{k}}_{q,(l)}\right) \cdot \left[\mathbf{s}\left(\hat{\mathbf{k}}_{(l)}\right) + \mathbf{A}\left(\hat{\mathbf{k}}_{q,(l)}\right)\right] \tag{9.79}$$

where $\mathbf{s}\left(\hat{\mathbf{k}}_{(l)}\right)$ comprises the transfers from all far groups on level $l$, and

$$\mathbf{A}\left(\hat{\mathbf{k}}_{(l)}\right) = \mathbf{W}_{(l-1,l)}^{\mathrm{T}} \cdot \left[\mathbf{s}\left(\hat{\mathbf{k}}_{(l-1)}\right)e^{-jk\hat{\mathbf{k}}_{(l-1)}\cdot\left(\mathbf{r}_{b_l}-\mathbf{r}_{b_{l-1}}\right)}\right] \tag{9.80}$$

is the anterpolated field from the parent group on level $l-1$, where the transpose of the interpolation matrix $\mathbf{W}_{(l-1,l)}$ has been used. If we now add an even coarser level (level $l-2$) to the tree, we would then update (9.80) to read

$$\mathbf{A}\left(\hat{\mathbf{k}}_{(l)}\right) = \mathbf{W}_{(l-1,l)}^{\mathrm{T}} \cdot \left[\left[\mathbf{s}\left(\hat{\mathbf{k}}_{(l-1)}\right) + \mathbf{A}\left(\hat{\mathbf{k}}_{(l-1)}\right)\right]e^{-jk\hat{\mathbf{k}}_{(l-1)}\cdot\left(\mathbf{r}_{b_l}-\mathbf{r}_{b_{l-1}}\right)}\right] \tag{9.81}$$

where

$$\mathbf{A}\left(\hat{\mathbf{k}}_{(l-1)}\right) = \mathbf{W}_{(l-2,l-1)}^{\mathrm{T}} \cdot \left[\mathbf{s}\left(\hat{\mathbf{k}}_{(l-2)}\right)e^{-jk\hat{\mathbf{k}}_{(l-2)}\cdot\left(\mathbf{r}_{b_{l-1}}-\mathbf{r}_{b_{l-2}}\right)}\right] \tag{9.82}$$

This is a recursive process which can now be applied in a straightforward way to a tree having any number of levels, yielding a downward pass which is as follows: We begin on the coarsest level (3) and perform transfers between all far groups, and as this is the coarsest level, there are no anterpolated fields. We now move to the next higher level, and perform the transfers between all far groups. To these fields we now compute and add the anterpolated fields from all parent groups. Once the transfers and anterpolations are computed on the finest level, the outermost sum in (9.78) is then performed for each receive function, ending the downward pass.

### 9.5.5 Interpolation Algorithms

The interpolations in the MLFMA consume a significant portion of the total time spent in the far multiply. Though we want these operations to be as fast as possible, we also need to minimize the interpolation error. In this section we will discuss several different interpolation algorithms and their implementation.

#### 9.5.5.1 Statement of the Problem

Given a function $f(\theta, \phi)$ on the unit sphere, let us compute $N_{k_1} = N_{\theta_1} \times N_{\phi_1}$ samples of this function and store them in vector $\mathbf{s}_1$. The interpolation problem involves computing an upsampled or downsampled version $f(\theta, \phi)$, using only the values stored in $\mathbf{s}_1$. If we want to compute a new vector $\mathbf{s}_2$ with $N_{k_2} = N_{\theta_2} \times N_{\phi_2}$ samples, this can be accomplished via the matrix-vector product

$$\mathbf{s}_2 = \mathbf{W}\mathbf{s}_1 \tag{9.83}$$

where $\mathbf{W}$ is an interpolation matrix of size $N_{k_2} \times N_{k_1}$. Individual samples of $\mathbf{s}_2$ are computed as

$$s_{2,p} = \sum_{q=1}^{N_{k_1}} W_{pq} s_{1,q} \tag{9.84}$$

If $N_{k_2} > N_{k_1}$, this operation *upsamples* $\mathbf{s}_1$, and if $N_{k_2} < N_{k_1}$ it *downsamples* $\mathbf{s}_1$. Interpolation methods fall into one of two general categories: *global* or *local*. In a global interpolation, the matrix $\mathbf{W}$ is full and each sample in $\mathbf{s}_2$ is computed using all the samples in $\mathbf{s}_1$. If $f(\theta, \phi)$ is bandlimited, this interpolation can be exact provided that it is sampled at or above the Nyquist rate. In a local interpolation, $\mathbf{W}$ is sparse and operates on subset of the original samples. Local interpolations do not usually produce an exact result, however they are usually much faster and their error can be controlled. In the MLFMA we will use a local interpolation, but we will consider global as well as local schemes.

### 9.5.5.2 Global Interpolation by Spherical Harmonics

As we discussed in Section 9.4.3.1, if $f(\theta, \phi)$ is bandlimited, it can be decomposed into a sum of spherical harmonics (9.58) with the coefficients $a_l^m$ defined in (9.61). Once the coefficients are calculated, an interpolated value at the angles $(\theta_p, \phi_p)$ can be computed as

$$f(\theta_p, \phi_p) = \sum_{l=0}^{L} \sum_{m=-l}^{l} a_l^m Y_l^m(\theta_p, \phi_p) \tag{9.85}$$

If we now substitute (9.61) into (9.85), this yields

$$f(\theta_p, \phi_p) = \sum_{l=0}^{L} \sum_{m=-l}^{l} Y_l^m(\theta_p, \phi_p) \sum_{q=1}^{N_{k_1}} w_q Y_l^{m*}(\theta_q, \phi_q) f(\theta_q, \phi_q) \tag{9.86}$$

where the original function $f(\theta, \phi)$ is sampled at $N_{k_1}$ points, and $w_q$ is the quadrature weight on the unit sphere. Exchanging the order of the summations allow us to write

$$f(\theta_p, \phi_p) = \sum_{q=1}^{N_{k_1}} w_q \sum_{l=0}^{L} \sum_{m=-l}^{l} Y_l^{m*}(\theta_p, \phi_p) Y_l^m(\theta_q, \phi_q) f(\theta_q, \phi_q) \tag{9.87}$$

which can be re-written following (9.84) as

$$f(\theta_p, \phi_p) = \sum_{q=1}^{N_{k_1}} W_{pq} f(\theta_q, \phi_q) \tag{9.88}$$

where the matrix elements $W_{pq}$ are

$$W_{pq} = w_q \sum_{l=0}^{L} \sum_{m=-l}^{l} Y_l^{m*}(\theta_p, \phi_p) Y_l^{m}(\theta_q, \phi_q) \tag{9.89}$$

Using this method, the matrix $\mathbf{W}$ is full, and to yield an efficient MLFMA it must be made sparse. A method for doing this was discussed at some length by Darve in [14], however this author found this method to be less efficient than the local method in Section 9.5.5.4.

### 9.5.5.3 Global Interpolation by FFT

For a bandlimited function, the Fast Fourier Transform (FFT) can be used to produce an upsampled version of $f(\theta, \phi)$ that is free from error. This can be accomplished by performing the forward FFT in each dimension, zero padding by the appropriate amount, and taking the inverse FFT [15]. Because the FFT requires equally spaced samples in each dimension, the integration in $\theta$ must then be changed to a quadrature rule appropriate for uniformly spaced samples. As this requires more sample points in $\theta$ than the Gauss-Legendre quadrature, one could instead use the Fourier interpolation in $\phi$ and the dense interpolation of Section 9.5.5.2 along $\theta$. The dense interpolation might then be improved using a spectral truncation method as in [16].

### 9.5.5.4 Local Interpolation by Lagrange Polynomials

Lagrange polynomials are an effective means of creating a sparse interpolator having local support. Given $N$ samples of a function $f(x)$, this method generates a unique polynomial of order $N - 1$ to interpolate those samples. New values $f(x')$ can then be written as

$$f(x') = \sum_{i=1}^{N} w_i(x') f(x_i) \tag{9.90}$$

where the weights are

$$w_i(x') = \frac{(x' - x_1) \cdots (x' - x_{i-1})(x' - x_{i+1}) \cdots (x' - x_{N+1})}{(x_i - x_1) \cdots (x_i - x_{k-1})(x_i - x_{i+1}) \cdots (x_i - x_{N+1})} \tag{9.91}$$

or

$$w_i(x') = \prod_{\substack{k=1 \\ k \neq i}}^{N} \frac{x' - x_k}{x_i - x_k} \tag{9.92}$$

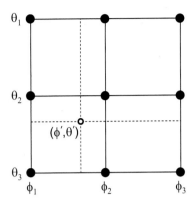

**FIGURE 9.10:** Interpolation Stencil

Interpolation on the unit sphere uses an $N \times N$ interpolation *stencil* as illustrated in Figure 9.10 for $N = 3$. In this case, the interpolation matrix $\mathbf{W}$ will have nine nonzero columns per row. This yields a very sparse interpolation, as in some problems there are hundreds of sample points on the unit sphere. The use of a quadratic or cubic interpolator ($N = 3$, $N = 4$) can provide reasonable results with small error provided that the function is oversampled by a small amount. An error analysis of this method is discussed in [17]. In this author's experience, a $3 \times 3$ Lagrange stencil is sufficient to solve most problems using the MLFMA, and more elaborate schemes such as those in Sections 9.5.5.2 and 9.5.5.3 are not necessary to obtain reasonable accuracy.

## 9.6 Preconditioners

We discussed preconditioning at a high level in Section 4.2.7, where the goal is to improve the convergence of the iterative solver on ill-conditioned problems. As only the near matrix elements are explicitly available in the FMM, any preconditioner must be tailored specifically toward sparse linear systems. Also, because the preconditioning step adds to the overall compute time per iteration, it should have close to the same computational cost as the FMM, or less if possible. In this section, we will discuss some commonly used sparse preconditioning schemes which are useful in the FMM.

### 9.6.1  Diagonal Preconditioner

The diagonal or *Jacobi* preconditioner is the simplest form of preconditioner and is useful for illustrating the concept, however in practice it results in a poor approximation of **A** and is not recommended for general use. The elements of the diagonal preconditioner matrix **M** are defined as

$$M_{ij} = A_{ij} \quad i = j \tag{9.93}$$

$$M_{ij} = 0 \quad i \neq j \tag{9.94}$$

The diagonal preconditioner requires no extra storage as the elements in $\mathbf{M}^{-1}$ can be computed on the fly during the preconditioning step.

### 9.6.2  Block Diagonal Preconditioner

Another simple preconditioner is the *block diagonal* preconditioner, where **M** is constructed from blocks of **A** that lie along the matrix diagonal. A conceptual block diagonal sparse matrix **A** can be written as

$$\mathbf{A} = \begin{bmatrix} A_{11} & A_{12} & 0 & 0 & 0 & 0 & 0 \\ A_{21} & A_{22} & 0 & 0 & 0 & 0 & 0 \\ 0 & 0 & A_{33} & A_{34} & A_{35} & 0 & 0 \\ 0 & 0 & A_{43} & A_{44} & A_{45} & 0 & 0 \\ 0 & 0 & A_{53} & A_{54} & A_{55} & 0 & 0 \\ 0 & 0 & 0 & 0 & 0 & A_{66} & A_{67} \\ 0 & 0 & 0 & 0 & 0 & A_{76} & A_{77} \end{bmatrix} \tag{9.95}$$

or

$$\mathbf{A} = \begin{bmatrix} \mathbf{A}_1 & 0 & 0 \\ 0 & \mathbf{A}_2 & 0 \\ 0 & 0 & \mathbf{A}_3 \end{bmatrix} \tag{9.96}$$

where **A** comprises the sub-matrix blocks $\mathbf{A}_1$, $\mathbf{A}_2$, and $\mathbf{A}_3$. In this case, $\mathbf{M}^{-1}$ will have the same block structure as **A**, and we can perform an LU factorization of each block independently. If each block is relatively small, the preconditioner setup will be quick and it will not cause much overhead per iteration. In implementing this preconditioner in the context of the FMM, consider that while the blocks of the FMM near matrix are not square, there is a square portion of these blocks that comprise the interactions of basis and testing functions within each group. These square sub-blocks can be extracted and used in building the preconditioner. This preconditioner was previously reported in [3, 18], where it was used on problems involving closed, conducting objects.

### 9.6.3 Incomplete LU (ILU) Preconditioners

An incomplete LU (ILU) factorization computes an upper and lower triangular sparse matrix pair $\mathbf{L}$ and $\mathbf{U}$ such that the residual matrix $\mathbf{R} = \mathbf{A} - \mathbf{L}\mathbf{U}$ satisfies certain criteria, such as a particular sparsity pattern. The simplest form of the ILU preconditioner is an incomplete LU factorization with no fill-in, denoted as ILU(0). In this factorization, the factor $\mathbf{L}$ has the same pattern as the lower part of $\mathbf{A}$ and the factor $\mathbf{U}$ has the same pattern as the upper part of $\mathbf{A}$. Thus, the memory requirements for the preconditioner are the same as the original matrix. The problem with this factorization is that the product $\mathbf{L}\mathbf{U}$ will have nonzero diagonals that are not present in $\mathbf{A}$. Despite this limitation, the ILU(0) factorization can perform well in some cases.

Improvements to the ILU(0) factorization can be made by allowing additional fill-ins outside the sparsity pattern of $\mathbf{A}$. There are schemes such as ILU(k), which allow $k$ fixed levels of fill-in, or the incomplete LU factorization with thresholding (ILUT). The ILUT algorithm generates the zero pattern of the factors on the fly by dropping elements based on their amplitude instead of their location. In ILUT, the user specifies a value for the drop tolerance, as well the maximum number elements per row in the factorization, typically a small fraction of $N$. The memory demands of ILUT may vary depending on the amount of fill-in, and can exceed the storage of the original near matrix depending on the settings. In the *Serenity* code, the ILUT fill-in is chosen to be a multiple of the average number of nonzero columns per row of the near-matrix ($L_{avg}$). For a more detailed discussion of these preconditioners, the reader is referred to Saad [7].

Experiments with the ILUT preconditioner have shown it to perform well when used in the FMM. Drawbacks to ILUT tend to be the memory requirements and long setup time, as efficient parallelization strategies have not yet been established and most implementations use a single processing thread in the factorization.

### 9.6.4 Sparse Approximate Inverse (SAI)

Another approach for building a sparse preconditioner is through minimizing the Frobenius norm of the residual matrix $\mathbf{I} - \mathbf{A}\mathbf{M}$ (for right-preconditioning), or

$$F(\mathbf{M}) = ||\mathbf{I} - \mathbf{A}\mathbf{M}||_F^2 \tag{9.97}$$

where $\mathbf{M}$ is an *approximate inverse*. The minimization problem can be decoupled into the sum of the squares of the 2-norms of the individual columns of the residual matrix as

$$||\mathbf{I} - \mathbf{A}\mathbf{M}||_F^2 = \sum_{n=1}^{N} ||\mathbf{e}_n - \mathbf{A}\mathbf{m}_n||_2^2 \tag{9.98}$$

where $\mathbf{e}_n$ and $\mathbf{m}_n$ are the columns of $\mathbf{I}$ and $\mathbf{M}$, respectively. The right-hand side of (9.98) comprises $N$ independent least-square sub-problems, which can be solved in parallel. In [7], Saad suggests solving for each column in (9.98) using a few iterations of a solver such as GMRES, which can be made fast by using sparse vectors and performing the matrix-vector product in a sparse-sparse mode. However, experiments by this author found this approach to be inefficient when used in the FMM, a conclusion also reached by the investigators in [19].

In building an SAI preconditioner from the FMM near matrix, one of the most important choices to make is the sparsity pattern of $\mathbf{M}$. Some treatments in the literature find this pattern through ad hoc or heuristic means, or iterative refinement. In other cases, $\mathbf{M}$ is assumed to have the same pattern as $\mathbf{A}$. In some cases, the elements in $\mathbf{A}$ are first down-selected (filtered) based on certain criteria (such as amplitude), yielding a further sparsified matrix $\tilde{\mathbf{A}}$ which is used to build $\mathbf{M}$. Approaches specifically tailored to the FMM problem such as those in [19, 20, 21] use the block structure of the near matrix to determine the pattern. We will discuss one such approach in the next section.

### 9.6.4.1 Dense QR Factorization

We will now describe a process of finding an approximate preconditioner matrix $\mathbf{M}$ which is assumed to have the same the sparsity pattern as $\mathbf{A}$. Following (9.98), this requires solving $N$ problems of the form

$$\mathbf{Am}_n = \mathbf{e}_n \qquad (9.99)$$

in the least-squares sense. To illustrate, consider the $9 \times 9$ sparse matrix

$$\mathbf{A} = \begin{bmatrix} A_{11} & A_{12} & 0 & 0 & A_{15} & 0 & 0 & 0 & 0 \\ A_{21} & A_{22} & 0 & A_{24} & 0 & 0 & 0 & A_{28} & 0 \\ 0 & 0 & A_{33} & 0 & 0 & A_{36} & 0 & 0 & A_{39} \\ 0 & A_{42} & 0 & A_{44} & 0 & 0 & A_{47} & 0 & 0 \\ A_{51} & 0 & 0 & 0 & A_{55} & 0 & 0 & 0 & 0 \\ 0 & 0 & A_{63} & 0 & 0 & A_{66} & 0 & A_{68} & 0 \\ 0 & 0 & 0 & A_{74} & 0 & 0 & A_{77} & 0 & 0 \\ 0 & A_{82} & 0 & 0 & 0 & A_{86} & 0 & A_{88} & A_{89} \\ 0 & 0 & A_{93} & 0 & 0 & 0 & 0 & A_{98} & A_{99} \end{bmatrix} \qquad (9.100)$$

which is a simple but realistic representation of an FMM near matrix. As each column of $\mathbf{M}$ is computed independently of the others, let us consider column 4 of $\mathbf{M}$. In this case, the nonzero rows of $\mathbf{m}_4$ are rows 2, 4, and 7, which operate on columns 2, 4, and 7 of $\mathbf{A}$ during the matrix-vector product in (9.99). Retaining only these rows of $\mathbf{m}_4$ and columns of $\mathbf{A}$ allows us to write the matrix-vector product as

$$
\begin{bmatrix}
A_{12} & 0 & 0 \\
A_{22} & A_{24} & 0 \\
0 & 0 & 0 \\
A_{42} & A_{44} & A_{47} \\
0 & 0 & 0 \\
0 & 0 & 0 \\
0 & A_{74} & A_{77} \\
A_{82} & 0 & 0 \\
0 & 0 & 0
\end{bmatrix}
\begin{bmatrix}
M_{24} \\
M_{44} \\
M_{74}
\end{bmatrix}
=
\begin{bmatrix}
0 \\
0 \\
0 \\
1 \\
0 \\
0 \\
0 \\
0 \\
0
\end{bmatrix}
\tag{9.101}
$$

We see that many of the rows of the matrix in (9.101) are zero. If we remove these rows from the left- and right-hand sides, we are then left with

$$
\begin{bmatrix}
A_{12} & 0 & 0 \\
A_{22} & A_{24} & 0 \\
A_{42} & A_{44} & A_{47} \\
0 & A_{74} & A_{77} \\
A_{82} & 0 & 0
\end{bmatrix}
\begin{bmatrix}
M_{24} \\
M_{44} \\
M_{74}
\end{bmatrix}
=
\begin{bmatrix}
0 \\
0 \\
1 \\
0 \\
0
\end{bmatrix}
\tag{9.102}
$$

which is a dense, overdetermined system. This can be solved in the least-squares sense using a QR factorization, such as the LAPACK function `cgels`, and the nonzero elements of $\mathbf{m}_4$ populated from the result vector.

### Near Matrix Block Structure

The $N$ least-square problems in (9.98) can be approached in a naive way by solving for each column of $\mathbf{M}$ independently, however this requires a total of $N$ QR factorizations, which will consume significant CPU time as the size of the problem grows. If instead we exploit the block structure of the near matrix, this process can be made much more efficient.

In single-region problems, the near matrix has a block structure described by the octree. Consider a cube $C_j$ on the finest level of the octree, which contains $N_j$ basis functions. These basis functions are tested by testing functions in the cubes near to $C_j$. Thus, for the columns $\mathbf{m}_n$ corresponding to the basis functions in $C_j$, the dense QR matrix is the same. Thus, if we process the columns of $\mathbf{M}$ on a *per cube* basis, only one QR factorization and $N_j$ right-hand side solves are required per cube.

In multi-region problems there are multiple octrees, and in general we cannot rely on them to determine the block structure. Instead, we can operate directly on the near matrix. Noting that the zero pattern of the near matrix is symmetric, we can locate blocks by finding rows of the CSR near matrix that have the same nonzero columns. Assuming that the CSR vector $\mathbf{j}$ is stored in memory as a contiguous array of integers, this can be done by hashing the portion of **j** corresponding to each row, and then sorting the hashes. Rows having identical hashes comprise a block, which can be processed together using a single QR factorization. This approach is also valid for single region problems.

## 9.7 Examples

In this section, we will compute the radar cross section of several test articles, using the MLFMA-accelerated iterative solver capability in the *Serenity* code. We again consider conducting, dielectric, and coated spheres, several EMCC benchmark targets, and the monoconic RV with dielectric nose. For each example, the triangle meshes are built so that there are at least $\lambda/10$ unknowns per wavelength in each dielectric region, and the CFIE is applied on closed conductors with $\alpha = 0.5$. Octree construction is halted when the cube size reaches $\lambda/4$ in each region. For the iterative solver, we use a non-restarted GMRES with residual tolerance of $10^{-4}$ per right-hand side, and an ILUT preconditioner with a maximum fill of $L_{avg}$ elements per row and a drop tolerance of $10^{-3}$.

### 9.7.1 Test System

The examples in this chapter were run on an SGI UV-100 system having eight six-core Intel Xeon (X7542) processors @ 2.66 GHz (a total of 48 cores) and 128 GB of RAM, running Red Hat Enterprise Linux Version 6.1. On this system, *Serenity* was compiled using the GNU C++ Compiler (Version 4.4.6). Though the test system described in Section 8.8.2 is still adequate in solving most of these examples, the increased memory capacity and thread count of the SGI system made it possible to solve many right-hand sides in parallel, which minimized the turnaround time.

### 9.7.2 Spheres

In this section we will consider the bistatic RCS of a series of conducting, fully dielectric, and coated spheres.

#### 9.7.2.1 Conducting Sphere

We first consider a conducting sphere with a radius of 0.5 meters at frequencies of 0.75, 1.5, and 3.0 GHz. The exact expressions for the scattered field of a conducting sphere were previously summarized in Section 7.9.1.1. At each frequency, we use facet models comprising 5120, 20480, and 81920 triangles, respectively. The bistatic RCS from the FMM is compared to the Mie series in Figures 9.11a–9.11f. The comparison is very good at all frequencies.

### 9.7.2.2 Dielectric Sphere

We next consider a dielectric sphere with a radius of 0.5 meters at frequencies of 0.75, 1.5, and 3.0 GHz. The exact expressions for the scattered field of a stratified sphere were previously summarized in Section 7.9.1.2. For the numerical simulation, we re-use the facet models from Section 9.7.2.1. In Figures 9.12a–9.12f, we compare the bistatic RCS of the FMM and Mie series for a lossless sphere ($\epsilon_r = 2.56$). The comparison is very good. In Figures 9.13a–9.13f is a similar comparison for a lossy sphere ($\epsilon_r = 2.56 - j.102$), where the comparison is again very good.

### 9.7.2.3 Coated Sphere

We next consider a conducting sphere with a thick dielectric coating at frequencies of 0.75, 1.5 and 3.0 GHz. The dimensions of the core and the coating are the same as the sphere in Section 8.8.3.3. At 0.75 GHz, the facet models representing the conducting core and the exterior of the coating comprise 1280 and 5120 triangles, respectively. At 1.5 GHz, they comprise 5120 and 20480 triangles, and at 3.0 GHz, they comprise 20480 and 81920 triangles, respectively. In Figures 9.14a-9.14f, we compare the bistatic RCS of the FMM and Mie series for a lossless coating ($\epsilon_r = 2.56$). The comparison is very good. In Figures 9.15a-9.15f is a similar comparison for a lossy coating ($\epsilon_r = 2.56 - j.102$), where the comparison is again very good.

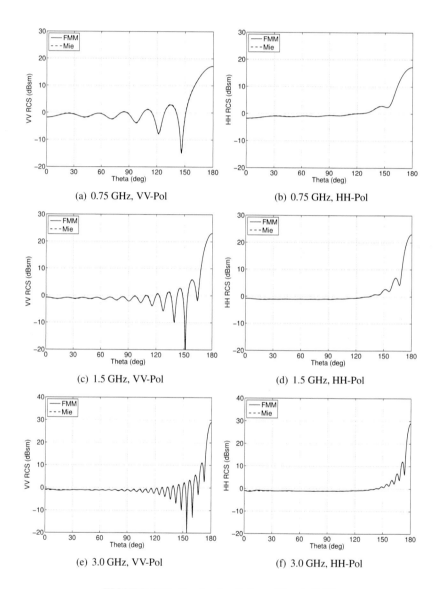

(a) 0.75 GHz, VV-Pol

(b) 0.75 GHz, HH-Pol

(c) 1.5 GHz, VV-Pol

(d) 1.5 GHz, HH-Pol

(e) 3.0 GHz, VV-Pol

(f) 3.0 GHz, HH-Pol

**FIGURE 9.11:** Conducting Sphere: Bistatic RCS

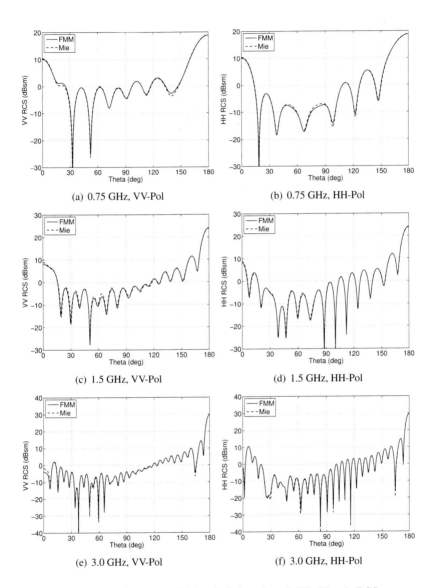

(a) 0.75 GHz, VV-Pol

(b) 0.75 GHz, HH-Pol

(c) 1.5 GHz, VV-Pol

(d) 1.5 GHz, HH-Pol

(e) 3.0 GHz, VV-Pol

(f) 3.0 GHz, HH-Pol

**FIGURE 9.12:** Lossless Dielectric Sphere ($\epsilon = 2.56$): Bistatic RCS

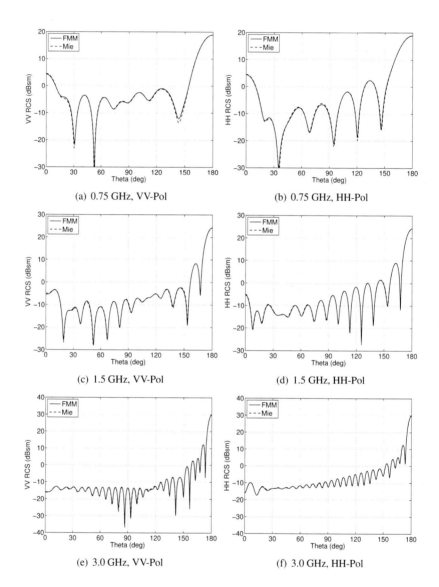

(a) 0.75 GHz, VV-Pol

(b) 0.75 GHz, HH-Pol

(c) 1.5 GHz, VV-Pol

(d) 1.5 GHz, HH-Pol

(e) 3.0 GHz, VV-Pol

(f) 3.0 GHz, HH-Pol

**FIGURE 9.13:** Lossy Dielectric Sphere ($\epsilon = 2.56 - j.102$): Bistatic RCS

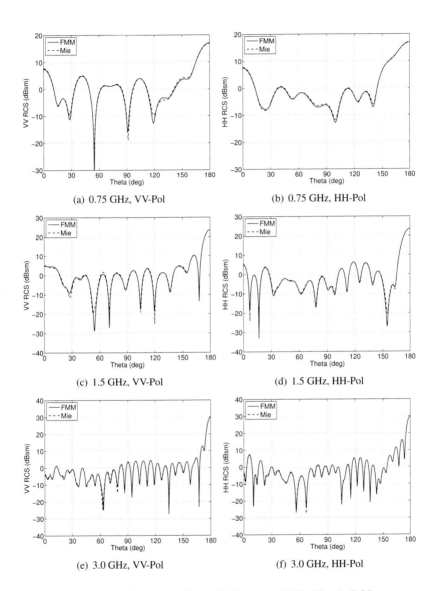

(a) 0.75 GHz, VV-Pol

(b) 0.75 GHz, HH-Pol

(c) 1.5 GHz, VV-Pol

(d) 1.5 GHz, HH-Pol

(e) 3.0 GHz, VV-Pol

(f) 3.0 GHz, HH-Pol

**FIGURE 9.14:** Lossless Coated Sphere ($\epsilon = 2.56$): Bistatic RCS

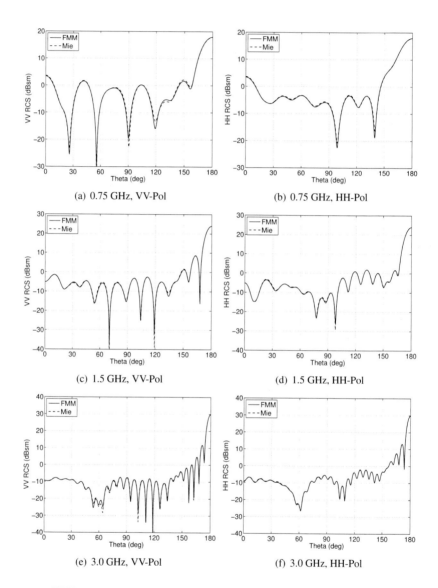

(a) 0.75 GHz, VV-Pol

(b) 0.75 GHz, HH-Pol

(c) 1.5 GHz, VV-Pol

(d) 1.5 GHz, HH-Pol

(e) 3.0 GHz, VV-Pol

(f) 3.0 GHz, HH-Pol

**FIGURE 9.15:** Lossy Coated Sphere ($\epsilon = 2.56 - j.102$): Bistatic RCS

### 9.7.3 EMCC Benchmark Targets

In this section, we will consider the monostatic RCS of the EMCC benchmark radar targets previously discussed in Section 7.9.2, as well as the NASA almond from [22].

#### 9.7.3.1 NASA Almond

The 9.936-inch NASA almond of Figure 9.16 is a target comprising a mostly smooth surface that supports specular and creeping wave scattering, and a pointed end that gives rise to tip diffraction effects. It can be expressed for $0 \leq \phi \leq 2\pi$ as

$$x = dt \qquad (9.103)$$

$$y(t) = 0.193333d\sqrt{1 - \left(\frac{t}{0.416667}\right)^2} \cos \phi \qquad (9.104)$$

$$z(t) = 0.64444d\sqrt{1 - \left(\frac{t}{0.416667}\right)^2} \sin \phi \qquad (9.105)$$

for $-0.41667 < t < 0$ inches, and

$$x = dt \qquad (9.106)$$

$$y(t) = 4.83345d\left[\sqrt{1 - \left(\frac{t}{2.08335}\right)^2} - 0.96\right] \cos \phi \qquad (9.107)$$

$$z(t) = 1.61115d\left[\sqrt{1 - \left(\frac{t}{2.08335}\right)^2} - 0.96\right] \sin \phi \qquad (9.108)$$

for $0 < t < 0.58333$, where $d = 9.936$ inches. The facet model for this example comprises 46984 triangles, and surface details of the pointed and rounded ends are shown in Figures 9.17a and 9.17b, respectively. Monostatic observations are made in the xy plane of Figure 9.16a. The RCS from the FMM is compared to the EMCC measurements at 7 GHz in Figures 9.18a and 9.18b for vertical and horizontal polarizations, respectively. The computed results were lower than the measurements by approximately 1.5 dB across a large range of angles, and were adjusted by that amount in each figure. After this adjustment, the comparison is very good. Results at 9.92 GHz are shown in Figures 9.19a and 9.19b for vertical and horizontal polarizations, respectively. A similar bias was also observed in this case, and after adjustment the comparison is again fairly good. However, the difference is much greater in this case, particularly in VV polarization. The reason for the difference is not known, though a similar difference was also noted between the measurements and the numerical results in [22]. The authors of that paper suggested it may have been due to the

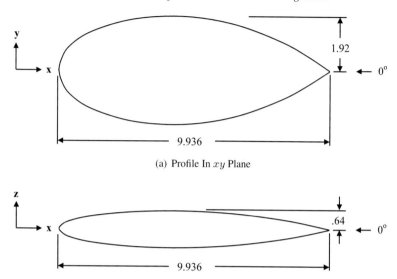

(a) Profile In $xy$ Plane

(b) Profile In $xz$ Plane

**FIGURE 9.16:** NASA Almond

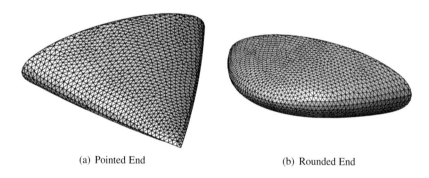

(a) Pointed End                           (b) Rounded End

**FIGURE 9.17:** Almond Facet Model Detail

density of the unknowns $(6/\lambda)$ in the CAD model that was used. As we use more than double this density in our model (approximately $13/\lambda$), this author believes the unknown density is not the reason for the discrepancy.

In Figure 9.20 are plotted maps of the surface current on the almond at several incident angles. The real component of the induced electric currents for vertical and horizontal polarizations are plotted at $\theta^i = 0$ in Figures 9.20a and 9.20b, respectively, and at $\theta^i = \pi/4$ and $\theta^i = \pi/2$ in Figures 9.20c–9.20f. The darker areas indicate the currents of higher amplitude.

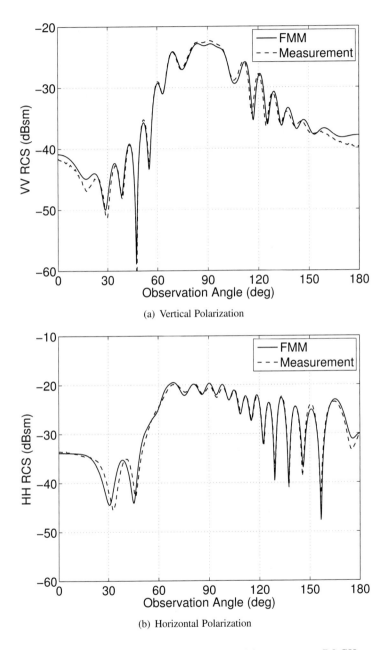

(a) Vertical Polarization

(b) Horizontal Polarization

**FIGURE 9.18:** NASA Almond: MLFMA vs. Measurement at 7.0 GHz

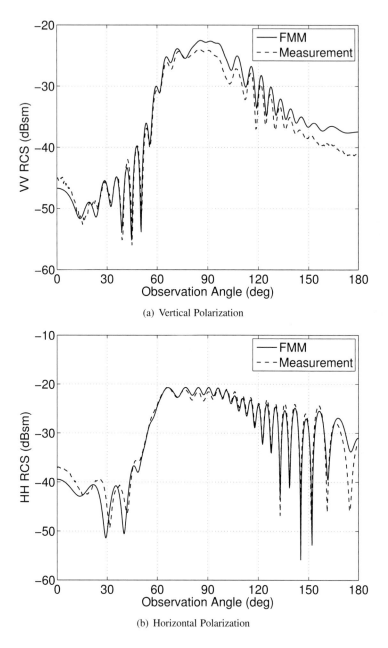

(a) Vertical Polarization

(b) Horizontal Polarization

**FIGURE 9.19:** NASA Almond: MLFMA vs. Measurement at 9.92 GHz

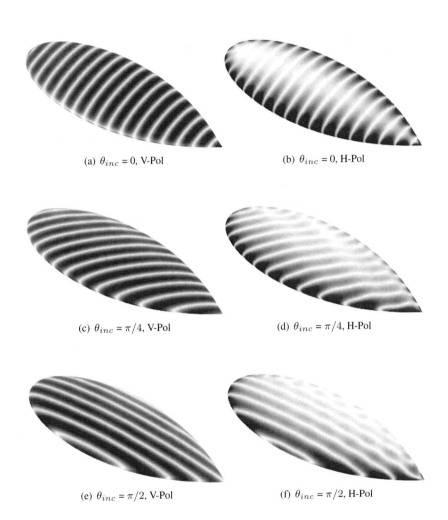

(a) $\theta_{inc} = 0$, V-Pol

(b) $\theta_{inc} = 0$, H-Pol

(c) $\theta_{inc} = \pi/4$, V-Pol

(d) $\theta_{inc} = \pi/4$, H-Pol

(e) $\theta_{inc} = \pi/2$, V-Pol

(f) $\theta_{inc} = \pi/2$, H-Pol

**FIGURE 9.20:** NASA Almond: Surface Currents at 9.92 GHz

### 9.7.3.2   EMCC Ogive

The EMCC ogive was previously described in Section 7.9.2.1, and the facet model used in this example comprises 21760 triangles. The numerical results are compared to the EMCC measurements at 9.0 GHz in Figures 9.21a and 9.21b for vertical and horizontal polarizations, respectively. The agreement is very good. Surface current maps for $\theta^i = \pi/4$ are shown in Figures 9.25a and 9.25b for vertical and horizontal polarizations, respectively.

### 9.7.3.3   EMCC Double Ogive

The EMCC double ogive was previously described in Section 7.9.2.2, and the facet model used in this example comprises 36240 triangles. The numerical results are compared to the EMCC measurements at 9.0 GHz in Figures 9.22a and 9.22b for vertical and horizontal polarizations, respectively. The agreement is again very good. Surface current maps on for $\theta^i = \pi/4$ are shown in Figures 9.25c and 9.25d for vertical and horizontal polarizations, respectively.

### 9.7.3.4   EMCC Cone-Sphere

The EMCC cone-sphere was previously described in Section 7.9.2.3, and the facet model used in this example comprises 79496 triangles. The numerical results are compared to the EMCC measurements at 9.0 GHz in Figures 9.23a and 9.23b for vertical and horizontal polarizations, respectively. The agreement is fairly good at all angles. Surface current maps on for $\theta^i = \pi/4$ are shown in Figures 9.25e and 9.25f for vertical and horizontal polarizations, respectively.

### 9.7.3.5   EMCC Cone-Sphere with Gap

The EMCC cone-sphere with gap was previously described in Section 7.9.2.4, and the facet model used in this example comprises 82528 triangles. The numerical results are compared to the EMCC measurements at 9.0 GHz in Figures 9.24a and 9.24b for vertical and horizontal polarizations, respectively. The agreement is fairly good at all angles. Surface current maps on for $\theta^i = \pi/4$ are shown in Figures 9.25g and 9.25h for vertical and horizontal polarizations, respectively.

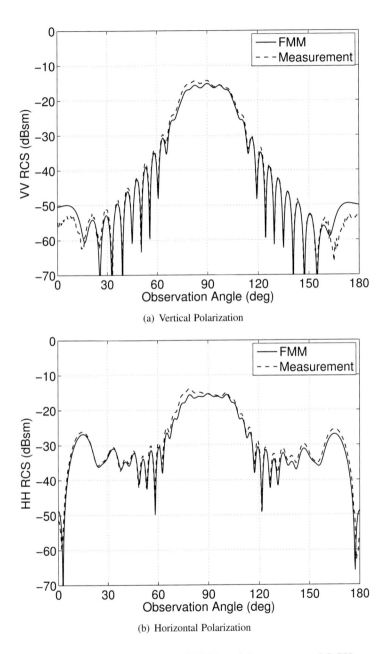

(a) Vertical Polarization

(b) Horizontal Polarization

**FIGURE 9.21:** EMCC Ogive: MLFMA vs. Measurement at 9.0 GHz

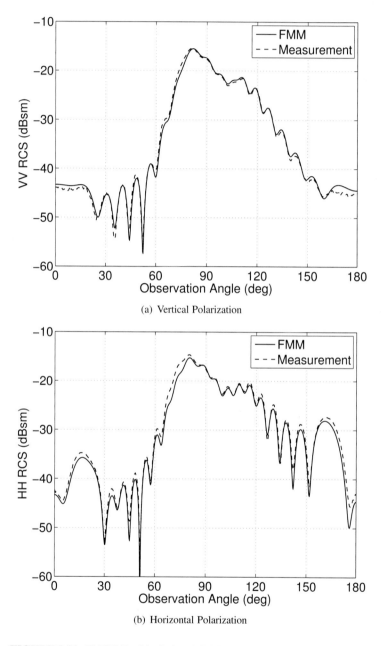

(a) Vertical Polarization

(b) Horizontal Polarization

**FIGURE 9.22:** EMCC Double Ogive: MLFMA vs. Measurement at 9.0 GHz

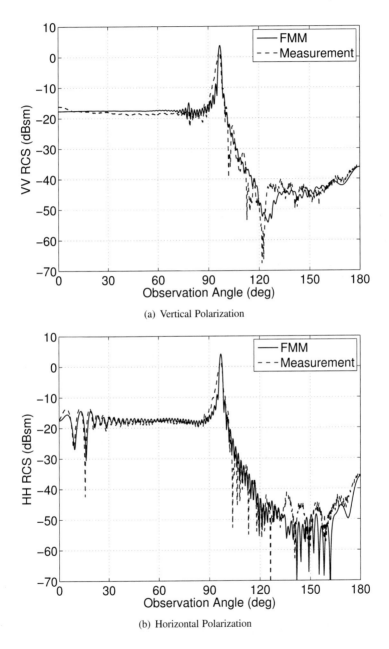

(a) Vertical Polarization

(b) Horizontal Polarization

**FIGURE 9.23:** EMCC Cone-Sphere: MLFMA vs. Measurement at 9.0 GHz

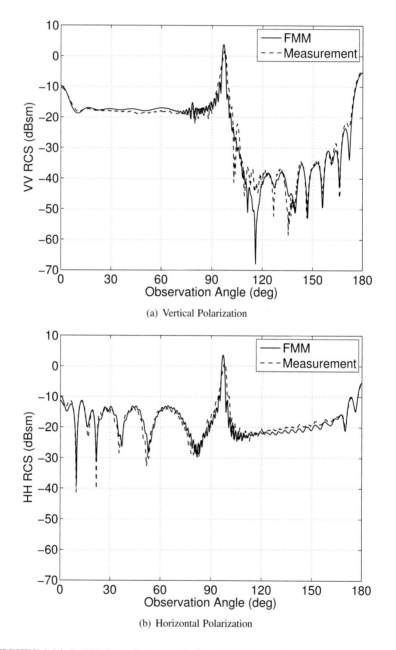

(a) Vertical Polarization

(b) Horizontal Polarization

**FIGURE 9.24:** EMCC Cone-Sphere with Gap: MLFMA vs. Measurement at 9.0 GHz

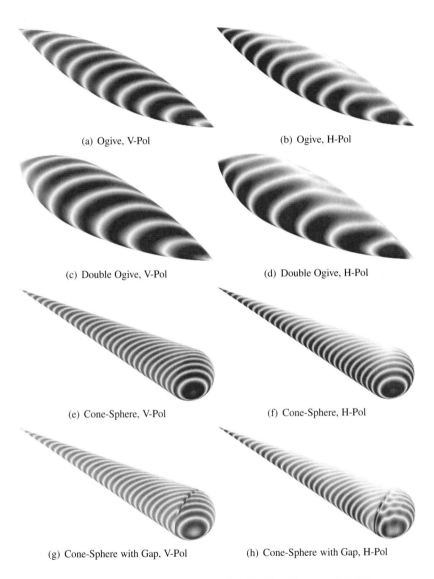

(a) Ogive, V-Pol

(b) Ogive, H-Pol

(c) Double Ogive, V-Pol

(d) Double Ogive, H-Pol

(e) Cone-Sphere, V-Pol

(f) Cone-Sphere, H-Pol

(g) Cone-Sphere with Gap, V-Pol

(h) Cone-Sphere with Gap, H-Pol

**FIGURE 9.25:** Surface Currents on EMCC Test Targets at 9.0 GHz

### 9.7.3.6   Monoconic Reentry Vehicle with Dielectric Nose

Finally, we consider the monoconic reentry vehicle with dielectric nose-tip, examined previously in Sections 7.10.2.4 and 8.8.8. In this case, we compute the monostatic RCS at frequencies of 2.0 and 5.0 GHz, and compare the FMM results to those from the *Galaxy* BOR code. At 2.0 GHz, the facet models of the nose, septum, and frustum comprise 1944, 472, and 15552 triangles, respectively, and at 5.0 GHz, they comprise 4216, 3312, and 99168 triangles. The results at 2.0 GHz for a lossless nose tip ($\epsilon = 5$) are compared in vertical and horizontal polarizations in Figures 9.26a and 9.26b, respectively. The comparison is very good. Similar results for a lossy nose tip ($\epsilon = 5 - j0.25$) are compared in Figures 9.27a and 9.27b, respectively, where the comparison is again good. Results at 5.0 GHz are compared in Figures 9.28 and 9.29, where the comparisons are again very good.

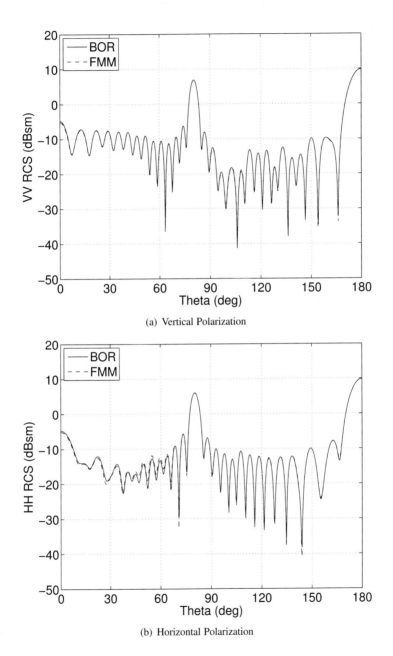

(a) Vertical Polarization

(b) Horizontal Polarization

**FIGURE 9.26:** Monoconic RV RCS: Lossless Nose, BOR vs. FMM at 2.0 GHz

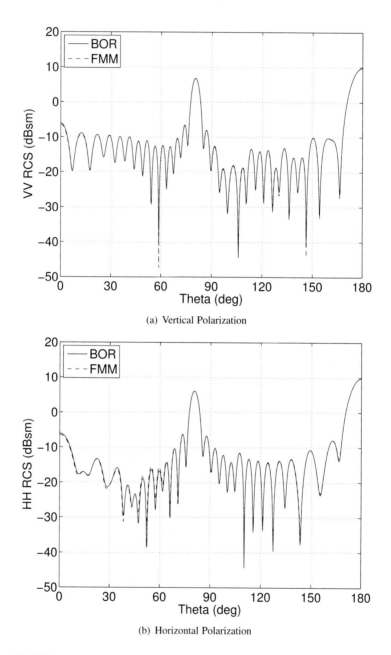

(a) Vertical Polarization

(b) Horizontal Polarization

**FIGURE 9.27:** Monoconic RV RCS: Lossy Nose, BOR vs. FMM at 2.0 GHz

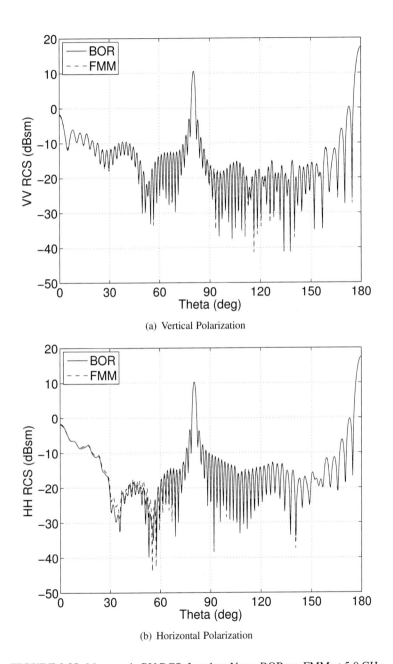

(a) Vertical Polarization

(b) Horizontal Polarization

**FIGURE 9.28:** Monoconic RV RCS: Lossless Nose, BOR vs. FMM at 5.0 GHz

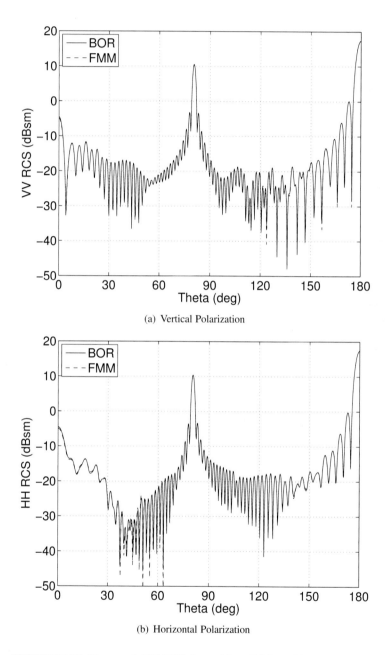

(a) Vertical Polarization

(b) Horizontal Polarization

**FIGURE 9.29:** Monoconic RV RCS: Lossy Nose, BOR vs. FMM at 5.0 GHz

### 9.7.4 Summary of Examples

Run metrics for the examples in this section are summarized in Table 8.5. Listed are the objects, operating frequency $f_0$ (in GHz), number of electric $N_J$ and magnetic $N_M$ basis functions, near matrix storage $M_{near}$, preconditioner storage $M_{pre}$, radiation and receive function storage $M_{rad}$ and $M_{rec}$, and number of levels in the FMM octree $N_O$. In cases where there are multiple dielectric regions, the number of FMM levels in each region is given, and storage is specified in megabytes. As the overall run-time is dependent on the software implementation strategy and optimizations, as well as the relative speed of the CPU, cache, and RAM, run-time information is not shown. Also, only the data for the lossless dielectrics are shown, as the metrics for the lossy dielectrics are identical except for preconditioner storage, which varied by less than one percent in all cases. We see from these examples the impact the number of FMM levels has on the size of the near matrix. This is most readily apparent in the runs involving the almond. Though a facet model having 46984 triangles and 70476 edges is used for both cases, the size of the near matrix at 7 GHz with a 5-level MLFMA (5280 MB) is nearly five times what it is at 9.92 GHz with a 6-level MLFMA (1083 MB). We also note the storage requirements of the ILUT preconditioner, which exceeds that of the near matrix in some cases. Though it greatly improved the convergence, the ILUT preconditioner was not strictly necessary in some cases. For example, although the block-diagonal preconditioner does not perform as well as ILUT, its storage

**TABLE 9.1:** Run Metrics

| Object | $f_0$ | $N_J$ | $N_M$ | $M_{near}$ | $M_{pre}$ | $M_{rad}$ | $M_{rec}$ | $N_O$ |
|---|---|---|---|---|---|---|---|---|
| PEC Sphere | 0.75 | 7680 | 0 | 35.2 | 54.3 | 15.9 | 15.9 | 4 |
| PEC Sphere | 1.5 | 30720 | 0 | 143.5 | 225.2 | 63.8 | 63.8 | 5 |
| PEC Sphere | 3.0 | 122880 | 0 | 571.5 | 905.0 | 255.0 | 255.0 | 6 |
| Dielectric Sphere | 0.75 | 7680 | 7680 | 140.9 | 254.0 | 63.8 | 127.5 | 4,5 |
| Dielectric Sphere | 1.5 | 30720 | 30720 | 573.9 | 997.5 | 255.0 | 510.0 | 5,6 |
| Dielectric Sphere | 3.0 | 122880 | 122880 | 2284.9 | 3756.5 | 1020.0 | 2040.0 | 6,7 |
| Coated Sphere | 0.75 | 9600 | 7680 | 2320.8 | 239.3 | 67.7 | 135.5 | 4,5 |
| Coated Sphere | 1.5 | 38400 | 30720 | 582.9 | 956.0 | 270.9 | 541.9 | 5,6 |
| Coated Sphere | 3.0 | 153600 | 122880 | 2320.8 | 3617.9 | 1083.8 | 2167.5 | 6,7 |
| Almond | 7.0 | 70476 | 0 | 5280.0 | 5280.0 | 146.3 | 146.3 | 5 |
| Almond | 9.92 | 70476 | 0 | 917.0 | 1082.9 | 113.0 | 113.0 | 6 |
| Ogive | 9.0 | 32640 | 0 | 1337.8 | 1337.8 | 85.2 | 85.2 | 5 |
| Double Ogive | 9.0 | 54360 | 0 | 2414.7 | 2414.7 | 112.8 | 112.8 | 5 |
| Cone-Sphere | 9.0 | 119244 | 0 | 1059.1 | 1490.8 | 247.5 | 247.5 | 7 |
| Cone-Sphere/Gap | 9.0 | 123792 | 0 | 1208.4 | 1658.7 | 256.9 | 256.9 | 7 |
| RV | 2.0 | 26932 | 2896 | 813.5 | 813.5 | 137.6 | 275.3 | 5,3 |
| RV | 5.0 | 159992 | 6272 | 1029.3 | 1514.5 | 501.4 | 1002.8 | 7,4 |

requirement is far less, and for some problems this can be tolerated on systems where memory is at a premium. We also note that in some cases, the density of the facet model was increased to better represent local surface curvatures, which resulted in more than $\lambda/10$ unknowns per wavelength. This increased the storage requirements of the near matrix and preconditioners in those cases, though this seemed reasonable given the larger amount of memory available in the test system.

### 9.7.5   Initial Guess in Iterative Solution

When using an iterative solver, there is often little in the way of *a priori* knowledge regarding the solution vector. Therefore, for an arbitrary right-hand side vector, the typical approach is to simply fill the solution vector with zeros. In radar cross section problems, where the right-hand side and solution vectors are similar between closely spaced incident angles, we can re-use the previous solution vector, reducing the number of iterations. Additionally, it was suggested in [3] that this approach can be improved even further by adjusting the phase of the solution vector between incident angles. To illustrate, consider the plane-wave incident vector $\hat{\mathbf{r}}_1$, where the phase term at a point $\mathbf{r}$ in the free space region $R_0$ is

$$\psi_1(\mathbf{r}) = e^{jk_0\hat{\mathbf{r}}_1\cdot\mathbf{r}} \tag{9.109}$$

Between angles, we can make following correction to the solution vector

$$\Delta\psi(\mathbf{r}) = e^{jk_0(\hat{\mathbf{r}}_2-\hat{\mathbf{r}}_1)\cdot\mathbf{r}} \tag{9.110}$$

When using RWG basis functions, this can be implemented by adjusting the phase of the solution vector elements corresponding to the electric and magnetic basis functions assigned to each edge in $R_0$. For a given edge, the correction is

$$\Delta\psi(\mathbf{r}_c) = e^{jk_0(\hat{\mathbf{r}}_2-\hat{\mathbf{r}}_1)\cdot\mathbf{r}_c} \tag{9.111}$$

where $\mathbf{r}_c$ is the center point of the edge. Let us now consider several examples of the improvements this phase correction offers. We first consider the cone-sphere of Section 9.7.3.4, where the run settings are the same except for the use of phase correction. In this case and those that follow, we solve for 721 incident angles over 180 degrees (a step size of 0.25 degrees). The number of iterations per right-hand side are compared at vertical and horizontal polarizations in Figures 9.30a and 9.30b, respectively. The phase correction reduces the number of iterations at all angles. A similar comparison is made for the almond at 9.92 GHz in Figures 9.30c and 9.30d. Though the number of iterations was already small, an improvement is still seen. A more dramatic improvement occurs in the case of the monoconic RV with lossless nose, as shown in Figures 9.30e and 9.30f.

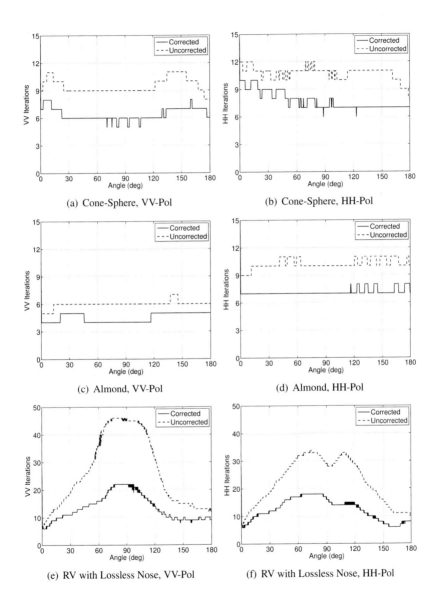

(a) Cone-Sphere, VV-Pol

(b) Cone-Sphere, HH-Pol

(c) Almond, VV-Pol

(d) Almond, HH-Pol

(e) RV with Lossless Nose, VV-Pol

(f) RV with Lossless Nose, HH-Pol

**FIGURE 9.30:** Phase Correction Impact on Iterative Solver Performance

### 9.7.6   Preconditioner Performance

We next consider the performance of the preconditioners discussed in Section 9.6 by comparing the number of iterations per right-hand side for two sample cases. For this comparison, the RCS of the almond (Section 9.7.3.1) and the cone-sphere with gap (Section 9.7.3.5) are computed using each preconditioner with identical run parameters. As in Section 9.7.5, we solve for 721 incident angles over 180 degrees (a step size of 0.25 degrees), and the phase correction (9.110) is applied between angles. The results for the almond at 9.92 GHz are compared at vertical and horizontal polarization in Figures 9.31a and 9.31b, respectively. The data in each figure has been thinned so that the markers are easily identified. Using no preconditioner yields the highest iteration count, as expected. Though it is only intended to illustrate the concept, the diagonal preconditioner actually decreases the iteration count in both polarizations, which was unexpected. The block-diagonal preconditioner improves the iteration count significantly. The SAI and ILUT preconditioners reduce the number of iterations even further, though at greatly increased cost in setup time and memory requirements. We next compare the results for the cone-sphere with gap at 9.0 GHz at vertical and horizontal polarization in Figures 9.32a and 9.32b, respectively. The SAI and ILUT preconditioners again show the best performance, with ILUT performing slightly better than the SAI.

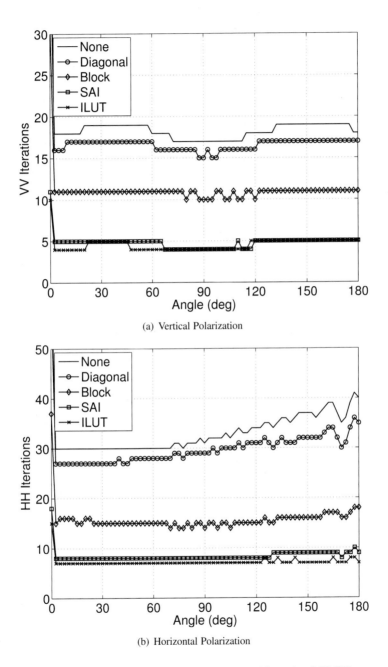

(a) Vertical Polarization

(b) Horizontal Polarization

**FIGURE 9.31:** Preconditioner Performance on Almond at 9.92 GHz

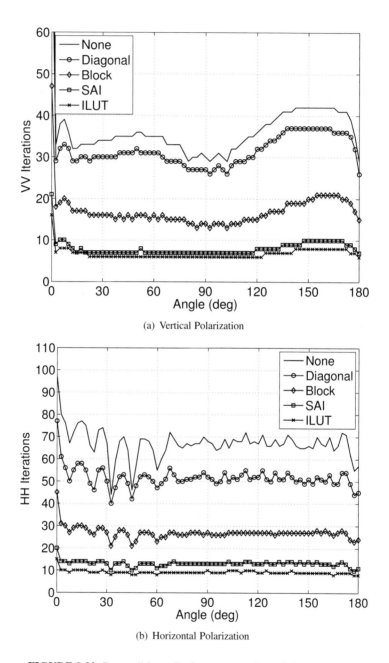

(a) Vertical Polarization

(b) Horizontal Polarization

**FIGURE 9.32:** Preconditioner Performance on Cone-Sphere with Gap

## 9.8 Notes on Software Implementation

The *Serenity* software used for the first edition of this book was written in the C programming language. This allowed for a high level of optimization in many areas, but added considerable complexity in the programming of several key algorithms, as well as the overall memory management and error handling. When *Serenity* was re-written in C++, many algorithms that were difficult to implement in C were very easy to write in C++ using classes and templates, and the total number of source code lines decreased significantly. Careful programming of the C++ version of *Serenity* has yielded a code which suffers no speed penalty versus the C-based version of the code on the same problems. This suggests that the notion of compiled, object-oriented (OO) programming languages such as C++ being unsuitable for implementing scientific software due to their overhead is a thing of the past, and any overhead that does exist is far outweighed by the simplified programming offered by these languages.

### 9.8.1 Parallelization

Parallelizing an FMM-accelerated Moment Method software is a challenging problem. We have parallelized some of the algorithms in our *Serenity* code using POSIX threads on shared memory systems. Computation of the radiation, receive functions, and transfer functions is parallelized over sample points on the unit sphere, and a mutex is not required. The near-matrix fill uses much of the same code as the full-matrix fill, and is parallelized over triangle pairs. In this case, a mutex is required when adding the results to the near matrix. Parallelization on distributed-memory systems is a problem currently receiving considerable attention at the academic level, and is beyond the scope of this text. The reader is referred to several approaches discussed in the literature [6, 23, 24, 25, 26, 27] for additional information.

#### 9.8.1.1 Single Right-Hand Side Solve

*Serenity* can solve a single right-hand in parallel by using multiple threads to compute the matrix-vector product. The aggregation step can be parallelized over groups on each level. As the aggregated field in a group only depends on the shifted and up-sampled field of its children, thread contention is provided that all threads are synchronized with a barrier moving to the next coarsest level. The disaggregation step can also be parallelized by group. As the local field in a group depends only on transfers from far groups and the shifted and anterpolated fields from its parent, contention is again avoided using a barrier before moving to the next finest level. As basis functions are unique to a group, a mutex is not required when adding to the result vector elements.

### 9.8.1.2 Multiple Right-Hand Side Solve

*Serenity* can also solve multiple right-hand sides simultaneously, with a single thread per right-hand side. This eliminates work imbalance and synchronization overhead in the matrix-vector product, but it requires separate scratch space for each thread. This mode is attractive only on systems having the memory to support it.

## 9.8.2   Vectorization

In the past, supercomputers had processors that operated on many data elements at once, and were referred to as single instruction multiple data (SIMD) or *vector* processors. Conversely, most consumer-level processors at that time were largely scalar, operating only on single data elements at a time. Modern consumer-level processor designs by manufacturers such as Intel and AMD now implement vector registers and instructions that can operate on multiple data elements in parallel. The Streaming SIMD Extensions 2 (SSE2) are one such implementation, introduced by Intel in the Pentium 4 CPU architecture. These instructions are particularly effective in vectorizing floating-point operations, and sections of code that operate on floating-point arrays become good candidates for vectorization. The aggregation, transfer, and disaggregation steps in the FMM far-multiply contain array operations that can be optimized using SSE2 instructions, in particular the `_mm_add_ps`, `_mm_sub_ps` and `_mm_mul_ps` compiler intrinsics, which are supported by most compilers on SSE2-capable x86 processors.

---

## References

[1] L. Greengard and V. Rokhlin, "A fast algorithm for particle simulations," *J. Comput. Phys.*, vol. 73, pp. 325–348, 1987.

[2] R. Coifman, V. Rokhlin, and S. Wandzura, "The fast multipole method for the wave equation: A pedestrian prescription," *IEEE Antennas Propagat. Magazine*, vol. 35, pp. 7–12, June 1993.

[3] W. C. Chew, J. M. Jin, E. Michielssen, and J. Song, *Fast and Efficient Algorithms in Computational Electromagnetics*. Artech House, 2001.

[4] J. Stratton, *Electromagnetic Theory*. McGraw-Hill, 1941.

[5] D. E. Amos, "A subroutine package for Bessel functions of a complex argument and nonnegative order," Tech. Rep. SAND85-1018, Sandia National Laboratory, May 1985.

[6] P. Havé, "A parallel implementation of the fast multipole method for Maxwell's equations," *Int. J. Numer. Meth. Fluids*, vol. 43, no. 8, pp. 839–864, 2003.

[7] Y. Saad, *Iterative Methods for Sparse Linear Systems*. PWS, 1st ed., 1996.

[8] J. M. Song and W. C. Chew, "Error analysis for the truncation of multipole expansion of vector Green's functions," *IEEE Microwave Wireless Components Lett.*, vol. 11, no. 7, pp. 311–313, 2001.

[9] O. M. Bucci and G. Franceschetti, "On the spatial bandwidth of scattered fields," *IEEE Trans. Antennas Propagat.*, vol. 35, pp. 1445–1455, December 1987.

[10] N. Geng, A. Sullivan, and L. Carin, "Fast multipole method for scattering from an arbitrary PEC target above or buried in a lossy half space," *IEEE Trans. Antennas Propagat.*, vol. 49, pp. 740–748, May 2001.

[11] O. M. Bucci, C. Gennarelli, and C. Savarese, "Optimal interpolation of radiated fields over a sphere," *IEEE Trans. Antennas Propagat.*, vol. 39, pp. 1633–1643, November 1991.

[12] M. Böhme and D. Potts, "A fast algorithm for filtering and wavelet decomposition on the sphere," *Trans. Numer. Anal*, vol. 16, pp. 70–93, 2002.

[13] J. Foley, A. van Dam, S. Feiner, and J. Hughes, *Computer Graphics: Principles and Practice*. Addison-Wesley, 1996.

[14] E. Darve, "The fast multipole method: Numerical implementation," *J. Comput. Phys.*, vol. 160, pp. 195–240, 2000.

[15] J. Sarvas, "Performing interpolation and anterpolation entirely by fast Fourier transform in the 3-D multilevel fast multipole algorithm," *SIAM J. Numer. Anal.*, vol. 41, no. 6, pp. 2180–2196, 2003.

[16] R. Jakob-Chien and B. Alpert, "A fast spherical filter with uniform resolution," *J. Comput. Phys.*, vol. 136, pp. 580–584, 1997.

[17] S. Koc, J. M. Song, and W. C. Chew, "Error analysis for the numerical evaluation of the diagonal forms of the spherical addition theorem," *SIAM J. Numer. Anal.*, vol. 36, no. 3, pp. 906–921, 1999.

[18] J. M. Song, C. C. Lu, and W. C. Chew, "Multilevel fast multipole algorithm for electromagnetic scattering by large complex objects," *IEEE Trans. Antennas Propagat.*, vol. 45, pp. 1488–1493, October 1997.

[19] T. Malas and L. Gurel, "Accelerating the multilevel fast multipole algorithm with the sparse-approximate-inverse (SAI) preconditioning," *SIAM J. Sci. Comput.*, vol. 31, no. 3, pp. 1968–1984, 2009.

[20] B. Carpentieri, I. Duff, L. Giraud, and G. Sylvand, "Combining fast multipole techniques and an approximate inverse preconditioner for large electromagnetism calculations," *SIAM J. Sci. Comput.*, vol. 27, no. 3, pp. 774–792, 2005.

[21] J. Lee, J. Zhang, and C. Lu, "Sparse inverse preconditioning of multilevel fast multipole algorithm for hybrid integral equations in electromagnetics," *IEEE Trans. Antennas Propagat.*, vol. 52, no. 9, pp. 2277–2287, 2004.

[22] A. C. Woo, H. T. G. Wang, M. J. Schuh, and M. L. Sanders, "Benchmark radar targets for the validation of computational electromagnetics programs," *IEEE Antennas Propagat. Magazine*, vol. 35, pp. 84–89, February 1993.

[23] B. Hariharan, S. Aluru, and B. Shanker, "A scalable parallel fast multipole method for analysis of scattering from perfect electrically conducting surfaces," in *Supercomputing '02: Proceedings of the 2002 ACM/IEEE conference on Supercomputing*, pp. 1–17, IEEE Computer Society Press, 2002.

[24] K. Donepudi, J. Jin, S. Velamparambil, J. Song, and W. Chew, "A higher-order parallelized multilevel fast multipole algorithm for 3-D scattering," *IEEE Trans. Antennas Propagat.*, vol. 49, pp. 1069–1078, July 2001.

[25] S. Velamparambil and W. C. Chew, "Parallelization of multilevel fast multipole algorithm on distributed memory computers," Tech. Rep. 13-01, Center for Computational Electromagnetics, University of Illinois at Urbana-Champaign, 2001.

[26] S. Velamparambil and W. C. Chew, "Analysis and performance of a distributed memory multilevel fast multipole algorithm," *IEEE Trans. Antennas Propagat.*, vol. 53, pp. 2719–2727, August 2005.

[27] S. Velamparambil, W. C. Chew, and J. Song, "10 million unknowns: Is it that big?," *IEEE Antennas Propagat. Mag.*, vol. 45, no. 2, pp. 43–58, 2003.

# Chapter 10

# *Integration*

In this book we have considered many integrals, where most operated in one or two dimensions. Though some of these can be evaluated analytically, most have no analytic solution and numerical integration (quadrature) must be used instead. When using the Moment Method, the accuracy of the solution depends heavily on the accuracy of the matrix elements, thus a reliable and computationally efficient quadrature scheme is vital. In this chapter we will consider several one-dimensional quadrature schemes, as well as quadratures designed specifically for two-dimensional planar triangles.

## 10.1   One-Dimensional Integration

For the one-dimensional definite integral given by

$$I = \int_a^b f(x) \, dx \tag{10.1}$$

a numerical quadrature generates an approximate solution using the sum

$$I = \int_a^b f(x) \, dx \approx \sum_{i=1}^N w(x_i) f(x_i) \tag{10.2}$$

where the function $f(x)$ is evaluated at $N$ unique locations $x_i$ and multiplied by the quadrature weights $w(x_i)$. The weights and locations are dependent on the type and order of quadrature, often referred to as the quadrature *rule*. We will discuss several one-dimensional quadrature schemes in this section.

### 10.1.1   Centroidal Approximation

Some of the integrations in this book were evaluated using a centroidal approximation. This method simply evaluates the function once and assumes that its value is constant within the limits of integration. The evaluation is typically done at the midpoint of the domain, resulting in an approximation given by

$$I = \int_a^b f(x)\,dx \approx (b-a) \cdot f([b+a]/2) \tag{10.3}$$

This "one-point" rule is useful in small domains where the integrand varies slowly, such as in computing the matrix elements between basis and testing function pairs that are far away from one another. In this case, the results from the centroidal rule often differ little from the results of a more accurate rule, but are much faster to compute. The error is often tolerable, since the magnitude of these elements are usually small compared to those closer to the matrix diagonal.

### 10.1.2 Rectangular Rule

The rectangular rule uses a stair-step approximation to $f(x)$ throughout the range of integration. We subdivide the interval $[a, b]$ into $N$ small segments of equal length $h = (b-a)/N$ and compute the values $f(x_1)$, $f(x_2)$, ..., $f(x_N)$, as shown in Figure 10.1. The value of the integral within each segment is approximated as

$$\int_{x_i}^{x_{i+1}} f(x)\,dx \approx h f(x_i) \tag{10.4}$$

Summing (10.4) across the entire interval, we derive the following expression

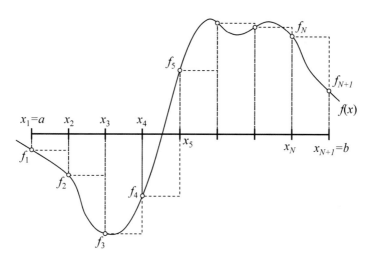

**FIGURE 10.1:** Rectangular (Stair-step) Approximation of $f(x)$

for the rectangular rule

$$I = \int_a^b f(x)\, dx \approx h \sum_{n=1}^{N} f(x_n) + E_R \tag{10.5}$$

where $E_R$ is the error. If $f''(x)$ is continuous within $[a, b]$, then the bound on the error is [1]

$$|E_R| \leq \frac{(b-a)h^2}{24} f''(\xi) \tag{10.6}$$

for some $\xi \in [a, b]$.

### 10.1.3 Trapezoidal Rule

The trapezoidal rule uses a piecewise linear approximation to $f(x)$ throughout the range of integration. We subdivide the interval $[a, b]$ into $N$ small segments of equal length $h = (b-a)/N$ and compute the values $f(x_1)$, $f(x_2), \ldots, f(x_N), f(x_{N+1})$, as shown in Figure 10.2. The value of the integral within each segment is approximated as

$$\int_{x_i}^{x_{i+1}} f(x)\, dx \approx \frac{h}{2}\left[ f_i + f_{i+1} \right] \tag{10.7}$$

Summing (10.7) across the entire interval, we derive the following expression

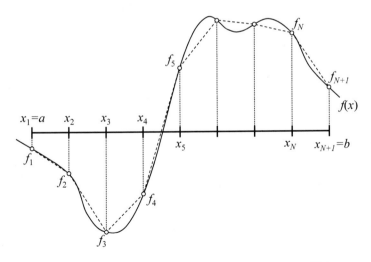

**FIGURE 10.2:** Piecewise Linear Approximation of $f(x)$

for the trapezoidal rule

$$I = \int_a^b f(x)\,dx \approx h\left[\frac{f_1}{2} + f_2 + \cdots + f_N + \frac{f_{N+1}}{2}\right] + E_T \qquad (10.8)$$

where $E_T$ is the error. If $f''(x)$ is continuous within $[a, b]$, then the bound on the error is [1]

$$|E_T| \leq \frac{(b-a)h^2}{12} f''(\xi) \qquad (10.9)$$

for some $\xi \in [a, b]$. If the result is not accurate enough, the number of segments can be doubled and the previous samples reused, requiring only $N$ additional evaluations of $f(x)$ for the new rule. This process can be repeated as necessary until the desired accuracy is reached.

### 10.1.3.1  Romberg Integration

Repeated applications of the Trapezoidal Rule results in an error comprising even powers of $1/N$ [2], which can be written as

$$I = I_{n,1} + \frac{A}{N^2} + \frac{B}{N^4} + \frac{C}{N^6} + \cdots \qquad (10.10)$$

We note that the second-order term in the above decreases by a factor of four if we double the number of quadrature points. Using two applications of the rule, we can therefore write

$$I \approx I_{n,1} + \frac{K}{N^2} \qquad (10.11)$$

and

$$I \approx I_{n+1,1} + \frac{K}{4N^2} \qquad (10.12)$$

If we solve the above two equations for $K$ we can eliminate the second order term and improve the previous solution to at least an $O(N^{-4})$ error. Doing so yields

$$I \approx I_{n+1,2} = \frac{4}{3}I_{n+1,1} - \frac{1}{3}I_{n,1} \qquad (10.13)$$

which is a Richardson's Extrapolation. Combining successive applications of the rule with higher orders of extrapolation can greatly increase p the accuracy of the solution and is known as *Romberg Integration*. In this method, higher-order extrapolates have the form

$$I_{n,m} = I_{n+1,m-1} + \frac{I_{n+1,m-1} - I_{n,m-1}}{4^{m-1} - 1} \qquad (10.14)$$

An example hierarchy of extrapolates is illustrated in Figure 10.3 for $n \leq 4$. Applying Romberg Integration using this hierarchy requires the results of the trapezoidal rule using 1, 2, 4, and 8 segments.

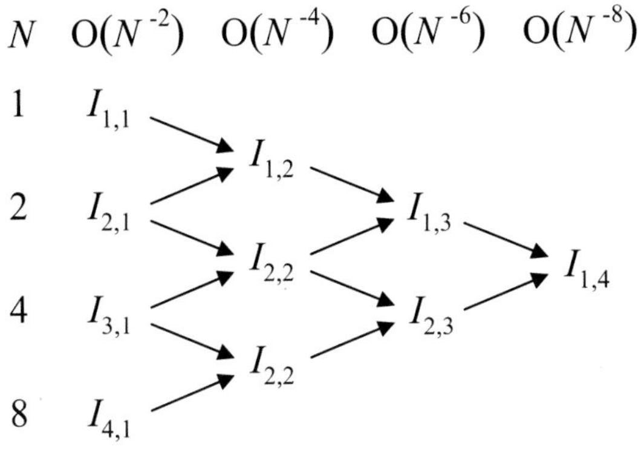

$$N \quad O(N^{-2}) \quad O(N^{-4}) \quad O(N^{-6}) \quad O(N^{-8})$$

$$1 \quad I_{1,1}$$
$$\qquad\qquad I_{1,2}$$
$$2 \quad I_{2,1}$$
$$\qquad\qquad I_{2,2} \qquad I_{1,3}$$
$$4 \quad I_{3,1} \qquad\qquad I_{2,3} \qquad I_{1,4}$$
$$\qquad\qquad I_{2,2}$$
$$8 \quad I_{4,1}$$

**FIGURE 10.3:** Hierarchy of Romberg Extrapolates

## 10.1.4 Simpson's Rule

Simpson's rule approximates $f(x)$ by a series of quadratic polynomials, as shown in Figure 10.4. The interval $[a, b]$ is divided into $N$ segments of equal length $h = (b - a)/N$, where $N \geq 2$ and even. Starting with the equation for a parabola

$$f(x) = Ax^2 + Bx + C \tag{10.15}$$

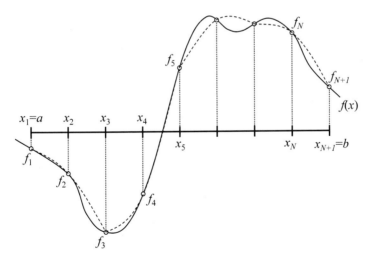

**FIGURE 10.4:** Piecewise Parabolic Approximation of $f(x)$

we perform an integration over a two-segment interval $[-h, h]$, yielding

$$\int_{-h}^{h} \left(Ax^2 + Bx + C\right) dx = \left[A\frac{x^3}{3} + B\frac{x^2}{2} + Cx\right]\Big|_{-h}^{h} = \frac{h}{3}\left(2Ah^2 + 6C\right)$$

$$(10.16)$$

Since the parabola passes through the points $(-h, y_1)$, $(0, y_2)$, and $(h, y_3)$, we can compute the values

$$f_1 = Ah^2 - Bh + C \tag{10.17}$$

$$f_2 = C \tag{10.18}$$

and

$$f_3 = Ah^2 + Bh + C \tag{10.19}$$

from which we obtain

$$2Ah^2 = f_1 + f_3 - 2f_2 \tag{10.20}$$

Inserting (10.18) and (10.20) into (10.16) yields

$$\int_{-h}^{h} \left(Ax^2 + Bx + C\right) dx = \frac{h}{3}\left[f_1 + 4f_2 + f_3\right] \tag{10.21}$$

Summing (10.21) across the entire interval, we derive the following expression known as *Simpson's Rule*

$$\int_{a}^{b} f(x)\, dx = \frac{h}{3}\left[f_1 + 4f_2 + 2f_3 + 4f_4 + \cdots + 2f_{N-1} + 4f_N + f_{N+1}\right] + E_S$$

$$(10.22)$$

where the term $E_S$ is the error. If $f^{(4)}(x)$ is continuous on $[a, b]$, then the bound on the error is [1]

$$|E_S| \leq \frac{b-a}{180} h^4 f^4(\xi) \tag{10.23}$$

for some $\xi \in [a, b]$. Note that the result of (10.21) is equal to that obtained by the extrapolation of the Trapezoidal Rule in (10.13).

### 10.1.4.1    Adaptive Simpson's Rule

The Simpson's Rule outlined thus far can be used in a recursive fashion similar to the Trapezoidal rule, where the number of segments is doubled and previous computations re-used. Writing the rule as

$$\int_{a}^{b} f(x)\, dx = S(a, b) + E_S \tag{10.24}$$

and starting at the coarsest level (2 segments), we have

$$S(a, b) = \frac{b-a}{6}\left[f(a) + 4f\left(\frac{a+b}{2}\right) + f(b)\right] \tag{10.25}$$

and

$$E_S^{(1)} \le \frac{b-a}{180} h^4 M \tag{10.26}$$

resulting in

$$I^{(1)} = S(a,b) + E_S^{(1)} \tag{10.27}$$

If we now divide the interval $[a,b]$ in half at $c = (a+b)/2$, we get

$$I^{(2)} = S(a,c) + S(c,b) + E_S^{(2)} \tag{10.28}$$

The error is now

$$E_S^{(2)} \le \frac{b-a}{180} \frac{h^4}{16} M = \frac{1}{16} E_S^{(1)} \tag{10.29}$$

hence

$$S^{(2)} - S^{(1)} = E_S^{(2)} - E_S^{(1)} = -15 E^2 \tag{10.30}$$

Using the above, we can stop the recursion once we reach a user-specified tolerance $\epsilon$ when

$$\frac{1}{15} \left| S^{(2)} - S^{(1)} \right| \le \epsilon \tag{10.31}$$

## 10.1.5   One-Dimensional Gaussian Quadrature

Using the Trapezoidal or Simpson's Rule, the interval $[a,b]$ is subdivided into equally spaced segments with fixed locations $x_i$. A more efficient integration rule can be developed if we allow the freedom to choose the optimal locations for integrating $f(x)$. Given the integral

$$\int_a^b W(x) f(x) \, dx \tag{10.32}$$

an $N$-point *Gaussian quadrature* integrates exactly a polynomial $f(x)$ of degree $2N - 1$ if the locations are chosen as the roots of the orthogonal polynomial for the same interval and *weighting function* $W(x)$. The most commonly used Gaussian quadrature is the *Gauss-Legendre* quadrature, of the form

$$\int_{-1}^1 f(x) \, dx = \sum_{i=1}^N w(x_i) f(x_i) \tag{10.33}$$

where $W(x) = 1$. The $x_i$ for this rule are the roots of the Legendre polynomial of order $N$. The weights $w(x_i)$ are [3]

$$w(x_i) = \frac{2}{(1-x_i)^2 \left[ P_N'(x_i) \right]^2} \tag{10.34}$$

Routines that compute an $N$-point Gauss-Legendre rule determine the locations of the zeros of $P_m(x)$ by a Newton-Raphson iteration, and then the associated weights via (10.34). There are many short software routines for this available in books and through the Internet. Many of these retain the above range $[-1, 1]$, while some instead map to the range $[0, 1]$. To use the rule, a change of variable should be used to map the interval $[a, b]$ to $[-1, 1]$ or $[0, 1]$. Note that in this rule, the spacing between the quadrature locations is not constant, thus previously computed values of $f(x)$ cannot be re-used if the number of quadrature points is changed. In some cases, Simpson's rule may be the better choice.

---

## 10.2   Integration over Triangles

Many of the integrals in Chapter 8 were performed over triangular subdomains, and therefore efficient analytic and numerical integration techniques for triangular elements are essential. In this section, we will consider these integrals in terms of *simplex coordinates*, also called *area coordinates* or *barycentric coordinates*. These coordinates comprise the transformation of a triangle of arbitrary shape to a canonical coordinate system. Analytic integrals are often easier to perform in simplex coordinates, and numerical quadrature rules are usually specified using them.

### 10.2.1   Simplex Coordinates

To develop the transformation to simplex coordinates, consider a triangle $T$ defined by the vertices $\mathbf{v}_1$, $\mathbf{v}_2$, and $\mathbf{v}_3$, with edges given by

$$\mathbf{e}_1 = \mathbf{v}_2 - \mathbf{v}_1 \qquad \mathbf{e}_2 = \mathbf{v}_3 - \mathbf{v}_2 \qquad \mathbf{e}_3 = \mathbf{v}_3 - \mathbf{v}_1 \qquad (10.35)$$

Any point $\mathbf{r}$ within the triangle can be written as a weighted sum of the three vertices as

$$\mathbf{r} = \gamma \mathbf{v}_1 + \alpha \mathbf{v}_2 + \beta \mathbf{v}_3 \qquad (10.36)$$

where $\alpha$, $\beta$, and $\gamma$ are the simplex coordinates

$$\alpha = \frac{A_1}{A}, \qquad \beta = \frac{A_2}{A}, \qquad \gamma = \frac{A_3}{A} \qquad (10.37)$$

$A$ is the area of $T$, and $A_1$, $A_2$ and $A_3$ are the areas of the sub-triangles, as illustrated in Figure 10.5a. These coordinates are subject to the constraint

$$\alpha + \beta + \gamma = 1 \qquad (10.38)$$

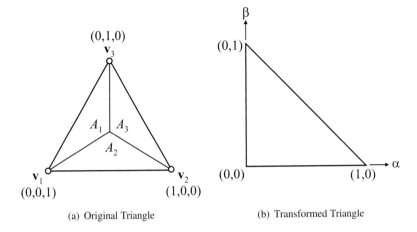

(a) Original Triangle          (b) Transformed Triangle

**FIGURE 10.5:** Simplex Coordinates for a Triangle

therefore,

$$\gamma = 1 - \alpha - \beta \tag{10.39}$$

and

$$\mathbf{r} = (1 - \alpha - \beta)\mathbf{v}_1 + \alpha\mathbf{v}_2 + \beta\mathbf{v}_3 \tag{10.40}$$

Note that (10.36) has the form of a linear interpolation. Indeed, simplex coordinates will also allow us to perform a linear interpolation of a function $f(x, y)$ at points inside the triangle if its values are known at the vertices. The resulting transformation of the integral to simplex coordinates is illustrated in Figure 10.5b, and is

$$\int_T f(\mathbf{r}) \, d\mathbf{r} = \int_{T'} f(\alpha, \beta)|J(\alpha, \beta)| \, d\alpha \, d\beta = 2A \int_0^1 \int_0^{1-\alpha} f(\alpha, \beta) \, d\beta \, d\alpha \tag{10.41}$$

Given a point $\mathbf{r}$ inside a triangle, it is also desirable to obtain the simplex coordinates $(\alpha, \beta, \gamma)$. We can write the barycentric expansion of $\mathbf{r} = (x, y, z)$ in terms of the components of the triangle vertices as

$$x = \gamma x_1 + \alpha x_2 + \beta x_3 \tag{10.42}$$
$$y = \gamma y_1 + \alpha y_2 + \beta y_3 \tag{10.43}$$
$$z = \gamma z_1 + \alpha z_2 + \beta z_3 \tag{10.44}$$

Substituting $\gamma = 1 - \alpha - \beta$ into the above gives

$$x = (1 - \alpha - \beta)x_1 + \alpha x_2 + \beta x_3 \tag{10.45}$$
$$y = (1 - \alpha - \beta)y_1 + \alpha y_2 + \beta y_3 \tag{10.46}$$
$$z = (1 - \alpha - \beta)z_1 + \alpha z_2 + \beta z_3 \tag{10.47}$$

Rearranging, this is

$$\alpha(x_2 - x_1) + \beta(x_3 - x_1) + x_1 - x = 0 \tag{10.48}$$
$$\alpha(y_2 - y_1) + \beta(y_3 - y_1) + y_1 - y = 0 \tag{10.49}$$
$$\alpha(z_2 - z_1) + \beta(z_3 - z_1) + z_1 - z = 0 \tag{10.50}$$

and solving for $\alpha$ and $\beta$ gives us

$$\alpha = \frac{B(F + I) - C(E + H)}{A(E + H) - B(D + G)} \tag{10.51}$$

and

$$\beta = \frac{A(F + I) - C(D + G)}{B(D + G) - A(E + H)} \tag{10.52}$$

where

$$A = x_2 - x_1 \tag{10.53}$$
$$B = x_3 - x_1 \tag{10.54}$$
$$C = x_1 - x \tag{10.55}$$
$$D = y_2 - y_1 \tag{10.56}$$
$$E = y_3 - y_1 \tag{10.57}$$
$$F = y_1 - y \tag{10.58}$$
$$G = z_2 - z_1 \tag{10.59}$$
$$H = z_3 - z_1 \tag{10.60}$$
$$I = z_1 - z \tag{10.61}$$

### 10.2.2 Radiation Integrals with a Constant Source

We will now consider radiation integrals having a constant-valued integrand, which can be written as

$$I = \int_S e^{j\mathbf{s}\cdot\mathbf{r}'} \, d\mathbf{r}' \tag{10.62}$$

Let us evaluate the integral over a flat, planar triangle $T$ of arbitrary shape. Using (10.41), this integral becomes

$$I = \int_T e^{j\mathbf{s}\cdot\mathbf{r}'} \, d\mathbf{r}' = 2A \int_0^1 \int_0^{1-\alpha} e^{j\mathbf{s}(\alpha,\beta)\cdot\mathbf{r}'(\alpha,\beta)} d\beta \, d\alpha \tag{10.63}$$

Expanding $\mathbf{r}'(\alpha, \beta)$ following (10.40), we can write (10.63) as

$$I = 2Ae^{j\mathbf{s}\cdot\mathbf{v}_1} \int_0^1 \int_0^{1-\alpha} e^{j\mathbf{s}\cdot\mathbf{e}_1\alpha} e^{j\mathbf{s}\cdot\mathbf{e}_3\beta} d\beta \, d\alpha \tag{10.64}$$

or

$$I = 2Ae^{js_1} \int_0^1 e^{js_2\alpha} \int_0^{1-\alpha} e^{js_3\beta} d\beta \, d\alpha \qquad (10.65)$$

where

$$s_1 = \mathbf{s} \cdot \mathbf{v}_1 \qquad s_2 = \mathbf{s} \cdot \mathbf{e}_1 \qquad s_3 = \mathbf{s} \cdot \mathbf{e}_3 \qquad (10.66)$$

Evaluating the integral versus $\beta$ yields

$$I = 2Ae^{js_1} \int_0^1 e^{js_2\alpha} \left( \frac{e^{js_3(1-\alpha)}}{js_3} - \frac{1}{js_3} \right) d\alpha \qquad (10.67)$$

which can be rewritten as

$$I = 2A\frac{e^{js_1}}{js_3} \int_0^1 \left( e^{js_3} e^{j\alpha(s_2-s_3)} - e^{j\alpha s_2} \right) d\alpha \qquad (10.68)$$

Integrating over $\alpha$ yields

$$I = 2A\frac{e^{js_1}}{js_3} \left( \frac{e^{js_2}}{j(s_2-s_3)} - \frac{e^{js_2}}{js_2} - \frac{e^{js_3}}{j(s_2-s_3)} + \frac{1}{js_2} \right) \qquad (10.69)$$

and after some tedious but straightforward algebra, we obtain

$$I = -\frac{2Ae^{js_1}}{(s_2-s_3)} \left( \frac{1-e^{js_3}}{s_3} - \frac{1-e^{js_2}}{s_2} \right) \qquad (10.70)$$

If we now define the quantities

$$a_1 = \mathbf{s} \cdot \mathbf{v}_1 \qquad a_2 = \mathbf{s} \cdot \mathbf{v}_2 \qquad a_3 = \mathbf{s} \cdot \mathbf{v}_3 \qquad (10.71)$$

we can then write

$$s_1 = a_1 \qquad s_2 = a_2 - a_1 \qquad s_3 = a_3 - a_1 \qquad (10.72)$$

Substitution of the above into (10.70) then yields

$$I = \frac{2Ae^{ja_1}}{(a_2-a_3)} \left( \frac{1-e^{ja_2}e^{-ja_1}}{a_2-a_1} - \frac{1-e^{ja_3}e^{-ja_1}}{a_3-a_1} \right) \qquad (10.73)$$

which after some manipulation becomes

$$I = \frac{2A}{(a_3-a_2)} \left( \frac{e^{ja_1}-e^{ja_2}}{a_1-a_2} - \frac{e^{ja_1}-e^{ja_3}}{a_1-a_3} \right) \qquad (10.74)$$

### 10.2.2.1   Special Cases

When **s** is perpendicular to any of the three edges, (10.74) has a singularity and must be rewritten. When $\mathbf{s} \cdot \mathbf{e}_1 = a_2 - a_1 = 0$, we use the formula

$$I = \frac{2A}{(a_3 - a_2)}\left(je^{ja_1} - \frac{e^{ja_1} - e^{ja_3}}{a_1 - a_3}\right) \tag{10.75}$$

Likewise, when $\mathbf{s} \cdot \mathbf{e}_3 = a_3 - a_1 = 0$,

$$I = \frac{2A}{(a_3 - a_2)}\left(\frac{e^{ja_1} - e^{ja_2}}{a_1 - a_2} - je^{ja_1}\right) \tag{10.76}$$

and when $\mathbf{s} \cdot \mathbf{e}_2 = a_3 - a_2 = 0$,

$$I = \frac{2A}{(a_1 - a_2)}\left(je^{ja_3} - \frac{e^{ja_1} - e^{ja_3}}{a_1 - a_3}\right) \tag{10.77}$$

When **s** is normal to the triangle, we are left with

$$I = Ae^{ja_1} \tag{10.78}$$

### 10.2.3   Radiation Integrals with a Linear Source

We next consider far-field radiation integrals with a linearly varying source on triangle $T$, such as the RWG function described in Section (8.2). These integrals can be written as

$$I = \int_T \rho(\mathbf{r}')e^{j\mathbf{s}\cdot\mathbf{r}'}\,d\mathbf{r}' \tag{10.79}$$

where the source function is defined as

$$\rho(\mathbf{r}') = \mathbf{r}' - \mathbf{v}_1 \tag{10.80}$$

This corresponds to the RWG function defined on the edge opposite vertex $\mathbf{v}_1$ on $T$. Using (10.41), we can write (10.79) as

$$I = 2A\int_0^1\int_0^{1-\alpha} e^{j\mathbf{s}(\alpha,\beta)\cdot\mathbf{r}'(\alpha,\beta)}\,d\beta\,d\alpha \tag{10.81}$$

and expanding $\mathbf{r}'(\alpha,\beta)$ following (10.40), we can write the above as

$$I = 2Ae^{j\mathbf{s}\cdot\mathbf{v}_1}\int_0^1\int_0^{1-\alpha}(\alpha\mathbf{e}_1 + \beta\mathbf{e}_3)e^{j\mathbf{s}\cdot\mathbf{e}_1\alpha}e^{j\mathbf{s}\cdot\mathbf{e}_3\beta}\,d\beta\,d\alpha \tag{10.82}$$

or

$$I = 2Ae^{js_1}\int_0^1\int_0^{1-\alpha}(\alpha\mathbf{e}_1 + \beta\mathbf{e}_3)e^{js_2\alpha}e^{js_3\beta}\,d\beta\,d\alpha \tag{10.83}$$

#### 10.2.3.1 General Case

For the general case, (10.83) evaluates to

$$I = -\frac{C}{s_3}\left[e^{js_2}\left(\frac{1-j(s_2-s_3)}{(s_2-s_3)^2}+\frac{js_2-1}{s_2^2}\right)-\frac{e^{js_3}}{(s_2-s_3)^2}+\frac{1}{s_2^2}\right]\mathbf{e}_1$$
$$-\frac{C}{s_2}\left[e^{js_3}\left(\frac{1-j(s_3-s_2)}{(s_3-s_2)^2}+\frac{js_3-1}{s_3^2}\right)-\frac{e^{js_2}}{(s_3-s_2)^2}+\frac{1}{s_3^2}\right]\mathbf{e}_3$$

$$(10.84)$$

where

$$C = j2Ae^{js_1} \qquad (10.85)$$

As in Section 10.2.2, this expression has singularities when **s** is perpendicular to any of the three edges. Therefore, we must also consider the special cases.

#### 10.2.3.2 Special Cases

When **s** is perpendicular to $\mathbf{e}_1$ ($s_2 = 0$), we can write (10.83) as

$$I = 2Ae^{js_1}\int_0^1\int_0^{1-\alpha}(\alpha\mathbf{e}_1+\beta\mathbf{e}_3)e^{js_3\beta}d\beta\,d\alpha \qquad (10.86)$$

which evaluates to

$$I = C\left[\frac{-1-js_3}{s_3^3}+\frac{1}{2s_3}+\frac{e^{js_3}}{s_3^3}\right]\mathbf{e}_1+C\left[e^{js_3}\left(\frac{js_3-2}{s_3^3}\right)+\frac{2}{s_3^3}+\frac{j}{s_3^2}\right]\mathbf{e}_3 \qquad (10.87)$$

When **s** is perpendicular to $\mathbf{e}_3$ ($s_3 = 0$), we can write (10.83) as

$$I = 2Ae^{js_1}\int_0^1\int_0^{1-\alpha}(\alpha\mathbf{e}_1+\beta\mathbf{e}_3)e^{js_2\alpha}d\beta\,d\alpha \qquad (10.88)$$

which evaluates to

$$I = C\left[e^{js_2}\left(\frac{js_2-2}{s_2^3}\right)+\frac{2}{s_2^3}+\frac{j}{s_2^2}\right]\mathbf{e}_1+C\left[\frac{-1-js_2}{s_2^3}+\frac{1}{2s_2}+\frac{e^{js_2}}{s_2^3}\right]\mathbf{e}_3 \qquad (10.89)$$

When $s_2 = s_3 = s, s \neq 0$, we can write (10.83) as

$$I = 2Ae^{js_1}\int_0^1\int_0^{1-\alpha}(\alpha\mathbf{e}_1+\beta\mathbf{e}_3)e^{js(\alpha+\beta)}d\beta\,d\alpha \qquad (10.90)$$

which evaluates to

$$I = C\left[e^{js}\left(\frac{1}{s^3}-\frac{1}{2s}-\frac{j}{s^2}\right)-\frac{1}{s^3}\right]\left[\mathbf{e}_1+\mathbf{e}_3\right] \qquad (10.91)$$

When **s** is normal to the triangle, 10.83 becomes

$$I = 2Ae^{j\mathbf{s}\cdot\mathbf{v}_1} \int_0^1 \int_0^{1-\alpha} (\alpha\mathbf{e}_1 + \beta\mathbf{e}_3)d\beta \, d\alpha \tag{10.92}$$

which evaluates to

$$I = \frac{A}{3}e^{js_1}(\mathbf{e}_1 + \mathbf{e}_3) \tag{10.93}$$

### 10.2.4   Gaussian Quadrature on Triangles

The most commonly referenced Gauss-Legendre locations and weights for triangles are the symmetric quadrature rules presented in [4]. In this reference are tables varying from degrees 1 up to 20 (79 quadrature points). Tables for orders 2–5 are reproduced in Figures 10.1, 10.2, 10.3, and 10.4, respectively. The weights in these tables are normalized with respect to triangle area, i.e.

$$\int_S f(x,y) \, dx \, dy \approx A \sum_{i=1}^N w(\alpha_i, \beta_i)f(\alpha_i, \beta_i) \tag{10.94}$$

It is this author's experience that the four- and seven-point quadrature rules work well for the example problems presented in Chapters 8 and 9.

**TABLE 10.1:** Three-Point Quadrature Rule ($n = 2$)

| i | $\alpha$ | $\beta$ | $\gamma$ | $w$ |
|---|----------|---------|----------|-----|
| 1 | 0.66666667 | 0.16666667 | 0.16666667 | 0.33333333 |
| 2 | 0.16666667 | 0.66666667 | 0.16666667 | 0.33333333 |
| 3 | 0.16666667 | 0.16666667 | 0.66666667 | 0.33333333 |

**TABLE 10.2:** Four-Point Quadrature Rule ($n = 3$)

| i | $\alpha$ | $\beta$ | $\gamma$ | $w$ |
|---|----------|---------|----------|-----|
| 1 | 0.33333333 | 0.33333333 | 0.33333333 | -0.56250000 |
| 2 | 0.60000000 | 0.20000000 | 0.20000000 | 0.52083333 |
| 3 | 0.20000000 | 0.60000000 | 0.20000000 | 0.52083333 |
| 4 | 0.20000000 | 0.20000000 | 0.60000000 | 0.52083333 |

**TABLE 10.3:** Six-Point Quadrature Rule ($n = 4$)

| i | $\alpha$ | $\beta$ | $\gamma$ | $w$ |
|---|---|---|---|---|
| 1 | 0.10810301 | 0.44594849 | 0.44594849 | 0.22338158 |
| 2 | 0.44594849 | 0.10810301 | 0.44594849 | 0.22338158 |
| 3 | 0.44594849 | 0.44594849 | 0.10810301 | 0.22338158 |
| 4 | 0.81684757 | 0.09157621 | 0.09157621 | 0.10995174 |
| 5 | 0.09157621 | 0.81684757 | 0.09157621 | 0.10995174 |
| 6 | 0.09157621 | 0.09157621 | 0.81684757 | 0.10995174 |

**TABLE 10.4:** Seven-Point Quadrature Rule ($n = 5$)

| i | $\alpha$ | $\beta$ | $\gamma$ | $w$ |
|---|---|---|---|---|
| 1 | 0.33333333 | 0.33333333 | 0.33333333 | 0.22500000 |
| 2 | 0.05971587 | 0.47014206 | 0.47014206 | 0.13239415 |
| 3 | 0.47014206 | 0.05971587 | 0.47014206 | 0.13239415 |
| 4 | 0.47014206 | 0.47014206 | 0.05971587 | 0.13239415 |
| 5 | 0.79742698 | 0.10128650 | 0.10128650 | 0.12593918 |
| 6 | 0.10128650 | 0.79742698 | 0.10128650 | 0.12593918 |
| 7 | 0.10128650 | 0.10128650 | 0.79742698 | 0.12593918 |

### 10.2.4.1  Comparison with Analytic Solution

The formulas presented in Sections 10.2.3.1 and 10.2.3.2 are advantageous as they are analytic solutions which are valid at all frequencies. By comparison, an accurate numerical quadrature requires an increasing number of integration points as the frequency increases, which is undesirable. Let us compare the analytic solution to (10.83) to that obtained using a 7-point Gaussian quadrature rule (Table 10.4). The source triangle is equilateral and lies in the $xy$ plane with vertices $\mathbf{v}_1 = (0,0)$, $\mathbf{v}_2 = (l,0)$, and $\mathbf{v}_3 = (l/2, l\sqrt{3}/2)$. The direction of radiation is along the spherical angles $(\theta, \phi) = (\pi/2, \pi/4)$. In Figure 10.6a we compare the magnitude of the $\hat{\mathbf{x}}$ components of (10.83) versus the triangle edge length $l$ in wavelengths. In Figure 10.6b we plot the difference between the results. We see that the analytical and numerical results compare very well to just over $l = 1\lambda$, at which point the numerical results start to diverge.

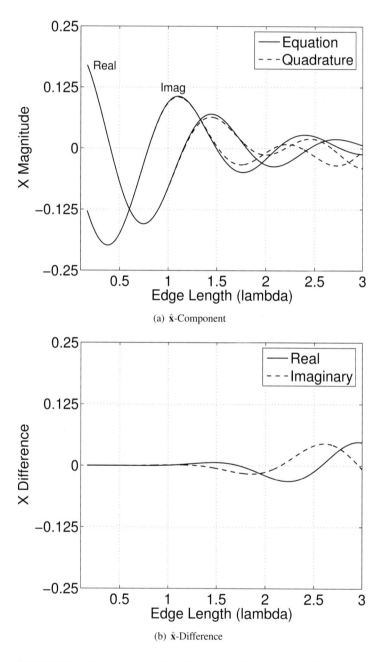

(a) $\hat{x}$-Component

(b) $\hat{x}$-Difference

**FIGURE 10.6:** Linear Source Radiation Integral: Equation vs. Quadrature

# References

[1] G. Thomas and R. Finney, *Calculus and Analytic Geometry*. Addison-Wesley, 1992.

[2] A. F. Peterson, S. L. Ray, and R. Mittra, *Computational Methods for Electromagnetics*. IEEE Press, 1998.

[3] W. H. Press, B. P. Flannery, S. A. Teukolsky, and W. T. Vetterling, *Numerical Recipes in C : The Art of Scientific Computing*. Cambridge University Press, 1992.

[4] D. Dunavant, "High degree efficient symmetrical Gaussian quadrature rules for the triangle," *Internat. J. Numer. Methods Engrg.*, vol. 21, pp. 1129–1148, 1985.

# Appendix A

## Scattering Using Physical Optics

We briefly discussed the Physical Optics approximation in Section 3.6.4, and we use the PO approximation for conducting interfaces (3.209) in several examples throughout this book. Because it is of vital importance in the treatment of electrically large objects, particularly in the area of radar cross section prediction, we discuss it in additional detail in this appendix. We will first consider the fields scattered from conducting surfaces, and then the fields scattered by dielectric interfaces.

### A.1   Field Scattered at a Conducting Interface

Recall again the surface equivalent problem of Figure 3.2c. Given an plane wave incident in Region $R_1$ with $\hat{\boldsymbol{\theta}}^i$ and $\hat{\boldsymbol{\phi}}^i$ components, the electric field in $R_1$ can be written as

$$\mathbf{E}_1^i(\mathbf{r}) = \left[ E_{1,\theta}^i \hat{\boldsymbol{\theta}}^i + E_{1,\phi}^i \hat{\boldsymbol{\phi}}^i \right] e^{jk_1 \hat{\mathbf{r}}^i \cdot \mathbf{r}} \tag{A.1}$$

and the incident magnetic field as

$$\mathbf{H}_1^i(\mathbf{r}) = -\frac{1}{\eta_1} \hat{\mathbf{r}} \times \mathbf{E}_1^i(\mathbf{r}) = \frac{1}{\eta_1} \left[ E_{1,\phi}^i \hat{\boldsymbol{\theta}}^i - E_{1,\theta}^i \hat{\boldsymbol{\phi}}^i \right] e^{jk_1 \hat{\mathbf{r}}^i \cdot \mathbf{r}} \tag{A.2}$$

For radar cross section problems, $R_1$ is the unbounded free space region $R_0$ where $k_1 = k_0 = w\sqrt{\mu_0 \epsilon_0}$ and $\eta_1 = \eta_0 = \sqrt{\mu_0/\epsilon_0}$. Using the PO approximation, the interface $S_{12} = S$ is assumed to be perfectly flat and of infinite extent. If $R_2$ is perfectly conducting, $\mathbf{M}_1(\mathbf{r}) = 0$, and the electric surface current is given by (3.209). Following Section 3.5.3, the total scattered far electric field $\mathbf{E}_1^s(\mathbf{r})$ is then

$$\mathbf{E}_1^s(\mathbf{r}) = -j\omega\mathbf{A}(\mathbf{r}) = -j\frac{k_0}{4\pi} \frac{e^{-jk_0 r}}{r} \int_S \eta_0 \mathbf{J}_1(\mathbf{r}') e^{jk_0 \hat{\mathbf{r}}^s \cdot \mathbf{r}'} \, d\mathbf{r}' \tag{A.3}$$

where $\mathbf{E}_1^s(\mathbf{r})$ has only $\hat{\boldsymbol{\theta}}$- and $\hat{\boldsymbol{\phi}}$-polarized components. Thus, the scattered field components are then related to the incident fields via the *scattering matrix*, written as

$$\begin{bmatrix} E_{1,\theta}^s \\ E_{1,\phi}^s \end{bmatrix} = \begin{bmatrix} S_{\theta\theta} & S_{\theta\phi} \\ S_{\phi\theta} & S_{\phi\phi} \end{bmatrix} \begin{bmatrix} E_{1,\theta}^i \\ E_{1,\phi}^i \end{bmatrix} S \tag{A.4}$$

Using (A.2) and (3.209), the scattering matrix elements can be written as

$$S_{\theta\theta} = [\hat{\boldsymbol{\phi}}^i \times \hat{\boldsymbol{\theta}}^s] \cdot \hat{\mathbf{n}}_1 \tag{A.5}$$

$$S_{\phi\theta} = [\hat{\boldsymbol{\phi}}^i \times \hat{\boldsymbol{\phi}}^s] \cdot \hat{\mathbf{n}}_1 \tag{A.6}$$

$$S_{\theta\phi} = [\hat{\boldsymbol{\theta}}^s \times \hat{\boldsymbol{\theta}}^i] \cdot \hat{\mathbf{n}}_1 \tag{A.7}$$

$$S_{\phi\phi} = [\hat{\boldsymbol{\phi}}^s \times \hat{\boldsymbol{\theta}}^i] \cdot \hat{\mathbf{n}}_1 \tag{A.8}$$

where

$$S = j \frac{k_0}{2\pi} \frac{e^{-jk_0 r}}{r} \int_S e^{jk_0(\hat{\mathbf{r}}^i + \hat{\mathbf{r}}^s) \cdot \mathbf{r}'} \, d\mathbf{r}' \tag{A.9}$$

We note that for monostatic observations

$$S_{\theta\theta} = S_{\phi\phi} \tag{A.10}$$

and

$$S_{\phi\theta} = S_{\theta\phi} = 0 \tag{A.11}$$

The integral (A.9) is dependent only on the shape of the subdomain over which the integral is performed, and is often referred to as a *shape function*. For planar triangles, an analytic solution was presented in Section 10.2.2, and for general $N$-sided polygons, one can refer to [1].

---

## A.2   Plane Wave Decomposition at a Planar Interface

At a planar interface between two regions, the incident electric field $\mathbf{E}_1^i$ can be decomposed into parallel ($\parallel$) and perpendicular ($\perp$) components, also known as vertical ($\hat{\mathbf{v}}$) and horizontal ($\hat{\mathbf{h}}$) components, respectively, as shown in Figure A.1. The decomposed components can be written as

$$\mathbf{E}_1^i = (\mathbf{E}_1^i \cdot \hat{\mathbf{v}}^i)\hat{\mathbf{v}}^i + (\mathbf{E}_1^i \cdot \hat{\mathbf{h}})\hat{\mathbf{h}} = E_{1,v}^i \hat{\mathbf{v}}^i + E_{1,h}^i \hat{\mathbf{h}} \tag{A.12}$$

The reflected field components can be written in terms of the incident field components and the parallel and perpendicular reflection coefficients ($R_\parallel$ and $R_\perp$, respectively) as

$$\mathbf{E}_1^r = R_\parallel E_{1,v}^i \hat{\mathbf{v}}^r + R_\perp E_{1,h}^i \hat{\mathbf{h}} \tag{A.13}$$

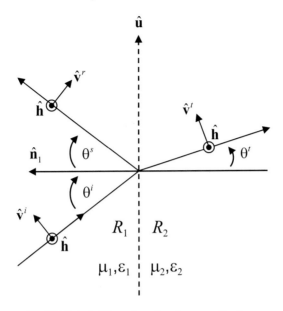

**FIGURE A.1:** Planar Interface between Regions

When $R_2$ is a dielectric, there may be fields transmitted into $R_2$. These can be written in terms of the incident field components and the parallel and perpendicular transmission coefficients ($T^{\parallel}$ and $T^{\perp}$, respectively) as

$$\mathbf{E}_2^t = T_{\parallel} E_{1,v}^i \hat{\mathbf{v}}^t + T_{\perp} E_{1,h}^i \hat{\mathbf{h}} \tag{A.14}$$

The angle of incidence $\theta^i$ is equal to the angle of reflection $\theta^s$ (Snell's Law), and the angle of transmission $\theta^t$ is related to $\theta^i$ via the relationship

$$k_1 \sin \theta^i = k_2 \sin \theta^t \tag{A.15}$$

For perpendicular polarization, the reflection coefficient $R_{\perp}$ is

$$R_{\perp} = \frac{\eta_2 \cos \theta^i - \eta_1 \cos \theta^t}{\eta_2 \cos \theta^i + \eta_1 \cos \theta^t} \tag{A.16}$$

and the transmission coefficient $T_{\perp}$ is

$$T_{\perp} = \frac{2\eta_2 \cos \theta^i}{\eta_2 \cos \theta^i + \eta_1 \cos \theta^t} \tag{A.17}$$

For parallel polarization, the reflection coefficient $R_{\parallel}$ is

$$R_{\parallel} = \frac{\eta_2 \cos \theta^t - \eta_1 \cos \theta^i}{\eta_1 \cos \theta^i + \eta_2 \cos \theta^t} \tag{A.18}$$

and the transmission coefficient $T_\|$ is

$$T_\| = \frac{\eta_2 \cos \theta^i}{\eta_1 \cos \theta^i + \eta_2 \cos \theta^t} \tag{A.19}$$

The expressions in (A.16)–(A.19) are known as the Fresnel coefficients for reflection and transmission [2]. Note that if region $R_2$ is a conductor, then

$$R_\perp = R_\| = -1 \tag{A.20}$$

and

$$T_\perp = T_\| = 0. \tag{A.21}$$

## A.3  Field Scattered at a Dielectric Interface

To compute the field scattered by a dielectric interface, we must find the equivalent currents $\mathbf{M}_1$ and $\mathbf{J}_1$ in (3.149) and (3.150). We note that only the components of the total electric and magnetic fields that lie in the plane of the interface contribute to the currents. Thus,

$$\mathbf{E} = E_{1,u}\hat{\mathbf{u}} + E_{1,h}\hat{\mathbf{h}} \tag{A.22}$$

$$\mathbf{H} = H_{1,u}\hat{\mathbf{u}} + H_{1,h}\hat{\mathbf{h}} \tag{A.23}$$

where $\hat{\mathbf{u}} = \hat{\mathbf{n}}_1 \times \hat{\mathbf{h}}$. The scalar components of the fields are

$$E_{1,h} = E_{1,h}^i(1 + R_\perp) \tag{A.24}$$

$$E_{1,u} = E_{1,u}^i(1 + R_\|) \tag{A.25}$$

$$H_{1,h} = H_{1,h}^i(1 - R_\|) \tag{A.26}$$

$$H_{1,u} = H_{1,u}^i(1 - R_\perp) \tag{A.27}$$

Substituting (A.22) and (A.23) into (3.149) and (3.150) yields

$$\mathbf{M}_1 = -\hat{\mathbf{n}}_1 \times \left[E_{1,u}^i\hat{\mathbf{u}} + E_{1,h}^i\hat{\mathbf{h}}\right] \tag{A.28}$$

$$\mathbf{J}_1 = \hat{\mathbf{n}}_1 \times \left[H_{1,u}^i\hat{\mathbf{u}} + H_{1,h}^i\hat{\mathbf{h}}\right] \tag{A.29}$$

which can be written as

$$\mathbf{M}_1 = -\hat{\mathbf{n}}_1 \times \mathbf{E}^i \cdot \mathbf{A} \tag{A.30}$$

$$\mathbf{J}_1 = \hat{\mathbf{n}}_1 \times \mathbf{H}^i \cdot \mathbf{B} \tag{A.31}$$

where

$$\mathbf{A} = \hat{\mathbf{h}}(1 + R_\perp)\hat{\mathbf{h}} + \hat{\mathbf{u}}(1 + R_\parallel)\hat{\mathbf{u}} \tag{A.32}$$

$$\mathbf{B} = \hat{\mathbf{h}}(1 - R_\parallel)\hat{\mathbf{h}} + \hat{\mathbf{u}}(1 - R_\perp)\hat{\mathbf{u}} \tag{A.33}$$

Following Section 3.5.3, the total scattered far electric field $\mathbf{E}_1^s(\mathbf{r})$ is

$$\mathbf{E}_1^s(\mathbf{r}) = -j\omega\left[\mathbf{A}(\mathbf{r}) - \eta_0\,\hat{\mathbf{r}}^s \times \mathbf{F}(\mathbf{r})\right] \tag{A.34}$$

or

$$\mathbf{E}_1^s(\mathbf{r}) = -j\frac{k_0}{4\pi}\frac{e^{-jk_0 r}}{r}\int_S \left[\eta_0\,\mathbf{J}_1(\mathbf{r}') - \hat{\mathbf{r}} \times \mathbf{M}_1(\mathbf{r}')\right]e^{jk_0\hat{\mathbf{r}}^s\cdot\mathbf{r}'}\,d\mathbf{r}' \tag{A.35}$$

Using vector identities and some algebra, we can write the scattered fields in terms of the incident fields following (A.4), where the matrix elements are

$$S_{\theta\theta} = \frac{1}{2}\left[\hat{\boldsymbol{\theta}}^i \cdot \mathbf{A}' \times \hat{\boldsymbol{\phi}}^s + \hat{\boldsymbol{\phi}}^i \cdot \mathbf{B}' \times \hat{\boldsymbol{\theta}}^s\right] \tag{A.36}$$

$$S_{\phi\theta} = \frac{1}{2}\left[\hat{\boldsymbol{\phi}}^i \cdot \mathbf{A}' \times \hat{\boldsymbol{\phi}}^s - \hat{\boldsymbol{\theta}}^i \cdot \mathbf{B}' \times \hat{\boldsymbol{\theta}}^s\right] \tag{A.37}$$

$$S_{\theta\phi} = \frac{1}{2}\left[-\hat{\boldsymbol{\theta}}^i \cdot \mathbf{A}' \times \hat{\boldsymbol{\theta}}^s + \hat{\boldsymbol{\phi}}^i \cdot \mathbf{B}' \times \hat{\boldsymbol{\phi}}^s\right] \tag{A.38}$$

$$S_{\phi\phi} = \frac{1}{2}\left[-\hat{\boldsymbol{\phi}}^i \cdot \mathbf{A}' \times \hat{\boldsymbol{\theta}}^s - \hat{\boldsymbol{\theta}}^i \cdot \mathbf{B}' \times \hat{\boldsymbol{\phi}}^s\right] \tag{A.39}$$

and

$$\mathbf{A}' = -\hat{\mathbf{u}}(1 + R_\perp)\hat{\mathbf{h}} + \hat{\mathbf{h}}(1 + R_\parallel)\hat{\mathbf{u}} \tag{A.40}$$

$$\mathbf{B}' = -\hat{\mathbf{u}}(1 - R_\parallel)\hat{\mathbf{h}} + \hat{\mathbf{h}}(1 - R_\perp)\hat{\mathbf{u}} \tag{A.41}$$

---

## A.4 Layered Dielectrics over Conductor

While interfaces between dielectric half-spaces are not realistic in most general three-dimensional problems, conductors with thin dielectric coatings are encountered often in practice. Examples are reentry vehicles or aircraft which are often coated with a heat shield or a thin layer of radar-absorbing material (RAM). Such a geometry is illustrated in Figure A.2, where a conductor is coated with $N$ thin dielectric layers, each having its own thickness $d$ and dielectric parameters $(\mu, \epsilon)$. For incident and scattered fields in $R_1$, this problem can be treated much the same way as in Section A.3, except that we compute *effective* reflection coefficients $(R_\perp, R_\parallel)$ at the outermost interface, taking into account the penetration and reflections inside of each layer. This

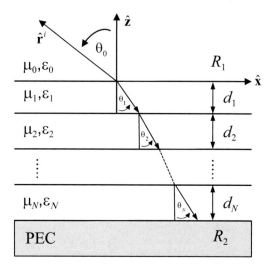

**FIGURE A.2:** Layered Dielectrics over Conductor

can be done by starting at the innermost interface (PEC) and propagating the reflection coefficients at each interface to the surface via the expression [3]

$$\tilde{R}_{i,i+1} = \frac{R_{i,i+1} + \tilde{R}_{i+1,i+2}e^{-2jk_{i+1,z}d_{i+1}}}{1 + R_{i,i+1}\tilde{R}_{i+1,i+2}e^{-2jk_{i+1,z}d_{i+1}}} \tag{A.42}$$

At each step, $\tilde{R}_{i,i+1}$ represents the generalized reflection coefficient at the interface between layers $i$ and $i+1$, taking into account reflections and transmissions in all layers below it, and $R_{i,i+1}$ are the Fresnel coefficients (A.16) or (A.18) at the interface between layers $i$ and $i+1$. The term $k_{i+1,z}$ comprises the $\hat{\mathbf{z}}$ component of the wavenumber in each layer, i.e.

$$k_{i,z} = k_i \cos\theta_i \tag{A.43}$$

We can determine $k_{i,z}$ by noting that in the free space exterior region,

$$\mathbf{k}_0 = k_{x,0}\hat{\mathbf{x}} + k_{z,0}\hat{\mathbf{z}} = k_0 \sin\theta_0\hat{\mathbf{x}} + k_0 \cos\theta_0\hat{\mathbf{z}} \tag{A.44}$$

and by Snell's law, $k_{x,0} = k_{x,1} = k_{x,2} \ldots k_{x,N} = k_x$, therefore

$$k_{i,z} = \sqrt{k_i^2 - k_x^2} \tag{A.45}$$

# References

[1] S. Lee and R. Mittra, "Fourier transform of a polygonal shape function and its application in electromagnetics," *IEEE Trans. Antennas Propagat.*, vol. 31, pp. 99–103, January 1983.

[2] C. A. Balanis, *Advanced Engineering Electromagnetics*. John Wiley and Sons, 1989.

[3] W. C. Chew, *Waves and Fields in Inhomogeneous Media*. IEEE Press, 1995.

# *Index*